Vib

and Applications

Vibration Theory
and Applications

William T. Thomson
Professor of Engineering
University of California

London
GEORGE ALLEN & UNWIN
Boston   Sydney

First published in Great Britain in 1966
Sixth impression 1978

GEORGE ALLEN & UNWIN LTD
40 Museum Street, London WC1A 1LU

© Prentice-Hall Inc., USA, 1965

ISBN 0 04 531003 3

Printed in Great Britain by
Biddles Ltd, Guildford, Surrey

Preface

This text had its beginnings in the elementary book, *Mechanical Vibrations*, by the author, which appeared in its second edition in 1953. In the intervening years many changes have taken place in technology and in the manner in which engineering problems are handled. For example the development of the machine computing facilities has greatly increased the sophistication to which engineering problems can be treated. New and more difficult concepts are being incorporated into engineering design, leading to considerable revision in the mode of engineering instruction. This book *Vibration Theory and Applications*, presents the author's views and experience in the practice and teaching of vibrations. The presentation stresses fundamentals of vibrations based on dynamical principles with applications serving to illustrate the theory.

Certain elementary considerations of the subject of vibrations remain unchanged. These are dealt with in the first three chapters of the text which take up the free, damped, and forced harmonic vibrations of the linear single degree of freedom system. Chapter 4 takes up the shock induced transient vibrations and the concept of the shock spectrum. Included is an elementary but adequate discussion of the electronic analog computer which today is an essential tool for vibration analysis. Chapter 5 is a brief introduction to the nonlinear single degree of freedom system with a discussion of the phase-plane and some of the more familiar analytical techniques.

All the concepts of the multidegree of freedom system can be demonstrated analytically, without undue algebraic difficulties, interms of systems of two degrees of freedom. Chapter 6 attempts to do this, the emphasis being on coordinates and normal modes which can be superimposed for transient or forced vibrations. The treatment is mainly in terms of matrix notation which adds to conceptual clarification.

The computing machine has greatly altered the method of treatment for the multidegree of freedom system. Chapter 7 treats multidegree of freedom systems with emphasis on formulation on matrix notation for efficient machine computation. Systems with repeating sections are included for analytical treatment by the difference equation. In spite of the emphasis for machine computation, the formal mathematical solution is still an important discipline which should not be slighted. The continuous system of Chapter 8 offers an excellent opportunity to discuss partial differential equations, boundary value problems, and eigenvalue solutions.

For advanced problems, Chapter 9 brings to light the power of the Lagrange's method. It leads to a clear understanding of the mode summation method which is one of the most important concepts in multidegree of freedom and continuous systems. The author believes that the value of the Lagrange's method lies in the formulation of the more complex problems and in the possible conceptual extraction of dynamical conclusions, and that the method should not be imposed on elementary problems.

During the past several years a somewhat new phenomenon of random vibrations has entered the field. These are vibrations induced by nondeterministic excitations which must be examined from a statistical viewpoint. Here the emphasis is on the introduction of the relatively new concepts, as applied to vibrations, of spectral density, correlation functions, and probabilities, rather than on computation. Computationally, even the simplest problem of random vibrations become tedious, and the task should be delegated to computing machines.

As to the compatibility of the text to the engineering curriculum, the first 5 chapters would represent adequate material for a first course of one semester, taught generally to seniors. The remainder of the book may then be covered in a second course which might be at the graduate level.

The author wishes to acknowledge his appreciation to the many colleagues and students who contributed to his education in the field of vibrations. The association with the Space Technology Laboratories has been particularly valuable. Appreciation is due also to my son Roger who typed the manuscript for the new book.

WILLIAM T. THOMSON

Contents

Chapter 4—Transient Vibrations   99

Chapter 5—Nonlinear Vibrations   130

1 | Undamped Free Vibration

1.1 Introduction

The subject of vibration deals with the oscillatory behavior of bodies. All bodies possessing mass and elasticity are capable of vibration. Most machines and engineering structures experience vibration in differing degrees, and their design generally requires consideration for their oscillatory behavior.

Oscillatory systems can be broadly characterized as linear or nonlinear. For linear systems the principle of superposition holds, and the mathematical techniques available for their treatment are well-developed. In contrast, techniques for the analysis of nonlinear systems are less well known and difficult to apply. Some knowledge of nonlinear systems is desirable since all systems tend to become nonlinear with increasing amplitude of oscillation.

In order to introduce certain terminology necessary for discussion, we will describe briefly the simplest form of oscillation, known as *harmonic motion*. Such motion may be described by a rotating vector, shown in Fig. 1.1-1, resulting in the trigonometric functions $A \sin \omega t$ or $A \cos \omega t$, which repeat themselves in equal intervals of time. The time elapsed while the motion repeats itself is called the *period*; the motion completed during the period is referred to as the *cycle*; and the number of complete cycles in a unit of time is designated as the *frequency* of vibration. The peak value of the motion is A, which is called the *amplitude*.

In the first four chapters we consider only linear systems described by linear differential equations. Vibrations of linear systems fall into two general classes, free and forced. *Free vibration* takes place when a system vibrates under the action of forces inherent in the system itself, and in the absence of external impressed forces. The system under free vibration will vibrate at one or more of its *natural frequencies* which are properties of the dynamical system.

Vibration that takes place under the excitation of external forces is called *forced vibration*. When the exciting force is harmonic, the forced vibration takes place at the frequency of the excitation, which is an arbitrary quantity independent of the natural frequencies of the system. When the frequency of the exciting force coincides with one of the natural frequencies of the system, a condition of *resonance* is encountered, and dangerously large amplitudes

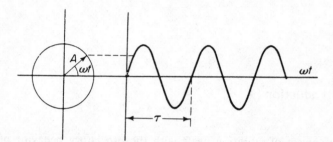

Fig. 1.1-1. Harmonic motion and its vector representation.

may result. Consequently, the calculation of natural frequencies is of interest in all types of vibrating systems.

Vibrating systems are all more or less subject to *damping* because energy is dissipated by friction and other resistances. Since no external energy is supplied in free vibration, the motion in free vibration will diminish with time, and is said to be *damped*. On the other hand, forced vibration may be maintained at constant amplitude, with the required energy supplied by an external force.

In describing the dynamical properties of any system the concept of the degrees of freedom must be introduced. The number of degrees of freedom is equal to the minimum number of independent coordinates necessary to describe the motion of the system. Thus a free particle undergoing general motion in space will have three degrees of freedom, while a finite body will have six degrees of freedom; i.e., three components of position and three angles defining its orientation. Furthermore, a continuous elastic body will require an infinite number of coordinates (three for each point on the body) to describe its motion; hence its degrees of freedom must be infinite. However, in many cases, parts of such bodies may be assumed to be rigid, and the system may be considered to be dynamically equivalent to one with finite degrees of freedom. In fact, a surprisingly large number of vibration problems can be treated with sufficient accuracy by reducing the system to one having a single degree of freedom.

In general, the free vibration of a multidegree-of-freedom system is periodic with several frequency components. However, among these there will be some simple harmonic motions called *principal modes of vibration*. These are characterized by a certain distribution of amplitude over the body, in which each point in the body undergoes harmonic motion of common frequency with all points passing through their equilibrium positions simultaneously. Later in the text it is shown that a system with n degrees of freedom will possess n principal modes of vibration with n natural frequencies, and that the more general type of vibratory motion for a linear system can be expressed by the superposition of principal modes.

The behavior of oscillatory systems may be examined in terms of the type of excitation to which the system is subjected. These forces of excitation may be divided into the headings of periodic, nonperiodic (pulse), and stochastic (random time functions), arranged in increasing order of difficulty.

1.2 Undamped Free Vibration

The simplest oscillatory system has a single degree of freedom, the motion of which can be described by a single coordinate x. A mass $m = W/g$ suspended from a spring of stiffness k lb./in. and negligible weight, as shown in Fig. 1.2-1, represents such a system.

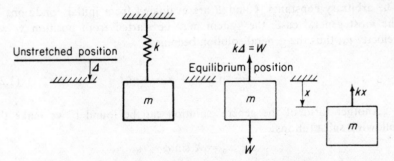

Fig. 1.2-1. Spring mass system and free-body diagram.

In the absence of impressed forces, free vibration at its natural frequency will take place when the mass is disturbed from its equilibrium position. Actually, such motion will gradually diminish in amplitude due to energy dissipated by the motion; however, natural frequency of the system is dependent mainly on its mass and stiffness and in most cases only slightly affected by the damping. Therefore, damping is generally neglected in the calculation of the natural frequency.

We take as reference the statical equilibrium position of the mass shown in Fig. 1.2-1. In this position the gravitational force W acting on the mass is balanced by the spring force $k\Delta$, where Δ is the statical deflection of the spring due to the weight W. It is evident then that by choosing the origin of coordinate x at the static equilibrium position, only forces due to displacement from this position need be considered for the equation of motion.

Letting all vector quantities in the downward direction be positive, the unbalanced force acting on the mass m in position x is $-kx$ and its motion is described by Newton's second law to be

$$m\ddot{x} = -kx \qquad (1.2\text{-}1)$$

Rearranging terms and introducing the quantity

$$\omega_n^2 = \frac{k}{m} \qquad (1.2\text{-}2)$$

Eq. (1.2-1) can be rewritten as

$$\ddot{x} + \omega_n^2 x = 0 \qquad (1.2\text{-}3)$$

Equation (1.2-3) is a homogeneous second order differential equation which has the following general solution

$$x = A \sin \omega_n t + B \cos \omega_n t \qquad (1.2\text{-}4)$$

The arbitrary constants A and B are evaluated from initial conditions. In the most general case, the system may be started from position x_0 with velocity v_0; thus the general solution becomes

$$x = \frac{v_0}{\omega_n} \sin \omega_n t + x_0 \cos \omega_n t \qquad (1.2\text{-}5)$$

Another form of the general solution can be found if we make the following substitutions:

$$x_0 = X \sin \phi$$

$$\frac{v_0}{\omega_n} = X \cos \phi$$

From these we obtain

$$X = \sqrt{x_0^2 + (v_0/\omega_n)^2}$$

$$\tan \phi = \frac{\omega_n x_0}{v_0}$$

and the alternative equation

$$x = X \sin (\omega_n t + \phi) \tag{1.2-6}$$

where X and ϕ are the two required arbitrary constants.

Equations (1.2-5) and (1.2-6) indicate that the motion of the spring mass system is harmonic and that the cycle of motion is repeated in the time $t = \tau$ where $\omega_n \tau = 2\pi$. Thus the period τ of the vibration is obtained as

$$\tau = 2\pi\sqrt{m/k} \tag{1.2-7}$$

The frequency of vibration is the number of complete cycles of motion in a unit of time, and is the reciprocal of the period. We then obtain

$$f_n = \frac{1}{2\pi} \sqrt{k/m} \tag{1.2-8}$$

which defines the natural frequency of the spring-mass system.

Equation 1.2-8 may be expressed in terms of the statical deflection by noting that $k\Delta = W = mg$, from which another expression for f_n becomes

$$f_n = \frac{1}{2\pi} \sqrt{g/\Delta} \tag{1.2-9}$$

The natural frequency is therefore found to be a function of the statical deflection Δ of the system. Using $g = 386$ in./sec.^{2}, and Δ in inches, the relationship between f_n and Δ may be given in the following two forms:

$$f_n = \frac{3.127}{\sqrt{\Delta}} \text{ c.p.s.} \tag{1.2-10}$$

$$f_n = \frac{187.6}{\sqrt{\Delta}} \text{ c.p.m.} \tag{1.2-11}$$

Figure 1.2-2 shows a logarithmic plot of Eq. (1.2-10).

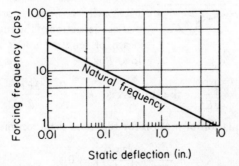

Fig. 1.2-2. Natural frequency versus static deflection.

EXAMPLE 1.2-1. A mass supported by a spring has a statical deflection of 0.020 in. Determine its natural frequency of vibration.

Solution. Using Eq. 1.2-9,

$$f_n = \frac{1}{2\pi}\sqrt{\frac{386}{0.02}} = 22.2 \text{ c.p.s.}$$
$$= 1330 \text{ c.p.m.}$$

This result can also be obtained by referring to Fig. 1.2-2.

EXAMPLE 1.2-2. An instrument weighing 2.5 lb. is set on three rubber mounts rated at $\frac{1}{16}$ in. deflection per pound each. Determine its natural frequency of vibration.

Solution. The stiffness of the three rubber mounts together is

$$k = 3 \times 16 = 48 \text{ lb./in. (See Sec. 1.6)}$$

The supported mass is

$$m = \frac{W}{g} = \frac{2.5}{386} \text{ lb. sec.}^2/\text{in.}$$

Using Eq. 1.2-8,

$$f_n = \frac{1}{2\pi}\sqrt{\frac{48 \times 386}{2.5}} = 13.7 \text{ c.p.s.} = 822 \text{ c.p.m.}$$

This frequency can also be determined from Fig. 1.2-2, and the statical deflection $\Delta = W/k = 2.5/48 = 0.052$.

EXAMPLE 1.2-3. What would be the frequency of vibration of a mass m attached to the end of a light cantilever beam of length l, and flexural stiffness EI, shown in Fig. 1.2-3?

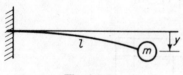

Fig. 1.2-3.

Solution. The spring force may be exerted by any elastic structure, not necessarily a helical spring. Since the statical deflection of the end of a cantilever beam of length l under a concentrated load P is

$$\Delta = \frac{Pl^3}{3EI}, \quad \text{the stiffness is} \quad k = \frac{P}{\Delta} = \frac{3EI}{l^3}$$

The differential equation of motion for the mass then becomes

$$m\frac{d^2y}{dt^2} + \left(\frac{3EI}{l^3}\right)y = 0$$

and the frequency of vibration is accordingly

$$f_n = \frac{1}{2\pi}\sqrt{\frac{3EI}{ml^3}}$$

EXAMPLE 1.2-4. A cylinder of Fig. 1.2-4 weighted to float vertically is D in. in diameter and weighs W lb. Neglecting any accompanying motion of water, determine the differential equation of motion and the period of oscillation of the cylinder if it is depressed slightly and released.

Solution. If the cylinder is depressed a vertical distance x, the buoyant force tending to restore the cylinder to its original position is

$$-\rho A x = m\frac{d^2x}{dt^2}$$

where ρ is the weight per unit volume of water and A is the cross-sectional area of the cylinder.

Rewriting the above equation in the following form,

$$\frac{d^2x}{dt^2} + \left(\frac{\rho A}{m}\right)x = 0$$

it is evident that the period of oscillation must be

$$\tau = 2\pi\sqrt{m/\rho A}$$

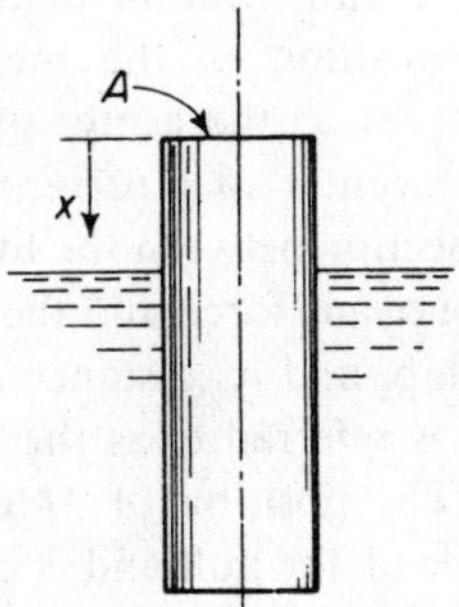

Fig. 1.2-4.

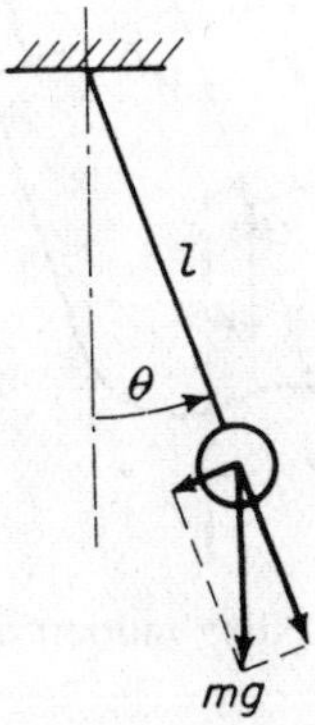

Fig. 1.2-5.

EXAMPLE 1.2-5. A mass m hangs by a string of length l as a simple pendulum, as shown in Fig. 1.2-5. Determine the equation of motion and the frequency of vibration when the amplitude is small.

Solution. Consider forces in the tangential direction. The angular displacement and force in the counterclockwise direction will be considered positive.

The tangential restoring force is $-mg \sin \theta$, and the tangential acceleration is $l(d^2\theta/dt^2)$. The equation of motion then becomes

$$ml \frac{d^2\theta}{dt^2} = -mg \sin \theta$$

For small amplitudes $\sin \theta$ may be assumed to be equal to θ, and the equation of motion reduces to

$$\frac{d^2\theta}{dt^2} + \left(\frac{g}{l}\right)\theta = 0$$

The frequency of vibration is therefore equal to

$$f_n = \frac{1}{2\pi}\sqrt{\frac{g}{l}}$$

EXAMPLE 1.2-6. Determine the oscillatory characteristics of a ship in rolling motion.

Solution. In Fig. 1.2-6, the weight W of the ship will act down through the center of gravity G of the ship. The buoyant force will be equal and opposite in direction to the weight W; however, it will act through the center of gravity of the displaced volume of water.

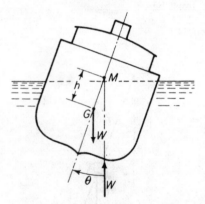

The oscillatory characteristics of the ship in rolling motion depend greatly on the position of the metacenter M with respect to the center of gravity G. The metacenter M represents the point of intersection between the line of ction of the buoyant force and the center line of the ship, and its distance h measured from G is referred to as the *metacentric height*. The position of M depends on the shape of the hull and is independent of the angular inclination θ of the ship for small values of θ. When M is above G, the couple formed by the two forces tends to restore the ship to its vertical position, and the oscillation is stable. When M lies below G, the couple tends to increase the angular inclination, and the system becomes unstable.

Fig. 1.2-6. Rolling motion of ships.

Letting J be the moment of inertia of the ship about the longitudinal rotational axis, the differential equation of motion for small θ becomes

$$J\ddot{\theta} = -Wh\theta, \qquad \ddot{\theta} + \left(\frac{Wh}{J}\right)\theta = 0$$

The period of oscillation can then be determined from the equation

$$\tau = 2\pi\sqrt{J/Wh}$$

In general, the position of the rotational axis is unknown, and J is obtained from the period of oscillation determined from a model test.

1.3 The Energy Method

When there is no dissipation of energy, the system is called conservative. At any instant the energy of a conservative system in free vibration is partly kinetic and partly potential. The kinetic energy is stored in the mass by virtue of its velocity, whereas the potential energy is stored in the form of work done in elastic deformation or work done against a force field such as gravity. The total energy, however, is constant, and the rate of change of the total energy must therefore be zero in a conservative system, as summarized by the following equations:

$$T + U = \text{constant}$$

$$\frac{d}{dt}(T + U) = 0 \qquad (1.3\text{-}1)$$

where T and U are the kinetic and potential energies, respectively. Furthermore, it is evident that the maximum kinetic energy must equal the maximum potential energy:

$$T_{\text{max}} = U_{\text{max}} \qquad (1.3\text{-}2)$$

The natural frequency of the system can then be determined if we assume that the vibrational motion is harmonic.

EXAMPLE 1.3-1. Consider the torsional pendulum of Fig. 1.3-1 consisting of a disk of mass moment of inertia J lb. in. sec.^{2}, restrained in rotation by a wire of torsional stiffness K in. lb./rad. Assuming the oscillatory motion to be harmonic and expressible by the equation

$$\theta = A \sin \omega_n t$$

the maximum kinetic and potential energies are,

$$T_{\text{max}} = \tfrac{1}{2}J\dot{\theta}^2_{\text{max}} = \tfrac{1}{2}J\omega_n^2 A^2$$
$$U_{\text{max}} = \tfrac{1}{2}K\theta^2_{\text{max}} = \tfrac{1}{2}KA^2$$

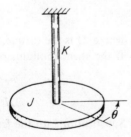

Fig. 1.3-1. A torsional pendulum.

Equating the two energies, we arrive at the expression for its natural frequency, which is

$$\omega_n = \sqrt{K/J}$$

EXAMPLE 1.3-2. A disk of unknown moment of inertia J, suspended at the end of a slender wire, executes 38 complete cycles in 1 min. To twist the wire 10 deg., a torque of 0.837 in. lb. was necessary. Determine the moment of inertia J of the disk.

Solution. The torsional stiffness of the wire is the torque necessary to twist it 1 radian.

$$K = \frac{0.837}{10°} \times 57.3° = 4.80 \text{ in. lb./rad.}$$

The period is $\tau = \frac{60}{38} = 1.58$ sec. The moment of inertia, from the frequency equation of Example 1.3-1 is

$$J = \frac{K\tau^2}{4\pi^2} = \frac{4.80 \times 1.58^2}{4\pi^2} = 0.303 \text{ lb. in. sec.}^2$$

EXAMPLE 1.3-3. A small generator is driven off the main engine through a $\frac{3}{8}$-in. steel shaft 8 in. long. If the moment of inertia of the generator rotor is 0.031 lb. in. sec.², determine its natural frequency in torsion.

Solution. The moment of inertia of the engine is so large compared to that of the generator rotor that the engine end of the shaft may be assumed to be fixed. The system therefore becomes a torsional pendulum composed of the rotor and shaft.

The angle of twist of the shaft is proportional to the torque and inversely proportional to the stiffness. For a circular shaft, θ can be expressed as

$$\theta = \frac{Ml}{\dfrac{\pi d^4}{32} G}$$

where M is the torque, l the length of shaft, d the diameter of the shaft, and G the shear modulus of elasticity. The torsional stiffness thus becomes

$$K = \frac{M}{\theta} = \frac{\pi d^4 G}{32 l} = \frac{\pi(\frac{3}{8})^4 \times 12 \times 10^6}{32 \times 8} = 2.91 \times 10^3 \text{ in. lb./rad.}$$

Substituting into the frequency equation,

$$f_n = \frac{1}{2\pi} \sqrt{\frac{2.91 \times 10^3}{0.031}} = 48.7 \text{ c.p.s.} = 2920 \text{ c.p.m.}$$

EXAMPLE 1.3-4. Determine the differential equation of motion and the natural period of oscillation of the fluid in a U tube shown in Fig. 1.3-2.

Solution. If the fluid in a U tube is set into oscillation, its motion is neither simple translation nor simple rotation about a fixed axis, and it becomes advantageous to use the energy method. Assuming all particles to have the same speed at any instant, the kinetic energy can be written as

$$T = \frac{1}{2} \rho \frac{Al}{g} \dot{x}^2$$

where l = length of fluid column, A = cross-sectional area of fluid, ρ = weight per unit volume of fluid.

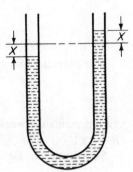

Fig. 1.3-2.

In considering the potential energy, it should be observed that the work done is the same as if the fluid column of length x had been transferred from the left side to the right side of the tube, leaving the remaining fluid undisturbed.

$$U = \rho A x^2$$

Substituting in Eq. (1.3-1) and dividing out $\dot{x}$, we obtain

$$\frac{\rho A l}{g} \ddot{x} + 2\rho A x = 0, \qquad \ddot{x} + \left(\frac{2g}{l}\right)x = 0$$

The period of oscillation is hence

$$.\tau = 2\pi\sqrt{l/2g}$$

The above equation indicates that the period of oscillation is independent of the shape of the tube, cross-sectional area of the tube, and the kind of fluid used.

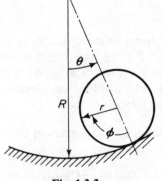

Fig. 1.3-3.

EXAMPLE 1.3-5. A cylinder of weight W and radius r rolls without slipping on a cylindrical surface of radius R, as shown in Fig. 1.3-3. Determine its natural frequency for small oscillations about the lowest point.

Solution. In determining the kinetic energy of the cylinder, it must be remembered that both translation and rotation take place. The translational velocity of the center of the cylinder is $(R - r)\dot{\theta}$, whereas the rotational velocity is $(\dot{\phi} - \dot{\theta})$. Since $R\theta = r\phi$, ϕ may be replaced by $R\theta/r$, and the kinetic energy may be written as

$$T = \frac{W}{2g}\,[(R - r)\dot{\theta}]^2 + \frac{1}{2}\left(\frac{Wr^2}{2g}\right)\left(\frac{R}{r} - 1\right)^2\dot{\theta}^2$$

where $Wr^2/2g$ is the moment of inertia of the cylinder about its center. The potential energy referred to its position at the lowest point is

$$U = W(R - r)(1 - \cos\theta)$$

and represents the work done in lifting the cylinder through the vertical height $(R - r)(1 - \cos\theta)$.

Substituting in Eq. (1.3-1),

$$\left[\frac{3W}{2g}\,(R - r)^2\ddot{\theta} + W(R - r)\sin\theta\right]\dot{\theta} = 0$$

$$\ddot{\theta} + \frac{2g}{3(R - r)}\,\theta = 0$$

where $\sin\theta$ was replaced by θ.

By inspection, the frequency of vibration in radians per second becomes

$$\omega_n = \sqrt{\frac{2g}{3(R - r)}}$$

EXAMPLE 1.3-6. Determine the natural frequency of the system shown in Fig. 1.3-4 for small oscillations, where the bar l is assumed to be rigid and weightless, and supported by the inextensible cord h.

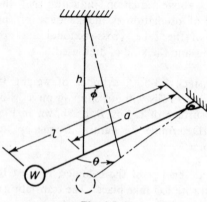

Solution. The maximum kinetic energy is

$$T_{\max} = \frac{W}{2g}\,(l\dot{\theta})^2_{\max}$$

Assuming harmonic motion, $\theta = \theta_0\cos\omega t$, $\dot{\theta}_{\max}$ can be replaced by $\omega\theta_0$:

$$T_{\max} = \frac{W}{2g}\,l^2\omega^2\theta_0^2$$

Fig. 1.3-4.

The potential energy in this case is equal to the work done in raising the weight W:

$$U_{max} = Wh(1 - \cos \phi_0)\frac{l}{a}$$

$$= Wh\left[1 - \left(1 - \frac{\phi_0^2}{2} + \cdots\right)\right]\frac{l}{a} \cong Wh\frac{\phi_0^2}{2}\frac{l}{a}$$

ϕ can be expressed in terms of θ by equating the arc $a\theta_0 = h\phi_0$ and the above equation can be rewritten as

$$U_{max} = \frac{Whl}{2a}\left(\frac{a}{h}\theta_0\right)^2$$

Equating T_{max} to U_{max},

$$\frac{W}{2g}l^2\omega^2\theta_0^2 = \frac{1}{2}Wl\left(\frac{a}{h}\right)\theta_0^2$$

from which the expression for the frequency becomes

$$\omega_n = 2\pi f_n = \sqrt{ga/lh}$$

1.4 Rayleigh's Method

In applying the energy method to systems with distributed masses, such as the beam, or a spring where the mass is not negligible, the deflection configuration of the system is necessary for the evaluation of the kinetic and potential energies. Rayleigh showed that the fundamental frequency (lowest frequency) of such systems can be determined with good accuracy by assuming any reasonable deflection curve. If the true deflection curve of the vibrating system is assumed, the fundamental frequency found by this method will be the correct frequency. For any other curve, the frequency determined will be higher than the correct frequency. This result may be explained from the fact that any deviation from the true curve requires additional constraints, a condition that implies greater stiffness and higher frequency. In general, the use of the static deflection curve of the elastic body results in a fairly accurate value of the frequency. If greater accuracy is desired, the approximate curve can be repeatedly improved.

Although Rayleigh's method is applied to systems of more than one degree of freedom, only the fundamental frequency can be found thereby. Also, the calculated frequency can never be lower than the true fundamental frequency of the system. In one procedure the energy method is used to lump the distributed mass at a particular point where the stiffness is known.

Thus, in effect, the distributed system is reduced to an equivalent single-degree-of-freedom system.

EXAMPLE 1.4-1. As an illustration of Rayleigh's method, let us consider the system shown in Fig. 1.4-1, where it is desired to determine the effect of the mass of the spring on the vibration frequency.

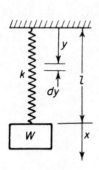

Fig. 1.4-1.

If the lower end of the spring has a maximum displacement of x_0, then any point a distance y from the fixed end can be assumed to have a displacement and velocity of

$$\frac{y}{l} x_0 \sin \omega t \quad \text{and} \quad \frac{y}{l} \omega x_0 \cos \omega t$$

where l is the length of the spring at static equilibrium.

The maximum kinetic energy of an element $w/g \, dy$ of the spring, a distance y from the fixed end, is then equal to

$$dT'_{\text{max}} = \frac{w}{2g} dy \left(\frac{y}{l} \omega x_0 \right)^2$$

where w is the weight per unit length of the spring. Integrating, the maximum kinetic energy of the spring becomes

$$T'_{\text{max}} = \frac{w}{2g} \left(\frac{\omega x_0}{l} \right)^2 \int_0^l y^2 \, dy = \frac{1}{2} \left(\frac{wl}{3g} \right) \omega^2 x_0^2$$

Adding this to the kinetic energy of the rigid mass, the total kinetic energy of the system becomes

$$T_{\text{max}} = \frac{1}{2} \left(\frac{W + \frac{1}{3}wl}{g} \right) \omega^2 x_0^2$$

The potential energy of the system is equal to

$$U_{\text{max}} = \int_0^{x_0} k(\Delta + x) \, dx - W x_0 = \frac{1}{2} k x_0^2$$

Equating the two, the natural frequency of the system becomes

$$f_n = \frac{\omega}{2\pi} = \frac{1}{2\pi} \sqrt{\frac{kg}{W + \frac{1}{3}wl}}$$

It is evident from this example that, for a spring-mass system of Fig. 1.4-1, the effect of the mass of the spring can be taken into account by adding one-third of its weight to the weight W of the rigid mass.

EXAMPLE 1.4-2. Fluid actuated by a movable element of a system, such as the one shown in Fig. 1.4-2, can have an effective mass which is greater than the actual mass. In order to determine the effective mass of the fluid we assume that the velocity distribution of the fluid in the tube is parabolic and calculate its kinetic energy. Letting v_0 be the maximum velocity at the center of the tube, the velocity at any radius r is

$$v = v_0\left[1 - \left(\frac{r}{R}\right)^2\right]$$

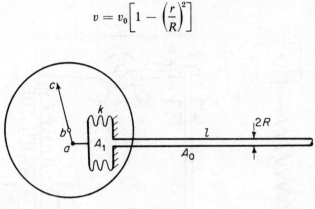

Fig. 1.4-2.

The kinetic energy of the fluid in the tube of length l is then

$$T = \tfrac{1}{2}\rho l \int_0^R 2\pi r v_0^2\left[1 - \left(\frac{r}{R}\right)^2\right]^2 dr = \tfrac{1}{2}(\rho A_0)\frac{v_0^2}{3}$$

Assuming next that the piston end of the bellows is given a velocity $\dot{x}$, we have from continuity

$$A_1\dot{x} = 2\pi \int_0^R r v_0\left[1 - \left(\frac{r}{R}\right)^2\right] dr = \tfrac{1}{2}v_0 A_0$$

$$v_0 = 2\left(\frac{A_1}{A_0}\right)\dot{x}$$

With this velocity the kinetic energy of the fluid becomes

$$T = \tfrac{1}{2}(\rho A_0 l)\frac{4}{3}\left(\frac{A_1}{A_0}\right)^2\dot{x}^2 = \tfrac{1}{2}M_{\text{eff}}\dot{x}^2$$

The effective mass referred to the piston end of the bellows is then

$$M_{\text{eff}} = \tfrac{4}{3}(\rho A_0 l)\left(\frac{A_1}{A_0}\right)^2$$

For liquids, M_{eff} is often many times the actual mass of the movable element and hence the natural frequency of the system is considerably reduced.

EXAMPLE 1.4-3. Oscillatory systems are often composed of levers, gears, and other linkages which seemingly complicate the analysis. The reduction of such a system to a simpler equivalent system is then generally desirable.

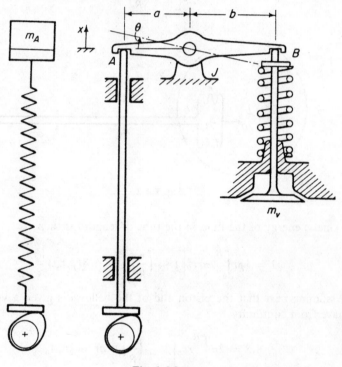

Fig. 1.4-3.

Levers. Consider the lever system of Fig. 1.4-3. The rocker arm of moment of inertia J and the valve of mass m_v can be reduced to a single mass at A by writing the kinetic energy as follows.

$$T = \tfrac{1}{2}J\dot{\theta}^2 + \tfrac{1}{2}m_v(b\dot{\theta})^2 = \tfrac{1}{2}(J + m_v b^2)\dot{\theta}^2 = \frac{1}{2}\left(\frac{J + m_v b^2}{a^2}\right)\dot{x}^2$$

The effective mass at A is then

$$m_A = \left(\frac{J + m_v b^2}{a^2}\right)$$

If the push rod is now reduced to a spring and an additional mass at the end, the entire system is reduced to a single spring and a mass.

1.5 Beam Vibration by Rayleigh's Method

In the previous section the energy method was used to reduce a distributed system into an equivalent spring-mass system. With the system vibrating harmonically, the procedure required an assumption as to the distribution of the amplitude for the evaluation of both the kinetic and potential energy.

In this section we wish to extend Rayleigh's method to beam vibrations. Letting m be the mass per unit length along the beam and y the amplitude of the assumed deflection curve, the kinetic energy is expressed by the equation

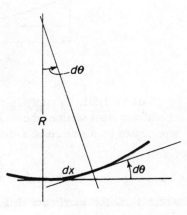

$$T_{max} = \tfrac{1}{2} \int \dot{y}^2 \, dm = \tfrac{1}{2}\omega^2 \int y^2 \, dm \quad (1.5\text{-}1)$$

where ω is the fundamental frequency in radians per second.

The potential energy of the beam is determined by the work done on the beam which is stored as elastic energy. Letting

Fig. 1.5-1.

M be the bending moment and θ the slope of the elastic curve, the work done is equal to

$$U = \tfrac{1}{2} \int M \, d\theta \qquad (1.5\text{-}2)$$

Since the deflection in beams is generally small, the following geometric relations are assumed to hold (see Fig. 1.5-1).

$$\theta = \frac{dy}{dx}, \qquad \frac{1}{R} = \frac{d\theta}{dx} = \frac{d^2 y}{dx^2} \qquad (1.5\text{-}3)$$

In addition to these relations we have, from the theory of beams, the flexure formula:

$$\frac{1}{R} = \frac{M}{EI} \qquad (1.5\text{-}4)$$

where EI is the flexural rigidity of the beam and R is the radius of curvature. Substituting for $d\theta$ and $1/R$, U may be written as

$$U_{max} = \frac{1}{2} \int \frac{M^2}{EI} \, dx = \frac{1}{2} \int EI \left(\frac{d^2 y}{dx^2}\right)^2 dx \qquad (1.5\text{-}5)$$

Equating the kinetic and potential energies, the fundamental frequency of the beam is determined from the equation

$$\omega^2 = \frac{\int EI(d^2y/dx^2)^2\, dx}{\int y^2\, dm} \tag{1.5-6}$$

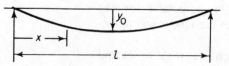

Fig. 1.5-2.

EXAMPLE 1.5-1. In applying this procedure to a simply supported beam of uniform cross section, shown in Fig. 1.5-2, we assume the deflection to be represented by a sine curve as follows:

$$y = \left(y_0 \sin \frac{\pi x}{l}\right) \sin \omega t$$

where y_0 is the maximum deflection at mid-span. The second derivative then becomes

$$\frac{d^2y}{dx^2} = -\left(\frac{\pi}{l}\right)^2 y_0 \sin \frac{\pi x}{l} \sin \omega t$$

Substituting into Eq. (1.5-6), we obtain

$$\omega^2 = \frac{EI\left(\dfrac{\pi}{l}\right)^4 \displaystyle\int_0^l \sin^2 \frac{\pi x}{l}\, dx}{\dfrac{w}{g} \displaystyle\int_0^l \sin^2 \frac{\pi x}{l}\, dx} = \pi^4 \frac{gEI}{wl^4}$$

The fundamental frequency is therefore found to be

$$\omega_1 = \pi^2 \sqrt{gEI/wl^4}$$

In this case the assumed curve happened to be the correct curve, and the exact frequency is obtained by Rayleigh's method. Any other curve assumed for the case will result in a constant greater than π^2 in the frequency equation.

EXAMPLE 1.5-2. If the distance between the ends of the beam of Fig. 1.5-2 is rigidly fixed, a tensile stress σ will be developed by the lateral deflection. Account for this additional strain energy in the frequency equation.

Solution. Due to the lateral deflection, the length dx of the beam is increased by an amount

$$[\sqrt{1 + (dy/dx)^2} - 1] \, dx \cong \frac{1}{2}\left(\frac{dy}{dx}\right)^2 dx$$

The additional strain energy in the element dx is

$$dU = \tfrac{1}{2}\sigma A \epsilon \, dx = \tfrac{1}{2} E A \epsilon^2 \, dx$$

where A is the cross-sectional area, σ the stress due to tension, and $\epsilon = \tfrac{1}{2}(dy/dx)^2$ is the unit strain.

Equating the kinetic energy to the total strain energy of bending and tension, we obtain,

$$\frac{1}{2}\omega^2 \int y^2 \, dm = \frac{1}{2} \int EI\left(\frac{d^2 y}{dx^2}\right)^2 dx + \frac{1}{2} \int \frac{EA}{4}\left(\frac{dy}{dx}\right)^4 dx$$

The above equation then leads to the frequency equation

$$\omega_1^2 = \frac{\displaystyle\int EI\left(\frac{d^2 y}{dx^2}\right)^2 dx + \int \frac{EA}{4}\left(\frac{dy}{dx}\right)^4 dx}{\displaystyle\int y^2 \, dm}$$

which contains an additional term due to the tension.

EXAMPLE 1.5-3. Consider next the cantilever beam shown in Fig. 1.5-3. We will assume here that the amplitude of the beam at any point x is given with sufficient accuracy by the statical deflection curve of a massless cantilever beam with a concentrated load at the end. Writing this equation in the form

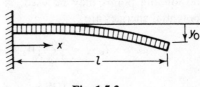

$$y = \tfrac{1}{2}y_0\left[3\left(\frac{x}{l}\right)^2 - \left(\frac{x}{l}\right)^3\right]$$

Fig. 1.5-3.

where $y_0 = Pl^3/3EI$ is the amplitude of the free end, the stiffness at the free end becomes $k = P/y_0 = 3EI/l^3$. The potential energy which is equal to the work done, is then,

$$U_{max}{}^* = \tfrac{1}{2}k y_0^2 = \frac{3EI}{2l^3} y_0^2$$

* This result can also be found from the equation $U_{max} = \tfrac{1}{2}\displaystyle\int_0^l EI\left(\frac{d^2 y}{dx^2}\right)^2 dx.$

The kinetic energy is next determined by integrating one half the product of the mass and the square of the velocity over the length of the beam

$$T_{\max} = \frac{w}{2g} \int_0^l (\omega y)^2 \, dx = \frac{w}{2g} \left(\frac{\omega y_0}{2}\right)^2 \int_0^l \left[3\left(\frac{x}{l}\right)^2 - \left(\frac{x}{l}\right)^3\right]^2 dx$$

$$= \frac{1}{2}\left(\frac{33wl}{140g}\right)\omega^2 y_0^2$$

The above equation indicates that for the assumed deflection curve, the continuous beam of w lb./ft. is equivalent in vibration characteristics to that of a weightless beam with a concentrated weight ($\frac{33}{140} wl$) at the end.

Equating the two energies, the fundamental frequency of vibration in radians per second becomes

$$\omega_1 = \sqrt{\frac{(3EI/l^3)g}{\frac{33}{140} wl}} = 3.56\sqrt{gEI/wl^4}$$

The exact solution for this case is

$$\omega_1 = 3.515\sqrt{gEI/\omega l^4}$$

In general, the deflection curve assumed for the problem should satisfy the boundary conditions of deflection, slope, shear, and moment. These conditions are satisfied by the static deflection curve which generally results in a frequency of acceptable accuracy.

If a beam is represented by a series of lumped weights $W_1, W_2, W_3, \cdots$, the maximum strain energy can be determined from the work done by these loads. As a first approximation, the static deflection $y_1, y_2, y_3, \cdots$ of corresponding points may be used, in which case the maximum kinetic and potential energies are

$$T_{\max} = \frac{1}{2}\frac{\omega^2}{g}[W_1 y_1^2 + W_2 y_2^2 + W_3 y_3^2 + \cdots] \qquad (1.5\text{-}7)$$

$$U_{\max} = \frac{1}{2}[W_1 y_1 + W_2 y_2 + W_3 y_3 + \cdots] \qquad (1.5\text{-}8)$$

By equating the two, the frequency equation is established as

$$\omega_1^2 = \frac{g \sum Wy}{\sum Wy^2} \qquad (1.5\text{-}9)$$

EXAMPLE 1.5-4. To illustrate the use of this equation, we will find the first approximation to the fundamental frequency of lateral vibration for the system shown in Fig. 1.5-4.

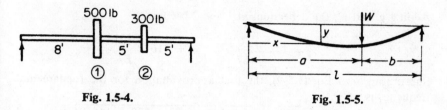

Fig. 1.5-4.   Fig. 1.5-5.

Referring to Fig. 1.5-5, the deflection at any point x, due to a concentrated load W a distance a and b from the ends, can be determined from the equation

$$y_x = \frac{Wbx}{6EIl}(l^2 - x^2 - b^2)$$

which can be found in any standard text on strength of materials. The deflections at the loads can be obtained from the super-position of the two loads, shown in Fig. 1.5-6:

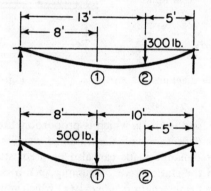

Fig. 1.5-6.

$$y_1' = \frac{300 \times 5 \times 8}{6 \times 18 \times EI}(18^2 - 8^2 - 5^2) \times 12^3 = \frac{45.2 \times 10^6}{EI} \text{ in.}$$

$$y_2' = \frac{300 \times 5 \times 13}{6 \times 18 \times EI}(18^2 - 13^2 - 5^2) \times 12^3 = \frac{40.7 \times 10^6}{EI} \text{ in.}$$

$$y_1'' = \frac{500 \times 8 \times 10}{6 \times 18 \times EI}(18^2 - 10^2 - 8^2) \times 12^3 = \frac{103. \times 10^6}{EI} \text{ in.}$$

$$y_2'' = \frac{500 \times 8 \times 5}{6 \times 1 \times EI}(18^2 - 5^2 - 8^2) \times 12^3 = \frac{75.3 \times 10^6}{EI} \text{ in.}$$

Adding y' and y'', the deflections at 1 and 2 become

$$y_1 = 148 \times \frac{10^6}{EI}, \qquad y_2 = 116 \times \frac{10^6}{EI}$$

Substituting into Eq. (1.5-9), the first approximation for the fundamental frequency is

$$\omega_1 = \sqrt{\frac{g \sum Wy}{\sum Wy^2}} = \sqrt{\frac{386(500 \times 148 + 300 \times 116)EI}{(500 \times 148^2 + 300 \times 116^2)10^6}}$$

$$= 0.0017\sqrt{EI} \text{ rad./sec.}$$

If further accuracy is desired, a better approximation to the dynamic curve can be made by using dynamic loads in place of the static weights.

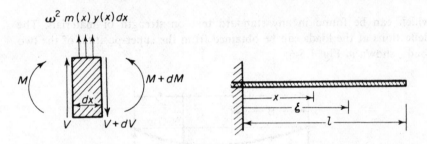

Fig. 1.5-7. Free-body diagram of   Fig. 1.5-8.
beam element.

Since the dynamic load is $m\omega^2 y$, which is proportional to the deflection, we can recalculate the deflection with the modified weights W_1 and $W_2(y_2/y_1)$.

The concept of dynamic loads can also be used, starting with a much simpler curve than the static curve. Assuming such a curve to be $y(x)$, the dynamic loading per unit length is $\omega^2 m(x)y(x)$ which must equal the change in shear along the beam

$$dV = \omega^2 m(x)y(x)\, dx \qquad (1.5\text{-}10)$$

as shown in Fig. 1.5-7. Since $dM = V\, dx$, the moment M can be found by integrating and substituting into the equation

$$U_{\max} = \frac{1}{2} \int \frac{M^2}{EI}\, dx \qquad (1.5\text{-}11)$$

which then is proportional to ω^4. Actually the equation for $T_{\max}$ is not so sensitive to the inaccuracies of the assumed curve, whereas the strain energy which depends on the curvature, could be very much in error and hence must be computed with care. Thus $T_{\max}$ can be computed from the inaccurate curve by using Eq. (1.5-1).

EXAMPLE 1.5-5. Determine the fundamental frequency of the uniform cantilever beam shown in Fig. 1.5-8, using the simple curve $y = cx^2$.

Solution. If we use Eq. (1.5-6), we would find the result to be very much in error since the above curve does not satisfy the boundary conditions at the free end. By using Eq. (1.5-6) we obtain

$$\omega = 4.47\sqrt{EI/ml^4}$$

whereas the exact value is

$$\omega_1 = 3.52\sqrt{EI/ml^4}$$

Acceptable results can be found using the given curve, by the procedure outlined in the previous section. From Eq. (1.5-10) we have

$$V(\xi) = \omega^2 \int_\xi^l mc\xi^2 \, d\xi = \frac{\omega^2 mc}{3}(l^3 - \xi^3)$$

and the bending moment becomes

$$M(x) = \int_x^l V(\xi) \, d\xi = \frac{\omega^2 mc}{3} \int_x^l (l^3 - \xi^3) \, d\xi$$

$$= \frac{\omega^2 mc}{12}(3l^4 - 4l^3 x + x^4)$$

The maximum strain energy is found by substituting $M(x)$ into Eq. (1.5-11):

$$U_{max} = \frac{1}{2EI}\left(\frac{\omega^2 mc}{12}\right)^2 \int_0^l (3l^4 - 4l^3 x + x^4)^2 \, dx$$

$$= \frac{\omega^4}{2EI}\frac{m^2 c^2}{144}\frac{312}{135} l^9$$

The maximum kinetic energy is

$$T_{max} = \frac{1}{2}\int_0^l \dot{y}^2 m \, dx = \frac{1}{2}c^2\omega^2 m \int_0^l x^4 \, dx = \frac{1}{2}c^2\omega^2 m \frac{l^5}{5}$$

By equating these results, we obtain

$$\omega_1 = \sqrt{12.47 \, EI/ml^4} = 3.53\sqrt{EI/ml^4}$$

which is very close to the exact result.

1.6 Stiffness of Spring Elements

The single-degree-of-freedom system is an idealization which in many cases is justified. Such approximations give excellent results when the mass of the elastic element is small compared with the lumped mass of the system, or when the distributed mass is weighted properly to give an equivalent lumped mass at a specified point. The fundamental frequency of the system can then be determined with good accuracy from the natural frequency equation of a single-degree-of-freedom system, provided the stiffness corresponding to the lumped mass position is known. The accompanying table of spring stiffness for various types of springs will be found convenient for reference.

EXAMPLE 1.6-1. Determine the spring constant for the system of springs shown in Fig. 1.6-1.

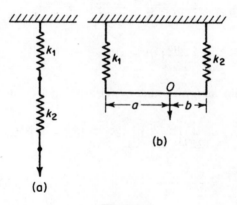

Fig. 1.6-1.

System (a). Applying a unit force at the end of the lower spring, each spring will stretch by an amount $1/k_1$ and $1/k_2$, and the total displacement of the end becomes $1/k_1 + 1/k_2$. The resultant spring constant which is by definition the force per unit deflection is then

$$k_0 = \frac{1}{\dfrac{1}{k_1} + \dfrac{1}{k_2}} = \frac{k_1 k_2}{k_1 + k_2}$$

System (b). The unit force applied at O is divided so that $b/(a + b)$ and $a/(a + b)$ act on springs k_1 and k_2 respectively. The deflections of k_1 and k_2 are $b/(a + b)k_1$ and $a/(a + b)k_2$, and that of point O is

$$\frac{b}{(a + b)k_1} + \left(\frac{a}{a + b}\right)\left[\frac{a}{(a + b)k_2} - \frac{b}{(a + b)k_1}\right] = \frac{1}{(a + b)^2}\left(\frac{a^2}{k_2} + \frac{b^2}{k_1}\right)$$

The resultant spring constant at O is then

$$k_0 = \frac{(a + b)^2}{\dfrac{a^2}{k_2} + \dfrac{b^2}{k_1}}$$

If $k_1 = k_2$ and $a = b$, the above equation reduces to $k_0 = 2k_1$.

Table of Spring Stiffness

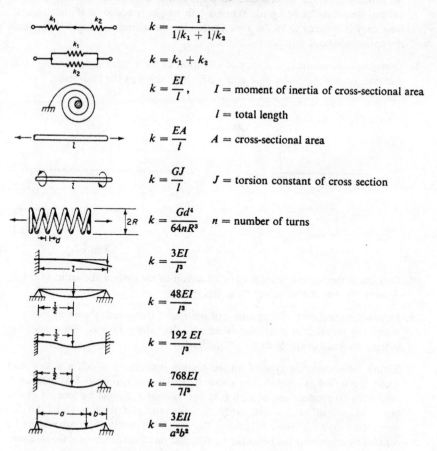

$$k = \frac{1}{1/k_1 + 1/k_2}$$

$$k = k_1 + k_2$$

$$k = \frac{EI}{l}, \qquad I = \text{moment of inertia of cross-sectional area}$$

$$l = \text{total length}$$

$$k = \frac{EA}{l} \qquad A = \text{cross-sectional area}$$

$$k = \frac{GJ}{l} \qquad J = \text{torsion constant of cross section}$$

$$k = \frac{Gd^4}{64nR^3} \qquad n = \text{number of turns}$$

$$k = \frac{3EI}{l^3}$$

$$k = \frac{48EI}{l^3}$$

$$k = \frac{192\,EI}{l^3}$$

$$k = \frac{768EI}{7l^3}$$

$$k = \frac{3EIl}{a^2b^2}$$

PROBLEMS

1. A 1-lb. weight attached to a light spring elongates it 0.31 in. Determine the natural frequency of the system.

2. What weight attached to the spring of Prob. 1 would result in a natural frequency of 100 cycles per minute?

3. In a spring-mass system k_1, m has a natural frequency of f_1. If a second spring k_2 is added in series with the first spring, the natural frequency is lowered to $\frac{1}{2}f_1$. Determine k_2 in terms of k_1.

4. A 10-lb. weight, attached to the lower end of a spring whose upper end is fixed, vibrates with a natural period of 0.45 sec. Determine the natural period when a 5-lb weight is attached to the mid-point of the same spring with the upper and lower ends fixed.

5. An unknown weight W lb. attached to the end of an unknown spring k has a natural frequency of 94 c.p.m. When a 1-lb weight is added to W, the natural frequency is lowered to 76.7 c.p.m. Determine the unknown weight W lb., and the spring constant k lb./in.

6. Determine the natural frequency of vibration of the system shown in Fig. 1-1. Assume the bar AB to be rigid and weightless with c as the mid-point.

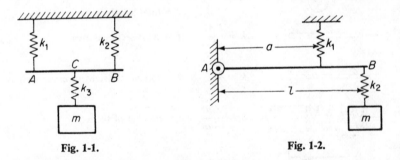

Fig. 1-1.   Fig. 1-2.

7. Determine the natural frequency of vibration of the system shown in Fig. 1-2. Assume the bar AB to be rigid and weightless.

8. Determine the differential equation of motion of the system shown in Fig. 1-3, where the moment of inertia of W and the bar about O is J_0. Show that the system becomes unstable when $b > ka^2/W$.

9. Figure 1-4 shows one type of impact-testing machine in which a heavy steel plate, supported as a pendulum about O, strikes the specimen S a distance L below O. To produce no reaction at O due to impact, S must be located at the center of percussion $L = k^2/\bar{r}$, where k is the radius of gyration and $\bar{r}$ the distance from O to the center of gravity. Show that the center of percussion can be located by measuring the period of free oscillation of the pendulum, the relationship being expressed by the equation $L = \tau^2 g/4\pi^2$.

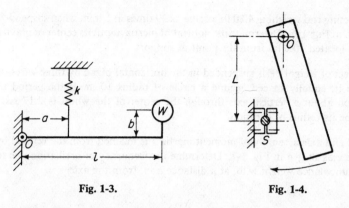

Fig. 1-3.   Fig. 1-4.

10. In the torsional pendulum of Fig. 1.3-1, explain how the natural frequency depends on (a) the length of wire, (b) the diameter of wire, (c) the material of the wire, (d) the suspended weight, and (e) the radius of gyration of the suspended weight.

11. A pulley weighing 60 lb. with radius of gyration of 8 in. is attached to a steel shaft 1 in. in diameter and 10 ft long. Determine the natural frequency of vibration if the pulley is clamped 6 and 4 ft. from the ends, which are fixed. (Neglect weight of shaft.)

12. A flywheel weighing 70 lb. was allowed to swing as a pendulum about a knife-edge at the inner side of the rim, as shown in Fig. 1-5. If the measured period of oscillation was 1.22 sec., determine the moment of inertia of the flywheel about its geometric axis.

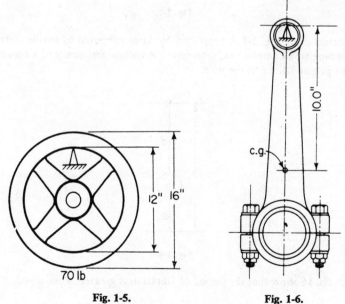

Fig. 1-5.   Fig. 1-6.

13. A connecting rod weighing 4.80 lb. oscillates 39 times in 1 min. when suspended as shown in Fig. 1-6. Determine its moment of inertia about its center of gravity which is located 10.0 in. from the point of support.

14. A flywheel of weight W is suspended in the horizontal plane by three wires of length 6 ft. equally spaced around a circle of radius 10 in. If the period of oscillation about a vertical axis through the center of the wheel is 2.17 sec., determine its radius of gyration.

15. A wheel and axle assembly of moment inertia J is inclined from the vertical by an angle α as shown in Fig. 1-7. Determine the frequency of oscillation due to a small unbalance weight w lb. at a distance a in. from the axle.

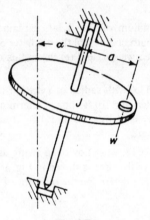

Fig. 1-7.

16. The mass m of Fig. 1-8 is supported by wires subjected to tensile force T lb. Assuming small amplitudes, determine the natural frequency of vibration in a plane perpendicular to the wire.

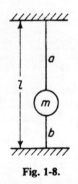

Fig. 1-8.

17. In Prob. 16 show that the period of vibration is greatest when $a = b$.

18. A chronograph is to be operated by a 2-sec. pendulum shown in Fig. 1-9. A platinum wire attached to the bob completes the electric timing circuit through a drop of mercury as it swings through the lowest point. (a) What should be the length L of the pendulum? (b) If the platinum wire is in contact with the mercury for $\frac{1}{8}$ in. of the swing, what must be the amplitude θ_0 to limit the duration of contact to 0.01 sec.? (Assume constant velocity during contact.)

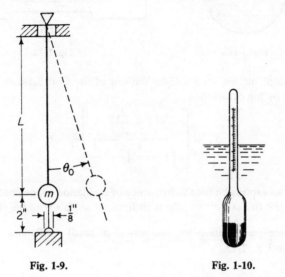

Fig. 1-9.   Fig. 1-10.

19. A spherical ball of radius R floats half submerged in water. If the ball is depressed slightly and released, write the differential equation of motion and determine the period of vibration. If $R = 2.0$ ft., what is the numerical value of the period?

20. Figure 1-10 shows a hydrometer float which is used to measure the specific gravity of liquids. The weight of the float is 0.082 lb., and the diameter of the cylindrical section protruding above the surface is $\frac{1}{4}$ in. Determine the period of vibration when the float is allowed to bob up and down in a fluid of specific gravity 1.20.

21. A spherical buoy 3 ft. in diameter is weighted to float half out of water, as shown in Fig. 1-11. The center of gravity of the buoy is 8 in. below its geometric center, and the period of oscillation in rolling motion is 1.30 sec. Determine the moment of inertia of the buoy about its rotational axis.

22. Two tanks of cross-sectional area A_1 and A_2 are connected by a pipe of cross-sectional area A_0 and length l, as shown in Fig. 1-12. Show that the differential

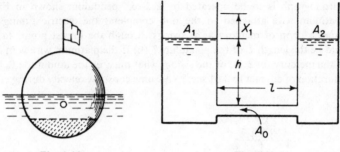

Fig. 1-11.   Fig. 1-12.

equation of motion for small oscillations of the fluid between the two tanks is given by the equation

$$\ddot{x}_1 + \left[\frac{g\left(1 + \dfrac{A_1}{A_2}\right)}{h\left(1 + \dfrac{A_1}{A_2}\right) + l\,\dfrac{A_1}{A_0}} \right] x_1 = 0$$

23. Obtain an expression for the frequency of oscillation of the fluid in an open U tube, when the axis of the tube is inclined an angle α with the vertical.

24. Determine the equation for the period of small oscillations of the system shown in Fig. 1-13.

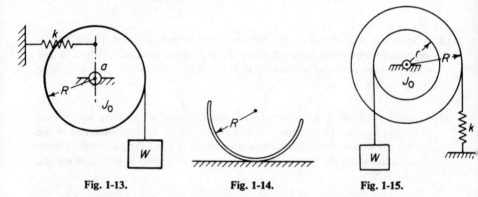

Fig. 1-13.   Fig. 1-14.   Fig. 1-15.

25. A thin rectangular plate is bent into a semicircular cylinder, as shown in Fig. 1-14. Determine its period of oscillation if it is allowed to rock on a horizontal surface.

26. Derive the expression for the frequency of oscillation of the system shown in Fig. 1-15. The two wheels are keyed together and their combined moment of inertia is J_0.

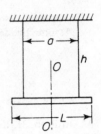

27. A uniform bar of length L and weight W is suspended symmetrically by two strings, as shown in Fig. 1-16. Set up the differential equation of motion for small angular oscillations of the bar about the vertical axis O-O, and determine its period.

Fig. 1-16.

28. A uniform bar of length L is suspended in the horizontal position by two vertical strings of equal length attached to the ends. If the period of oscillation in the plane of the bar and strings is t_1, and the period of oscillation about a vertical line through the center of gravity of the bar is t_2, show that the radius of gyration of the bar about the center of gravity is given by the expression

$$k = \left(\frac{t_2}{t_1}\right) \frac{L}{2}$$

29. A uniform bar of radius of gyration k about its center of gravity is suspended horizontally by two vertical strings of length h, at distances a and b from the mass center. Prove that the bar will oscillate about the vertical line through the mass center and determine the frequency of oscillation.

30. A steel shaft 50 in. long and $1\frac{1}{2}$ in. in diameter is used as a torsion spring for the wheels of a light automobile as shown in Fig. 1-17. Determine the natural frequency of the system if the weight of the wheel and tire assembly is 38 lb. and its radius of gyration about its axle is 9.0 in.

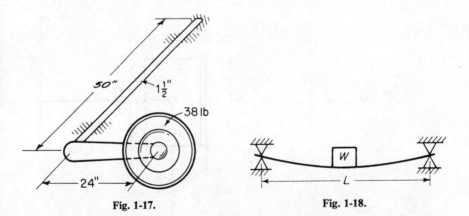

Fig. 1-17.   **Fig. 1-18.**

31. Determine the natural frequency of vibration of a weight W attached to the center of a beam of length L, as shown in Fig. 1-18. The beam is simply supported, its weight and lateral stiffness being w lb. and EI lb. in.^{2}, respectively. Assume the deflection to be equal to the static curve for concentrated load

$$y = y_{max}\left[3\left(\frac{x}{l}\right) - 4\left(\frac{x}{l}\right)^3\right], \qquad 0 < x < l/2$$

where $y_{max} = Pl^3/48EI$.

32. In Prob. 31, if $W = 15$ lb. and the beam is steel of dimensions 5 ft. $\times$ 3 in. $\times$ $\frac{3}{8}$ in., determine the numerical value of the natural frequency. ($E = 29 \times 10^6$ lb. in.$^{-2}$)

33. A steel cantilever beam 6 ft. long has a rectangular cross section $1 \times \frac{1}{2}$ in. with the 1-in. side in the vertical direction. Determine the natural frequency for oscillation in the vertical and horizontal planes.

34. The free end of the beam of Prob. 33 is given an initial displacement $\frac{1}{4}$ in. down and $\frac{1}{2}$ in. to the left. Determine the path described by a point on the end face.

35. Tachometers are a reed type of frequency-measuring instrument consisting of small cantilever beams with weights attached at the ends. When the frequency of vibration corresponds to the natural frequency of one of the reeds, it will vibrate, thereby indicating the frequency. How large a weight must be placed on the end of a reed made of spring steel 0.04 in. thick, 0.25 in. wide, and 3.50 in. long, for a natural frequency of 20 c.p.s.?

36. In Fig. 1-19, the mass which is restricted to move in the vertical direction has attached to it a second spring k_2 with initial tension T. Show that if x/l is small, the natural frequency is given by the equation

$$\omega = \sqrt{\frac{k_1 + (T/l)}{m}}$$

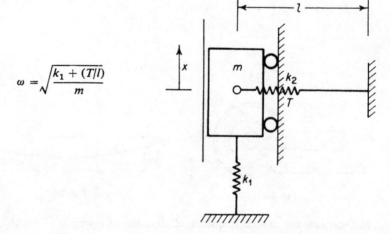

Fig. 1-19.

37. Derive the differential equation of motion of the system of Fig. 1-19 when x/l is not small.

38. The system of Fig. 1-20 rotates about the vertical axis with speed Ω rad./sec. The cantilever spring is stiff in the plane of rotation. Using the result of Prob. 36, show that the natural frequency is given by the equation

$$\omega = \omega_0 \sqrt{1 + (\Omega/\omega_0)^2(1 + r/l)}, \qquad \omega_0 = \sqrt{3EI/ml^3}$$

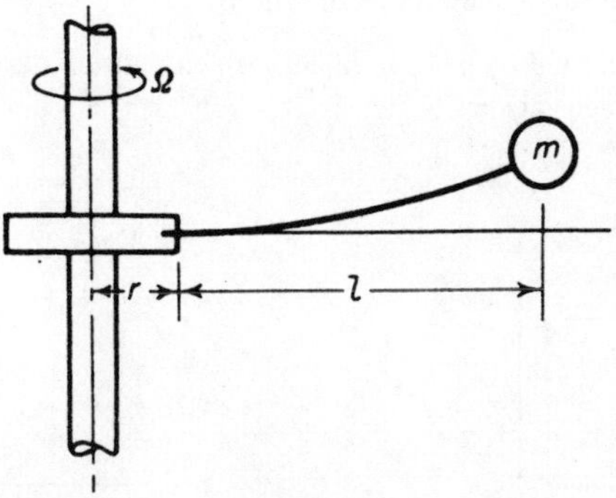

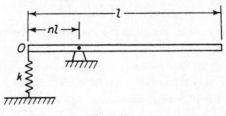

Fig. 1-20.

39. Using the results of Example 1.5-2 and the curve $y = y_0 \sin(\pi x/l) \sin \omega t$, determine the fundamental frequency of the simply supported beam of Fig. 1.5-2, including the effect of the axial tension.

40. Determine the effective mass at point O of a uniform rod of mass m and length l pivoted at a distance nl from O, as shown in Fig. 1-21.

Fig. 1-21.

41. The liquid-filled pressure transducer of Fig. 1.4-2 has the following specifications:

weight of translating parts at end of bellows = 0.098 lb.
weight of bellows = 0.13 lb.
weight of indicator needle = 0.0041 lb.
stiffness of bellows k = 0.12 lb./in.

Assume the needle to be a uniform rod of length 5 in. with $ab = \frac{1}{2}$ in. Determine the moment of inertia of the needle about the pivot and the effective mass of the needle at a. What is the natural frequency of the instrument when actuated by air in the tube?

If filled with oil of specific gravity 0.80, determine the kinetic energy of the moving fluid when the displacement of the end of the bellows is harmonic and equal to x. Assume the tube to be $\frac{1}{4}$ in. inside dia. and 10 ft. long; bellows to be 2 in. dia.; and the velocity distribution in the connecting tube to be parabolic. Determine the effective mass of the fluid at a and the natural frequency of the instrument including the connecting fluid.

42. The vibrometer of Fig. 1-22 is filled with oil. Show that its natural frequency is given by the equation

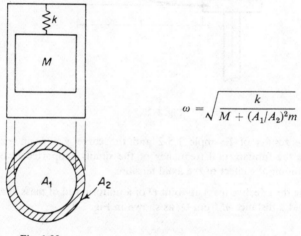

$$\omega = \sqrt{\frac{k}{M + (A_1/A_2)^2 m}}$$

Fig. 1-22.

where m is the mass of the oil in the annular space between the seismic mass and the case.

43. Assume in Prob. 42 that the two diameters are $d_1 = 1.0$ in. and $d_2 = 1.020$ in. and that the density ratio of the mass M to that of the fluid is 8. Show that the natural frequency of the instrument is $\omega = \sqrt{k/4.09M}$.

44. Using Rayleigh's method, estimate the fundamental frequency of the lumped mass system shown in Fig. 1-23.

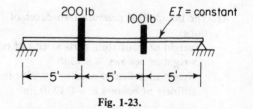

Fig. 1-23.

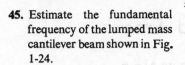

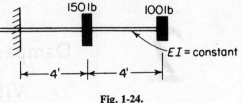

45. Estimate the fundamental frequency of the lumped mass cantilever beam shown in Fig. 1-24.

Fig. 1-24.

46. For a valve system similar to that of Fig. 1.4-3, the following data are given: $J = 0.32$ lb. in.^{2}; $m_v = 0.26$ lb.; $a = 2.2$ in.; $b = 3.1$ in. Determine the effective mass at A.

47. A uniform beam of mass M and stiffness $K = EI/l^3$, shown in Fig. 1-25, is supported on equal springs with total vertical stiffness of k lb./in. Using Rayleigh's method with the deflection $y_{max} = \sin(\pi x/l) + b$ show that the frequency equation becomes

$$\omega^2 = \frac{k}{M}\left(\frac{\dfrac{K}{k}\dfrac{\pi^4}{4} + \dfrac{b^2}{2}}{\dfrac{1}{2} + \dfrac{4b}{\pi} + b^2}\right)$$

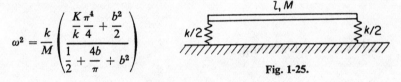

Fig. 1-25.

By $\partial \omega^2/\partial b = 0$, show that the lowest frequency results when

$$b = -\frac{\pi}{4}\left(\frac{1}{2} - \frac{K\pi^4}{2k}\right) \pm \sqrt{\left[\frac{\pi}{4}\left(\frac{1}{2} - \frac{K\pi^4}{2k}\right)\right]^2 + \frac{\pi^4 K}{2k}}$$

2

Damped Free
Vibration

2.1 Viscous Damping

During vibration, energy is dissipated in one form or another, and steady amplitude cannot be maintained without its continuous replacement. The actual description of the damping force associated with the dissipation of energy is difficult. It may be a function of the displacement, velocity, stress or other factors. However, ideal damping models can be conceived which will often permit a satisfactory approximation. Of these the viscous damping force, proportional to the first power of the velocity, leads to the simplest mathematical treatment.

Viscous damping force can be expressed by the equation

$$F = -c\dot{x} \qquad (2.1\text{-}1)$$

where c is a constant of proportionality and $\dot{x}$ the velocity. Symbolically, it is designated by a dashpot, as shown in Fig. 2.1-1. When such a system undergoes free vibration the equation of motion becomes

$$m\ddot{x} = -c\dot{x} - kx \qquad (2.1\text{-}2)$$

which, on rearranging, takes the form

$$m\ddot{x} + c\dot{x} + kx = 0 \qquad (2.1\text{-}3)$$

Equation (2.1-3), being a homogeneous second-order differential equation, can be solved by assuming a solution of the form

$$x = e^{st} \qquad (2.1\text{-}4)$$

36

where s is a constant to be determined. Upon substitution of Eq. (2.1-4) into (2.1-3), we obtain the equation

$$\left(s^2 + \frac{c}{m}s + \frac{k}{m}\right)e^{st} = 0 \qquad (2.1-5)$$

which is satisfied for all values of t when

$$s^2 + \frac{c}{m}s + \frac{k}{m} = 0 \qquad (2.1-6)$$

Equation (2.1-6), which is known as the characteristic equation, has two roots

$$s_{1,2} = -\frac{c}{2m} \pm \sqrt{\left(\frac{c}{2m}\right)^2 - \frac{k}{m}} \qquad (2.1-7)$$

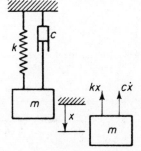

and hence the general solution for the damped free vibration as described by Eq. 2.1-3 is

$$x = Ae^{s_1 t} + Be^{s_2 t} \qquad (2.1-8)$$

where A and B are arbitrary constants depending on how the motion is started.

Critical Damping. The behavior of the damped system of Fig. 2.1-1 depends on the numerical value of the radical of Eq. (2.1-7). As a reference

Fig. 2.1-1. Free vibration with viscous damping.

quantity, we define critical damping as the value of c which reduces this radical to zero, or

$$\frac{c_c}{2m} = \sqrt{k/m} = \omega_n \qquad (2.1-9)$$

$$c_c = 2\sqrt{km} = 2m\omega_n$$

The actual damping of the system can then be specified in terms of the critical damping c_c by the nondimensional ratio

$$\zeta = \frac{c}{c_c} \qquad (2.1-10)$$

which is referred to as the damping ratio. The quantity $c/2m$ can now be expressed in terms of ζ:

$$\frac{c}{2m} = \zeta\frac{c_c}{2m} = \zeta\omega_n \qquad (2.1-11)$$

and Eq. (2.1-7) reduces to

$$s_{1,2} = (-\zeta \pm \sqrt{\zeta^2 - 1})\omega_n \qquad (2.1-12)$$

The cases of interest will now depend on whether ζ is greater, equal to, or less than unity.

2.2 Root Locus Study of Damping

We will introduce here a root locus procedure for the study of Eq. (2.1-12). In this study we vary ζ from zero to some large number and examine the roots s_1 and s_2.

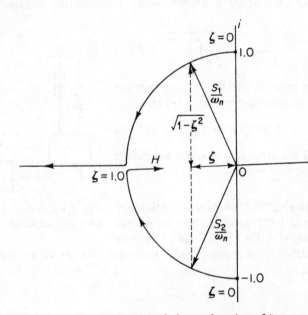

Fig. 2.2-1. Plot of s/ω_n as function of ζ.

If $\zeta = 0$, the roots of Eq. (2.1-12) can be written as

$$s_{1,2} = \pm i\omega_n$$

where $i = \sqrt{-1}$.

For ζ between zero and unity, Eq. (2.1-12) can be written as

$$\frac{s_{1,2}}{\omega_n} = -\zeta \pm i\sqrt{1 - \zeta^2} \qquad (2.2\text{-}1)$$

By plotting the real and imaginary parts of this equation along horizontal and vertical axes, respectively, the locus of all such points for $0 \leq \zeta \leq 1.0$ is found to be a half circle, as shown in Fig. 2.2-1. The two starting points on the vertical axis correspond to the undamped case, $\zeta = 0$. As ζ is increased from zero, the conjugate complex roots move toward the point -1.0 on the horizontal axis. As ζ is increased beyond unity, the two roots separate along the horizontal axis. With this diagram in mind, we are now ready to examine the solution given by Eq. (2.1-8).

(1) **Damping less than critical** $\zeta < 1.0$. Substituting Eq. (2.2-1) into Eq. (2.1-8), the general solution becomes

$$x = e^{-\zeta\omega_n t}(Ae^{i\sqrt{1-\zeta^2}\omega_n t} + Be^{-i\sqrt{1-\zeta^2}\omega_n t})$$

$$= Xe^{-\zeta\omega_n t}\sin(\sqrt{1-\zeta^2}\,\omega_n t + \phi) \qquad (2.2\text{-}2)$$

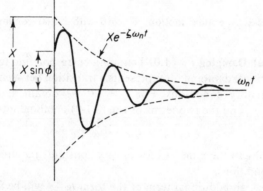

Fig. 2.2-2. Damped oscillations: $\zeta < 1$.

and the motion is oscillatory with diminishing amplitude, as shown in Fig. 2.2-2. The frequency of the damped oscillation

$$\omega_d = \sqrt{1-\zeta^2}\,\omega_n \qquad (2.2\text{-}3)$$

is represented by the ordinate of the root locus plot of Fig. 2.2-1, whereas the horizontal projection ζ is the damping ratio which determines the rate of decay of the amplitude.

(2) **Damping greater than critical** $\zeta > 1.0$. When $\zeta = 1.0$ the two roots of Eq. (2.1-12) approach the point -1.0 on the real axis of Fig. 2.2-1. As ζ exceeds unity, the two roots remain on the real axis but separate, one increasing and the other decreasing. The equation for the general solution then becomes

$$x = Ae^{(-\zeta+\sqrt{\zeta^2-1})\omega_n t} + Be^{(-\zeta-\sqrt{\zeta^2-1})\omega_n t} \qquad (2.2\text{-}4)$$

and the motion is no longer oscillatory, being exponentially decreasing with time, as shown in Fig. 2.2-3. Such motion is referred to as aperiodic.

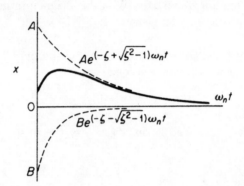

Fig. 2.2-3. Aperiodic motion: $\zeta > 1.0$ with initial conditions x_0 and v_0.

(3) **Critical Damping** $\zeta = 1.0$. Damping corresponding to $\zeta = 1.0$ is referred to as critical damping. It represents a transition between the oscillatory and nonoscillatory conditions. For $\zeta = 1.0$ we obtain a double root at $s_1 = s_2 = -\omega_n$, and the two terms of Eq. (2.1-8) combine into a single term

$$x = (A + B)e^{-\omega_n t} = Ce^{-\omega_n t}$$

which is lacking in the number of arbitrary constants required to satisfy the two initial conditions.

In this case, an additional term of the form $te^{-\omega_n t}$ will be found to satisfy the differential equation of motion, and the general solution

$$x = (A + Bt)e^{-\omega_n t} \quad (2.2\text{-}5)$$

retains the necessary number of arbitrary constants in conformity with the order of the differential equation. For a motion initiated with velocity v_0 from position x_0, the solution becomes

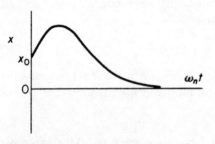

Fig. 2.2-4. Critically damped motion: $\zeta = 1.0$ with initial conditions x_0 and v_0.

$$x = \left[x_0 + \left(\frac{v_0}{\omega_n} + x_0 \right) \omega_n t \right] e^{-\omega_n t}$$

$$(2.2\text{-}6)$$

a graphical representation of which is shown in Fig. 2.2-4. It is evident that the motion is similar to that of aperiodic motion. However, critical damping represents the limit of aperiodic damping, and hence the motion returns to rest in the shortest time without oscillation. The moving parts of many electrical meters and instruments are critically damped to take advantage of this property.

EXAMPLE 2.2-1. A spring-mass system with viscous damping is displaced a distance x_0 and released. Determine the equation of motion when (a) $\zeta = 2$, (b) $\zeta = 0.20$, and (c) $\zeta = 1.0$, and compare the three cases by plotting the results in nondimensional form.

Solution. Part (a). $\zeta = 2$ leads to aperiodic motion which is represented by Eq. (2.2-4):

$$x = Ae^{(-2+\sqrt{3})\omega_n t} + Be^{(-2-\sqrt{3})\omega_n t}$$

The arbitrary constants A and B are determined from initial conditions, which are as follows:

$$(x)_{t=0} = A + B = x_0$$
$$(\dot{x})_{t=0} = (-2 + \sqrt{3})\omega_n A + (-2 - \sqrt{3})\omega_n B = 0$$

Solving for A and B,

$$A = \left(\frac{2 + \sqrt{3}}{2\sqrt{3}}\right)x_0 = 1.08x_0$$

$$B = \left(\frac{-2 + \sqrt{3}}{2\sqrt{3}}\right)x_0 = -0.0774x_0$$

The equation of motion then becomes

$$\frac{x}{x_0} = 1.08e^{-0.268\omega_n t} - 0.0774e^{-3.73\omega_n t}$$

Part (b). The motion for $\zeta = 0.20$ is oscillatory and represented by Eq. (2.2-2):

$$x = Xe^{-0.2\omega_n t}\sin\left(\sqrt{0.96}\,\omega_n t + \phi\right)$$

X and ϕ are determined from initial conditions, as before:

$$(x)_{t=0} = X\sin\phi = x_0$$

$$(\dot{x})_{t=0} = X[\sqrt{0.96}\,\omega_n \cos\phi - 0.2\omega_n \sin\phi] = 0$$

From the second equation we obtain

$$\phi = \tan^{-1}\frac{\sqrt{0.96}}{0.20} = 78°28'$$

Sin ϕ must then equal $\sqrt{0.96}$, from which we obtain

$$X = \frac{x_0}{\sqrt{0.96}}$$

The equation of motion can then be written in the following nondimensional form:

$$\frac{x}{x_0} = \frac{1}{\sqrt{0.96}} e^{-0.2\omega_n t} \sin(\sqrt{0.96}\,\omega_n t + 78°28')$$

Part (c). $\zeta = 1.0$ represents a critically damped case, and Eq. 2.2-5 becomes applicable:

$$x = (A + Bt)e^{-\omega_n t}$$

Substituting the initial conditions,

$$(x)_{t=0} = A = x_0, \qquad (\dot{x})_{t=0} = -\omega_n A + B = 0$$

from which we obtain

$$A = x_0, \qquad B = \omega_n x_0$$

The equation of motion can then be written as

$$\frac{x}{x_0} = (1 + \omega_n t)e^{-\omega_n t}$$

which could have been deduced from Eq. (2.2-6) with $v_0 = 0$.

The equation of motion obtained for each value of ζ is plotted in Fig. 2.2-5, with x/x_0 as ordinate and $\omega_n t$ as abscissa.

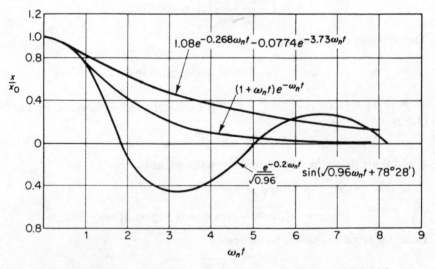

Fig. 2.2-5. Free vibration with $\zeta = 2.0$, 1.0, and 0.20, with initial conditions x_0, and $v_0 = 0$.

EXAMPLE 2.2-2. Large guns are designed so that, on firing, the barrel recoils against a spring. At the end of the recoil, a dashpot is engaged that allows the barrel to return to its initial position in the minimum time without oscillation. Determine the proper spring constant and dashpot damping coefficient for a barrel weighing 1610 lb. if the initial recoil velocity at the instant of firing is 80 ft./sec., and the recoil distance is 5 ft.

Solution. Since the dashpot is not operative during the recoil, the initial kinetic energy is equal to the work done on the spring:

$$\frac{1}{2} \cdot \frac{1610}{32.2} \cdot 80^2 = \frac{1}{2} \cdot k \cdot 5^2$$

$$k = 12,800 \text{ lb./ft.}$$

To return the barrel to its original position in the least time without oscillation, the dashpot should be designed for critical damping:

$$c_c = 2m\omega_n = 2\sqrt{mk} = 2\sqrt{\frac{1610}{32.2} \cdot 12,800} = 1600 \text{ lb./ft. per sec.}$$

2.3 Logarithmic Decrement

A convenient way to determine the amount of damping present in a system is to measure the rate of decay of oscillation. This is conveniently expressed by a term called *logarithmic decrement*, which is defined as the natural logarithm of the ratio of any two successive amplitudes.

Consider a damped vibration expressed by Eq. (2.2-2) and represented graphically in Fig. 2.3-1.

$$x = Xe^{-\zeta\omega_n t} \sin\left(\sqrt{1 - \zeta^2}\, \omega_n t + \phi\right)$$

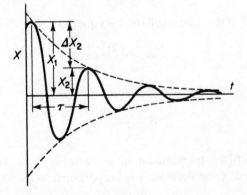

Fig. 2.3-1. Rate of decay of oscillation is measured by the logarithmic decrement.

When $\sin (\sqrt{1 - \zeta^2}\,\omega_n t + \phi) = 1.0$, the curve is tangent to the exponential envelope $Xe^{-\zeta\omega_n t}$; however, the tangents are not horizontal, and the points of tangency appear slightly to the right of the point of maximum amplitude. Generally this discrepancy is negligible, and the amplitude at the point of tangency may be taken equal to the maximum amplitude.

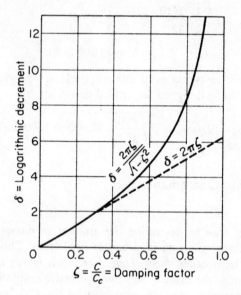

Fig. 2.3-2. Logarithmic decrement as function of ζ.

The logarithmic decrement δ is then expressed mathematically as

$$\delta = \ln \frac{x_1}{x_2} = \ln \frac{e^{-\zeta\omega_n t_1}}{e^{-\zeta\omega_n(t_1 + \tau)}} = \ln e^{\zeta\omega_n \tau} = \zeta\omega_n \tau \,. \qquad (2.3\text{-}1)$$

Since the period of damped oscillation is equal to

$$\tau = \frac{2\pi}{\omega_n \sqrt{1 - \zeta^2}} \qquad (2.3\text{-}2)$$

the decrement can also be written in terms of ζ as

$$\delta = \frac{2\pi\zeta}{\sqrt{1 - \zeta^2}} \qquad (2.3\text{-}3)$$

δ as a function of ζ is represented graphically in Fig. 2.3-2. It is evident that for small values of ζ the decrement is proportional to ζ and is given by

$$\delta \cong 2\pi\zeta \qquad (2.3\text{-}4)$$

EXAMPLE 2.3-1. The following data are given for a vibrating system with viscous damping: $w = 10$ lb., $k = 30$ lb./in., and $c = 0.12$ lb./in. per sec. Determine the logarithmic decrement and the ratio of any two successive amplitudes.

Solution. The undamped natural frequency of the system in radians per second is

$$\omega_n = \sqrt{\frac{k}{m}} = \sqrt{\frac{30 \times 386}{10}} = 34.0 \text{ rad./sec.}$$

The critical damping coefficient c_c and damping factor ζ are

$$c_c = 2m\omega_n = 2 \times \frac{10}{386} \times 34.0 = 1.76 \text{ lb./in. per sec.}$$

$$\zeta = \frac{c}{c_c} = \frac{0.12}{1.76} = 0.0681$$

The logarithmic decrement, from Eq. (2.3-3), is

$$\delta = \frac{2\pi\zeta}{\sqrt{1 - \zeta^2}} = \frac{2\pi \times 0.0681}{\sqrt{1 - 0.0681^2}} = 0.429$$

The amplitude ratio for any two consecutive cycles is

$$\frac{x_1}{x_2} = e^\delta = e^{0.429} = 1.54$$

EXAMPLE 2.3-2. Show that the logarithmic decrement is also given by the equation

$$\delta = \frac{1}{n} \ln \frac{x_0}{x_n}.$$

where x_n represents the amplitude after n cycles have elapsed. Plot a curve giving the number of cycles elapsed against ζ for the amplitude to diminish by 50 per cent.

Solution. The amplitude ratio for any two consecutive amplitudes is

$$\frac{x_0}{x_1} = \frac{x_1}{x_2} = \frac{x_2}{x_3} = \cdots \frac{x_{n-1}}{x_n} = e^\delta$$

The ratio x_0/x_n can be written as:

$$\frac{x_0}{x_n} = \left(\frac{x_0}{x_1}\right)\left(\frac{x_1}{x_2}\right)\left(\frac{x_2}{x_3}\right) \cdots \left(\frac{x_{n-1}}{x_n}\right) = (e^\delta)^n = e^{n\delta}$$

from which the required equation is obtained as

$$\delta = \frac{1}{n} \ln \frac{x_0}{x_n}$$

To determine the number of cycles elapsed for 50 per cent reduction in amplitude, we obtain the following relation from the above equation:

$$\delta \cong 2\pi\zeta = \frac{1}{n} \ln 2 = \frac{0.693}{n}$$

$$n\zeta = \frac{0.693}{2\pi} = 0.110$$

The last equation is that of a rectangular hyberbola, and is plotted in Fig. 2.3-3.

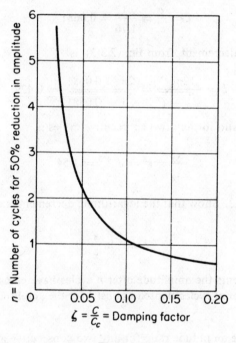

Fig. 2.3-3.

EXAMPLE 2.3-3. For small damping, show that the logarithmic decrement is expressible in terms of the vibrational energy U and the energy dissipated per cycle ΔU.

Solution. Figure 2.3-4 shows a damped vibration with consecutive amplitudes $x_1, x_2, x_3, \cdots$. From the definition of the logarithmic decrement

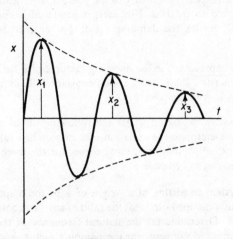

Fig. 2.3-4.

$\delta = \ln x_1/x_2$, we can write the ratio of amplitudes in exponential fo.m:

$$\frac{x_2}{x_1} = e^{-\delta} = 1 - \delta + \frac{\delta^2}{2!} - \cdots$$

The vibrational energy of the system is that stored in the spring at maximum displacement, or

$$U_1 = \tfrac{1}{2}kx_1^2, \qquad U_2 = \tfrac{1}{2}kx_2^2$$

The loss of energy divided by the original energy is

$$\frac{U_1 - U_2}{U_1} = 1 - \frac{U_2}{U_1} = 1 - \left(\frac{x_2}{x_1}\right)^2 = 1 - e^{-2\delta} = 2\delta - \frac{(2\delta)^2}{2!} + \cdots$$

Thus for small δ we obtain the relationship

$$\frac{\Delta U}{U} = 2\delta$$

PROBLEMS

1. A weight of 2 lb. is attached to the end of a spring with a stiffness of 4 lb./in. Determine the critical damping coefficient.

2. If in Prob. 1 the damping coefficient $c = 0.144$ lb./in. per sec., determine the damped natural frequency and compare it with the natural frequency of the undamped system.

3. To calibrate a dashpot, the velocity of the plunger was measured when a given force was applied to it. If a $\frac{1}{2}$-lb. weight produced a constant velocity of 1.20 in./sec., determine the damping factor ζ when used with the system of Prob. 1.

4. Determine the expression for the damping factor ζ of a torsional system with a moment of inertia J lb. in. sec.^{2}, spring constant K lb. in./rad., and damping coefficient η lb. in./rad. per sec.

5. A vibrating system is started under the following initial conditions: $x = 0$, $\dot{x} = v_0$, $t = 0$. Determine the equation of motion when (a) $\zeta = 2.0$, (b) $\zeta = 0.50$, (c) $\zeta = 1.0$. Plot nondimensional curves for the three cases with $\omega_n t$ as abscissa and $x\omega_n/v_0$ as ordinate.

6. A vibrating system consisting of a weight of 5 lb. and a spring of stiffness 10 lb./in. is viscously damped such that the ratio of any two consecutive amplitudes is 1.00 to 0.98. Determine (a) the natural frequency of the damped system, (b) the logarithmic decrement, (c) the damping factor, and (d) the damping coefficient.

7. A body vibrating in a viscous medium has a period of 0.20 sec. and an initial amplitude of 1.0 in. Determine the logarithmic decrement if the amplitude after 10 cycles is 0.02 in.

8. Plot curves similar to that of Fig. 2.3-3 for the number of cycles elapsed for the amplitude to decay to (a) 20 per cent of the initial value, (b) 80 per cent of the initial value.

9. A vibrating system consists of a weight of 10 lb., a spring of stiffness 20 lb./in., and a dashpot with a damping coefficient of 0.071 lb./in. per sec. Find (a) the damping factor, (b) the logarithmic decrement, and (c) the ratio of any two consecutive amplitudes.

10. A 12.2-lb. weight is viscously damped and is suspended from a spring of stiffness 5 lb./in. If the amplitude of vibration dies down to one-half the initial value in 10 sec., determine the damping factor.

11. A vibrating system has the following constants: $w = 38.6$ lb., $k = 40$ lb./in., and $c = 0.40$ lb./in. per sec. Determine (a) the damping factor, (b) the natural frequency of damped oscillation, (c) the logarithmic decrement, and (d) the ratio of any two consecutive amplitudes.

12. Set up the differential equation of motion for the system shown in Fig. 2-1. Determine the expression for (a) the critical damping coefficient, and (b) the natural frequency of damped oscillation.

13. Write the differential equation of motion for the system shown in Fig. 2-2, and determine the natural frequency of damped oscillation and the critical damping coefficient.

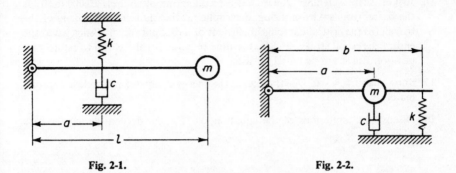

Fig. 2-1.   Fig. 2-2.

14. A thin plate of area A and weight W is attached to the end of a spring and is allowed to oscillate in a viscous fluid, as shown in Fig. 2-3. If τ_1 is the natural period of undamped oscillation, that is, with the system oscillating in air, and τ_2 the damped period with the plate immersed in the fluid, show that

$$\mu = \frac{2\pi W}{g A \tau_1 \tau_2} \sqrt{\tau_2^2 - \tau_1^2}$$

where the damping force on the plate is $F_d = \mu 2Av$, $2A$ is the total surface area of the plate, and v is its velocity.

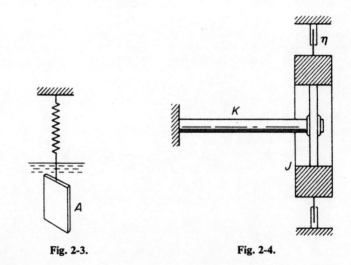

Fig. 2-3.   Fig. 2-4.

15. A torsional system J, K of Fig. 2-4 having a natural frequency f_1 in air was immersed in oil wherein the natural frequency was lowered to f_2. Determine the damping coefficient η lb. in./rad. per sec. of the system.

16. A gun barrel weighing 1200 lb. has a recoil spring of stiffness 20,000 lb./ft. If the barrel recoils 4 ft. on firing, determine (a) the initial recoil velocity of the barrel, (b) the critical damping coefficient of a dashpot which is engaged at the end of the recoil stroke, and (c) the time required for the barrel to return to a position 2 in. from its initial position.

17. Show that if $\zeta = 0.02$, the fraction of the energy dissipated per cycle is approximately $\frac{1}{4}$.

18. Discuss the limitations of the equation $\Delta U/U = 2\delta$ by considering the case where $x_2/x_1 = \frac{1}{2}$.

3

Forced Vibration with Harmonic Excitation

3.1 Steady State Solution with Viscous Damping

We consider here a viscously damped spring-mass system excited by a harmonic force $F_0 \sin \omega t$ as shown in Fig. 3.1-1. The differential equation of motion can then be written as

$$m\ddot{x} + c\dot{x} + kx = F_0 \sin \omega t \qquad (3.1\text{-}1)$$

where ω is the frequency of the harmonic excitation.

The solution of the above equation can be considered in two parts, the homogeneous solution and the particular solution. Because of the initial conditions at $t = 0$, the homogeneous equation, which has already been studied in the previous chapter, results in damped free vibrations at its natural frequency. This part of the solution soon disappears, leaving only the particular solution which is a steady state harmonic oscillation at the frequency of the exciting force with the displacement vector lagging the force vector by a phase angle ϕ. We can therefore assume the particular solution to be

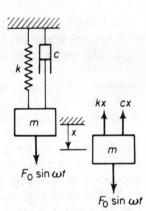

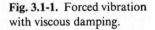

Fig. 3.1-1. Forced vibration with viscous damping.

$$x = X \sin (\omega t - \phi) \qquad (3.1\text{-}2)$$

The amplitude X and the phase angle ϕ are found by substituting the above solution into the differential equation of motion. Shifting all the terms of the differential equation to the right side of the equation, and substituting from the above, we obtain the following vector relation:

Inertia force + damping force + spring force + impressed force $= 0$

$$m\omega^2 X \sin(\omega t - \phi) - c\omega X \sin\left(\omega t - \phi + \frac{\pi}{2}\right)$$

$$- kX \sin(\omega t - \phi) + F_0 \sin \omega t = 0$$

This vector relation, depicted graphically in Fig. 3.1-2, shows that:

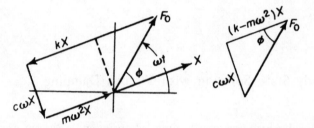

Fig. 3.1-2. Vector representation of forced vibration with viscous damping.

1. The displacement lags the impressed force by the angle ϕ, which can vary between 0 and 180 deg.

2. The spring force is always opposite in direction to the displacement.

3. The damping force lags the displacement by 90 deg. and hence is opposite in direction to the velocity.

4. The inertia force is in phase with the displacement and opposite in direction to the acceleration.

The above information is thus in agreement with the physical interpretation of harmonic motion. The vectors remain fixed with respect to each other and rotate together with angular velocity ω.

To determine expressions for X and ϕ, we consider the right triangle of Fig. 3.1-2 with F_0 as the hypotenuse, from which we obtain

$$X = \frac{F_0}{\sqrt{(k - m\omega^2)^2 + (c\omega)^2}} \qquad (3.1\text{-}3)$$

$$\tan \phi = \frac{c\omega}{k - m\omega^2} \qquad (3.1\text{-}4)$$

The complete solution, including the transient term, is then given by the following equation:

$$x = X_1 e^{-\zeta \omega_n t} \sin (\sqrt{1 - \zeta^2}\, \omega_n t + \phi_1) + \frac{F_0 \sin (\omega t - \phi)}{\sqrt{(k - m\omega^2)^2 + (c\omega)^2}} \quad (3.1\text{-}5)$$

For convenience of discussion, the steady-state solution may be reduced to nondimensional form. Dividing the numerator and denominator by k, we obtain

$$X = \frac{F_0/k}{\sqrt{\left(1 - \dfrac{m\omega^2}{k}\right)^2 + \left(\dfrac{c\omega}{k}\right)^2}} \quad (3.1\text{-}6)$$

$$\tan \phi = \frac{c\omega/k}{1 - \dfrac{m\omega^2}{k}} \quad (3.1\text{-}7)$$

The terms in the equations above may be further reduced in terms of the following quantities:

$\omega_n = \sqrt{k/m}$ = natural frequency of undamped oscillation in radians per second

$\zeta = c/c_c$ = damping factor

$c_c = 2m\omega_n$ = critical damping coefficient

$X_0 = F_0/k$ = zero frequency deflection of the spring-mass system under the action of a steady force F_0 (not to be confused with statical deflection $\Delta = W/k$).

The nondimensional form of X and ϕ then becomes

$$\frac{X}{X_0} = \frac{1}{\sqrt{[1 - (\omega/\omega_n)^2]^2 + [2\zeta(\omega/\omega_n)]^2}} \quad (3.1\text{-}8)$$

$$\tan \phi = \frac{2\zeta(\omega/\omega_n)}{1 - (\omega/\omega_n)^2} \quad (3.1\text{-}9)$$

The term X/X_0 called the *magnification factor*, represents the factor by which the zero frequency deflection must be multiplied to determine the amplitude X. These equations indicate that X/X_0 and ϕ are functions only of the frequency ratio ω/ω_n and the damping factor ζ, and can be plotted as shown in Fig. 3.1-3.

These curves indicate that the damping factor has a large influence on the amplitude and phase angle in the region of resonance. Further understanding

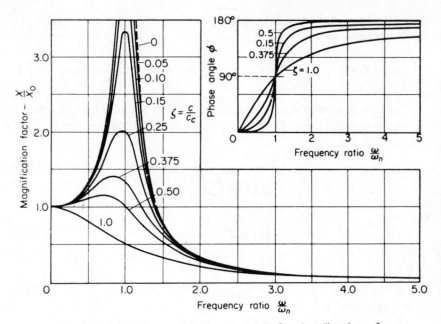

Fig. 3.1-3. Plot of Eqs. (3.1-8) and (3.1-9) for the vibration of a viscously damped system.

of the behavior of the system may be obtained by studying the vector force diagram in the three regions of ω/ω_n.

For small values of $\omega/\omega_n \ll 1.0$, both the inertia and damping terms are small, which results in a small phase angle ϕ. The magnitude of the impressed force is then nearly equal to the spring force, as shown in Fig. 3.1-4.

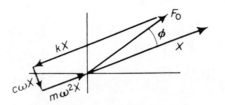

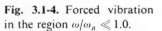

Fig. 3.1-4. Forced vibration in the region $\omega/\omega_n \ll 1.0$.

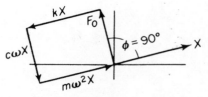

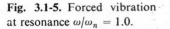

Fig. 3.1-5. Forced vibration at resonance $\omega/\omega_n = 1.0$.

For $\omega/\omega_n = 1.0$, the phase angle is 90 deg. and the vector diagram appears as shown in Fig. 3.1-5. The inertia force, which is now larger, is balanced by the spring force, whereas the impressed force overcomes the damping force. The amplitude at resonance can be obtained, either from Eqs. (3.1-3) and (3.1-8) or from the vector diagram of Fig. 3.1-5, as

$$X = \frac{F_0}{c\omega_n} = \frac{X_0}{2\zeta} \qquad (3.1\text{-}10)$$

At large values of $\omega/\omega_n \gg 1.0$, ϕ approaches 180 deg., and the impressed force is expended almost entirely in overcoming the large inertia force, as shown in Fig. 3.1-6.

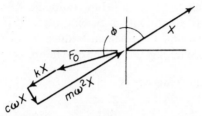

Fig. 3.1-6. Forced vibration in the region $\omega/\omega_n \gg 1.0$.

EXAMPLE 3.1-1. The amplitude and phase of the transient term depend on how the motion is started, whereas the amplitude and phase of the steady-state term are independent of the initial conditions. For instance, if the motion is started with zero displacement and velocity, and with the impressed frequency equal to $\omega = \omega_n = \sqrt{k/m}$, we obtain

$$\phi = \tan^{-1} \infty = \pi/2$$

The expression for the velocity is obtained by differentiating Eq. (3.1-5), for the condition $\omega = \omega_n$:

$$\dot{x} = X_1 e^{-\zeta\omega_n t}[\omega_n\sqrt{1 - \zeta^2} \cos(\sqrt{1 - \zeta^2}\, \omega_n t + \phi_1)$$
$$- \omega_n\zeta \sin(\sqrt{1 - \zeta^2}\, \omega_n t + \phi_1)] + \frac{F_0 \sin \omega_n t}{c}$$

Since the initial velocity is equal to zero,

$$(\dot{x})_{t=0} = \omega_n X_1[\sqrt{1 - \zeta^2} \cos \phi_1 - \zeta \sin \phi_1] = 0$$

$$\tan \phi_1 = \frac{\sqrt{1 - \zeta^2}}{\zeta}, \qquad \sin \phi_1 = \sqrt{1 - \zeta^2}$$

(See Fig. 3.1-7.) Setting the initial displacement equal to zero,

$$(x)_{t=0} = X_1 \sin \phi_1 - \frac{F_0}{c\omega_n} = 0, \qquad X_1 = \frac{F_0}{c\omega_n\sqrt{1 - \zeta^2}}$$

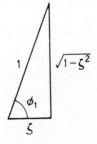

Fig. 3.1-7.

The final solution satisfying the given initial conditions and $\omega = \omega_n$ is then

$$x = \frac{F_0}{c\omega_n\sqrt{1-\zeta^2}}\, e^{-\zeta\omega_n t} \sin\left(\sqrt{1-\zeta^2}\,\omega_n t + \sin^{-1}\sqrt{1-\zeta^2}\right) - \frac{F_0\cos\omega_n t}{c\omega_n}$$

The last equation represents an oscillation that starts with zero amplitude and velocity and builds up to a steady oscillation of $(F_0\cos\omega_n t)/c\omega_n$.

EXAMPLE 3.1-2. A machine weighing 193 lb. is supported on springs of total stiffness 200 lb. per in. If a harmonic disturbing force of magnitude 10 lb. acts on the machine, determine the resonant frequency and resonant amplitude, assuming a viscous damping coefficient of 6 lb./in. per sec.

Solution. The resonant frequency is

$$f_1 = \frac{1}{2\pi}\sqrt{\frac{k}{m}} = \frac{1}{2\pi}\sqrt{\frac{200 \times 386}{193}} = 3.18 \text{ c.p.s.}$$

From Eq. (3.1-10), the resonant amplitude is

$$X = \frac{F}{c\omega_n} = \frac{10}{6 \times 2\pi \times 3.18} = 0.0836 \text{ in.}$$

EXAMPLE 3.1-3. Derive the expressions for the peak amplitude and the phase angle corresponding to the peak amplitude. Determine numerical values for these quantities for the system of Example 3.1-2.

Solution. The frequency ratio corresponding to the peak amplitude can be obtained from Eq. (3.1-8) by minimizing the denominator, which gives

$$\frac{\omega}{\omega_n} = \sqrt{1 - 2\zeta^2}$$

Substituting into Eqs. (3.1-8) and (3.1-9), the peak amplitude and the phase angle can be expressed in terms of the damping factor ζ as

$$\frac{X}{X_0} = \frac{1}{2\zeta\sqrt{1-\zeta^2}}, \qquad \tan\phi = \frac{1}{\zeta}\sqrt{1 - 2\zeta^2}$$

For the system of Example 3.1-2, the damping factor is determined as follows:

$$c_c = 2m\omega_n = 2 \times \tfrac{193}{386} \times 2\pi \times 3.18 = 20 \text{ lb./in. per sec.}$$

$$\zeta = \frac{c}{c_c} = \frac{6}{20} = 0.30$$

Substituting into the derived equations, we obtain

$$X = \frac{1}{2 \times 0.30\sqrt{1 - 0.30^2}} \times \frac{10}{200} = 0.0875 \text{ in.}$$

$$\phi = \tan^{-1} \frac{\sqrt{1 - 2 \times 0.30^2}}{0.30} = \tan^{-1} 3.02 = 71°40'$$

3.2 Method of Complex Algebra

The use of complex algebra often simplifies the procedure for solving the differential equation. In forced vibration problems where the impressed force is harmonic, the steady-state solution is also harmonic with the same frequency. As shown in Fig. 3.1-2, the impressed force and the resulting displacement are found to be vectors F_0 and X, the latter lagging the former by the angle φ, and the two rotating together with a common angular speed ω. If such vectors are represented by the exponential function, we can write

$$F = F_0 e^{i\omega t} \tag{3.2-1}$$

$$x = X e^{i(\omega t - \varphi)} = X e^{-i\varphi} e^{i\omega t} \tag{3.2-2}$$

In these equations F_0 and X are absolute values equal to the length of the vectors. It is also possible to write Eq. (3.2-2) as

$$x = \overline{X} e^{i\omega t} \tag{3.2-3}$$

where $\overline{X} = X e^{-i\varphi}$ is the complex amplitude designating its angular position with respect to F_0. This latter form is found to be convenient in establishing the amplitude and phase of the steady-state solution in many oscillatory systems.

Applying this type of analysis to the forced vibration problem of the preceding section, we can write the differential equation in the form

$$m\ddot{x} + c\dot{x} + kx = F_0 e^{i\omega t} \tag{3.2-4}$$

Letting $x = \overline{X} e^{i\omega t}$, the preceding equation becomes

$$(-m\omega^2 + i\omega c + k)\overline{X} = F_0 \tag{3.2-5}$$

The complex amplitude is then determined as

$$\overline{X} = \frac{F_0}{(k - m\omega^2) + ic\omega} = \frac{F_0 e^{-i\varphi}}{\sqrt{(k - m\omega^2)^2 + (\omega c)^2}} \tag{3.2-6}$$

from which the amplitude and phase are found to be

$$X = \frac{F_0}{\sqrt{(k - m\omega^2)^2 + (\omega c)^2}}$$
(3.2-7)

$$\varphi = \tan^{-1}\frac{\omega c}{k - m\omega^2}$$
(3.2-8)

3.3 Reciprocating and Rotating Unbalance

A very common source of periodically varying impressed force is the reciprocating or the rotating machine. Electric motors, turbines, and automobile engines are a few examples for the origin of such forces.

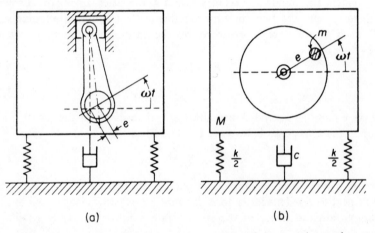

Fig. 3.3-1. Periodic disturbing force resulting from a reciprocating or rotating unbalance.

Figure 3.3-1 shows two elastically mounted systems with similar exciting force. In (a) the excitation is supplied by the reciprocating motion of the piston, whereas in (b) the rotation of an unbalanced disk or wheel becomes the source for the exciting force. Both sources of excitation are of the inertia type and result in the same form of impressed force on the system.

Consider the system represented by Fig. 3.3-1(b), where the exciting force is supplied by the rotation of an eccentric mass m with eccentricity e. If m rotates with angular velocity ω, its vertical displacement becomes

$$x + e \sin \omega t$$

where x represents the motion of the spring-supported mass. Letting M represent the total mass including m, the equation of motion for the system may be written as follows:

$$(M - m)\frac{d^2x}{dt^2} + m\frac{d^2}{dt^2}(x + e \sin \omega t) = -kx - c\frac{dx}{dt} \qquad (3.3\text{-}1)$$

Rearranging terms, Eq. (3.3-1) reduces to

$$M\frac{d^2x}{dt^2} + c\frac{dx}{dt} + kx = (me\omega^2)\sin \omega t \qquad (3.3\text{-}2)$$

Equation (3.3-2) is identical to Eq. (3.1-1) when the force amplitude F_0 is replaced by the inertia force $(me\omega^2)$. The solution is thus of the same form as Eqs. (3.1-3) and (3.1-4) and may be written as

$$X = \frac{me\omega^2}{\sqrt{(k - M\omega^2)^2 + (c\omega)^2}} \qquad (3.3\text{-}3)$$

$$\tan \phi = \frac{c\omega}{k - M\omega^2} \qquad (3.3\text{-}4)$$

If the numerator and denominator of the above equations are divided by k, and $\sqrt{k/M}$ is replaced by ω_n, the natural frequency of the system, Eqs. (3.3-3) and (3.3-4) may be rewritten as

$$X = \frac{\dfrac{m}{M}e\left(\dfrac{\omega}{\omega_n}\right)^2}{\sqrt{\left[1 - \left(\dfrac{\omega}{\omega_n}\right)^2\right]^2 + \left[2\zeta\dfrac{\omega}{\omega_n}\right]^2}} \qquad (3.3\text{-}5)$$

$$\tan \phi = \frac{2\zeta\dfrac{\omega}{\omega_n}}{1 - \left(\dfrac{\omega}{\omega_n}\right)^2} \qquad (3.3\text{-}6)$$

In representing Eqs. (3.3-5) and (3.3-6) graphically, it is convenient to plot MX/me and ϕ against ω/ω_n for various values of the damping factor ζ, as shown in Fig. 3.3-2. At low speeds, the exciting force $me\omega^2$ is small, and the curve starts at zero. When $\omega/\omega_n = 1.0$, $MX/me = 1/2\zeta$, and the amplitude is limited only by the damping present. When ω/ω_n is very large, the ratio MX/me approaches 1.0, and the mass $(M - m)$ has an amplitude $X = me/M$ 180 deg. out of phase with m. Since me/M is the distance to the center of gravity of the total mass M, at high speeds the spring-supported mass moves in such a way that its center of gravity remains stationary.

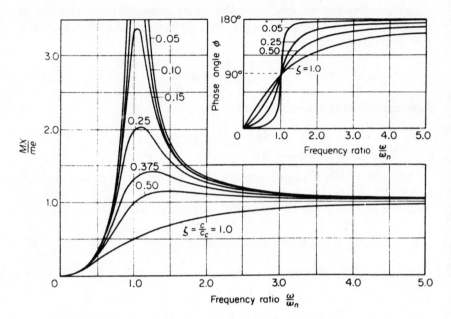

Fig. 3.3-2. Plot of Eqs. (3.3-5) and (3.3-6) for the forced vibration of a viscously damped system with a reciprocating or rotating unbalance.

EXAMPLE 3.3-1. A counterrotating eccentric weight exciter is used to produce forced oscillation of a spring-supported mass, as shown in Fig. 3.3-3. By varying the speed of rotation, a resonant amplitude of 0.60 in. was recorded. When the speed of rotation was increased considerably beyond the resonant frequency, the amplitude appeared to approach a fixed value of 0.08 in. Determine the damping factor of the system.

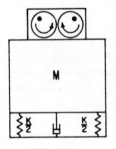

Fig. 3.3-3.

Solution. From Eq. (3.3-5), the resonant amplitude is

$$X = \frac{me/M}{2\zeta} = 0.60 \text{ in.}$$

When ω is very much greater than ω_n, the same equation becomes

$$X = \frac{me}{M} = 0.08 \text{ in.}$$

Solving the two equations simultaneously, the damping factor of the system is

$$\zeta = \frac{0.08}{2 \times 0.60} = 0.0666$$

3.4 Base Excitation

A vibratory system is sometimes excited by a prescribed motion of some point in the system. If we let y be the motion of the support point of the

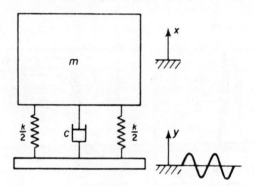

Fig. 3.4-1. Base excitation of a spring-mass system.

system of Fig. 3.4-1, the differential equation becomes

$$m\ddot{x} = -k(x - y) - c(\dot{x} - \dot{y})$$

which may be rearranged to

$$m\ddot{x} + c\dot{x} + kx = ky + c\dot{y} \tag{3.4-1}$$

Using the method of complex algebra, we let

$$y = Ye^{i\omega t} \qquad x = Xe^{i(\omega t - \phi)} = Xe^{-i\phi}e^{i\omega t} \tag{3.4-2}$$

Substituting these into the differential equation, we obtain

$$(-m\omega^2 + i\omega c + k)Xe^{-i\phi} = (k + i\omega c)Y \tag{3.4-3}$$

from which the amplitude ratio is

$$\frac{Xe^{-i\phi}}{Y} = \frac{k + i\omega c}{(k - m\omega^2) + i\omega c} \tag{3.4-4}$$

The absolute value of the amplitude ratio is then

$$\left|\frac{X}{Y}\right| = \sqrt{\frac{k^2 - (c\omega)^2}{(k - m\omega^2)^2 + (c\omega)^2}} = \sqrt{\frac{1 + (2\zeta\,\omega/\omega_n)^2}{[1 - (\omega/\omega_n)^2]^2 + (2\zeta\,\omega/\omega_n)^2}} \quad (3.4\text{-}5)$$

To find the phase angle ϕ, we put $e^{-i\phi} = \cos\phi - i\sin\phi$, and equate the real and imaginary parts of Eq. (3.4-3) to determine $\sin\phi$ and $\cos\phi$. The ratio then results in the equation for the phase angle, which is

$$\tan\phi = \frac{mc\omega^3}{k^2[1 - (\omega/\omega_n)^2] + (c\omega)^2} = \frac{2\zeta(\omega/\omega_n)^3}{1 - (\omega/\omega_n)^2 + (2\zeta\,\omega/\omega_n)^2} \quad (3.4\text{-}6)$$

Equations (3.4-5) and (3.4-6) are plotted in Fig. 3.4-2.

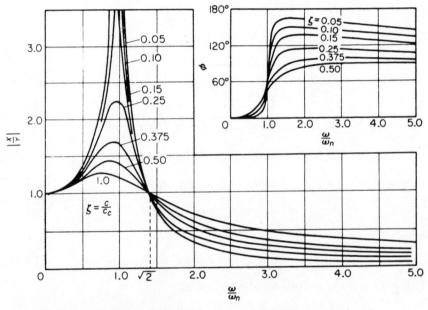

Fig. 3.4-2. Plot of Eqs. (3.4-5) and (3.4-6) for base excitation.

3.5 Vibration Isolation

Vibratory forces generated by machines and engines are often unavoidable; however, they can be reduced substantially by properly designed springs, which we can refer to as isolators.

In Fig. 3.5-1, let $F_0 \sin \omega t$ be the exciting source to be isolated by the springs. The transmitted force through the springs and damper, shown in Fig. 3.5-2, is

$$F_T = \sqrt{(kX)^2 + (c\omega X)^2} = kX\sqrt{1 + \left(\frac{c\omega}{k}\right)^2} \qquad (3.5\text{-}1)$$

Since the amplitude X developed under the force $F_0 \sin \omega t$ is given by Eq.

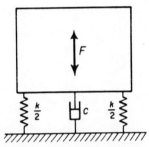

Fig. 3.5-1. Force is transmitted through springs and damper.

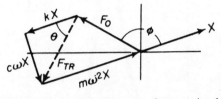

Fig. 3.5-2. Vector diagram of transmitted force.

(3.1-6), the above equation reduces to

$$F_T = \frac{F_0\sqrt{1 + \left(\dfrac{c\omega}{k}\right)^2}}{\sqrt{\left[1 - \dfrac{m\omega^2}{k}\right]^2 + \left(\dfrac{c\omega}{k}\right)^2}} = \frac{F_0\sqrt{1 + \left(2\zeta\dfrac{\omega}{\omega_n}\right)^2}}{\sqrt{\left[1 - \left(\dfrac{\omega}{\omega_n}\right)^2\right]^2 + \left(2\zeta\dfrac{\omega}{\omega_n}\right)^2}} \qquad (3.5\text{-}2)$$

Comparison of Eqs. (3.5-2) and (3.4-5) indicates that $|F_T/F_0|$ is identical to $|X/Y| = |\omega^2 X/\omega^2 Y|$. Thus the problem of isolating a mass from a base motion is identical to that of isolating disturbing forces. Each of these ratios is referred to as transmissibility and is plotted in Fig. 3.5-3 for various values of ζ. These curves show that the transmissibility is less than unity only for $\omega/\omega_n > \sqrt{2}$, thereby establishing the fact that vibration isolation is possible only for $\omega/\omega_n > \sqrt{2}$. The results also indicate that in the region $\omega/\omega_n > \sqrt{2}$ an undamped spring is superior to a damped spring in reducing the transmissibility; however, some damping may be desirable when it is necessary for ω to pass through the resonant region, although the large amplitude at resonance can be limited by stops.

When the damping is negligible the transmissibility equation reduces to the following

$$TR = \frac{1}{(\omega/\omega_n)^2 - 1} \tag{3.5-3}$$

where it is understood that the value of ω/ω_n to be used in the above equation is always greater than $\sqrt{2}$. Equation (3.5-3) is represented in Fig. 3.5-3 by the curve for $\zeta = 0$.

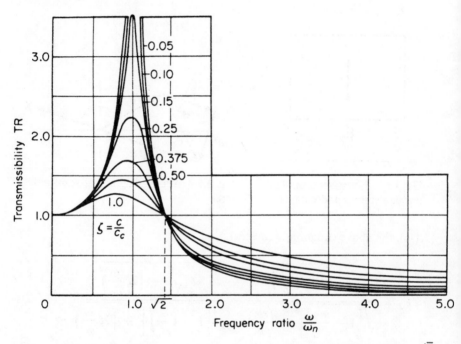

Fig. 3.5-3. Transmissibility is less than 1.0 for frequency ratio $\omega/\omega_n > \sqrt{2}$.

In Fig. 3.5-4 is shown another curve that is useful in problems of isolation. It is obtained from Eq. (3.5-3) by replacing ω_n^2 with g/Δ, where Δ is the statical deflection of the system. The transmissibility then becomes

$$TR = \frac{1}{\dfrac{(2\pi f)^2 \Delta}{g} - 1} \tag{3.5-4}$$

Solving for f in cycles per minute, we obtain the following equations for the curves:

$$f = 188\sqrt{\frac{1}{\Delta}\left(\frac{1}{TR} + 1\right)}, \quad f = 188\sqrt{\frac{1}{\Delta}\left(\frac{2 - R}{1 - R}\right)} \quad (3.5\text{-}5)$$

where the per cent reduction in the transmitted vibration is defined as $R = (1 - TR)$.

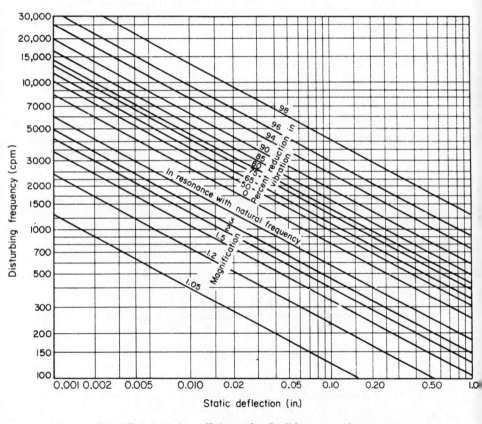

Fig. 3.5-4. Isolation efficiency for flexibly mounted system.

The discussion so far has been limited to bodies with translation along a single coordinate. In general, a rigid body has six degrees of freedom; namely, translation along the three perpendicular coordinate axes, and rotation about them. Each of the six degrees of freedom should thus be examined in terms of Eq. (3.5-3), which applies for either translation or rotation. In rotation the transmissibility is the ratio of the transmitted torque to the disturbing torque.

Materials most commonly used for isolators are steel springs, rubber, cork, and felt. Characteristics of the last three vary with the shape, size, and grade of material, and recommended design practices for them are specified by commercial firms dealing with such material.

EXAMPLE 3.5-1. A machine with a disturbing frequency of 800 c.p.m. is mounted on rubber pads with statical deflection of 0.30 in. Determine the per cent reduction of the transmitted vibration.

Solution. Referring to Fig. 3.5-4, the reduction in vibration is 77 per cent.

EXAMPLE 3.5-2. A machine weighing 200 lb. and supported on springs of total stiffness 4000 lb./in. has an unbalanced rotating element which results in a disturbing force of 80 lb. at a speed of 3000 r.p.m. Assuming a damping factor of $\zeta = 0.20$, determine (a) its amplitude of motion due to the unbalance, (b) the transmissibility, and (c) the transmitted force.

Solution. The statical deflection of the system is $\frac{200}{4000} = 0.05$ in., and its frequency is

$$f_n = \frac{60}{2\pi}\sqrt{\frac{386}{0.05}} = 841 \text{ c.p.m.}$$

(a) Substituting into Eqs. (3.1-8), the amplitude of vibration becomes

$$X = \frac{\left(\frac{80}{4000}\right)}{\sqrt{[1 - (\frac{3000}{841})^2]^2 + [2 \times 0.20 \times \frac{3000}{841}]^2}} = 0.00169 \text{ in.}$$

(b) The transmissibility, from Eq. (3.5-2), is

$$TR = \frac{\sqrt{1 + (2 \times 0.20 \times \frac{3000}{841})^2}}{\sqrt{[1 - (\frac{3000}{841})^2]^2 + (2 \times 0.20 \times \frac{3000}{841})^2}} = 0.148$$

(c) The transmitted force is the disturbing force multiplied by the transmissibility

$$F_{TR} = 80 \times 0.148 = 11.8 \text{ lb.}$$

3.6 Air Springs*

It was pointed out in the previous section that vibration isolation is possible only in the frequency range $\omega/\omega_n > \sqrt{2}$. To isolate against low frequency disturbances, this requirement necessitates a suspension of very large statical deflection. For example, to reduce by 80 per cent a disturbance

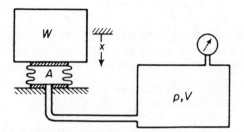

Fig. 3.6-1. Isolation by air springs.

of 1.0 c.p.s., it would require a suspension of natural frequency 0.408 c.p.s. with a statical deflection of 4.8 ft. It is evident then that practical difficulties would be encountered in designing the above suspension with conventional springs.

Fortunately for such problems, the characteristics of the air spring are found to be ideal. The air spring enables a system to have a very low natural frequency with zero statical deflection.

With this type of spring the body to be isolated is supported by air pressure, as shown in Fig. 3.6-1. In many cases, three such springs are used for leveling.

The behavior of the air spring can be determined by assuming the bellows to be a piston and cylinder of area A. By letting the pressure and volume of the air in the equilibrium and displaced positions be P_0, V_0 and P, V, respectively, the following gas law for the adiabatic change holds:

$$PV^{\gamma} = P_0 V_0^{\gamma} \tag{3.6-1}$$

where $\gamma = 1.4$ is the ratio of specific heats for air.

* B. Sussholz, "Forced and Free Motion of a Mass on an Air Spring." *Jour. of Applied Mechanics*, June, 1944, p. 101.

Differentiating with respect to the displacement x, we obtain the equation

$$\frac{dP}{dx} = -\gamma P_0 V_0^{\gamma} V^{-(\gamma+1)} \frac{dV}{dx} \qquad (3.6\text{-}2)$$

Since the volume corresponding to any x is

$$V = (V_0 - Ax), \qquad \frac{dV}{dx} = -A \qquad (3.6\text{-}3)$$

Remembering that the spring stiffness k is, by definition, equal to the force per unit displacement,

$$k = A \frac{dP}{dx} = \frac{\gamma P_0 A^2}{V_0} \left(1 - \frac{Ax}{V_0}\right)^{-(\gamma+1)} \qquad (3.6\text{-}4)$$

This equation indicates that the stiffness of the air spring is dependent on the displacement and hence is nonlinear. However the oscillatory deflection x is generally quite small and the stiffness is approximated by the equation

$$k \cong \frac{\gamma P_0 A^2}{V_0} = \frac{\gamma W A}{V_0} \qquad (3.6\text{-}5)$$

Equation (3.6-5) indicates that a very soft spring is possible by increasing the volume V_0 of the system by means of a connecting tank. The pressure P_0 required is established by the weight W and the area A of each piston. If damping is desired, a throttle valve may be placed in the air line. The equation for the natural frequency of the system can now be written as

$$\omega = \sqrt{k/m} = \sqrt{\gamma A g / V_0} \qquad (3.6\text{-}6)$$

3.7 Energy Dissipated by Damping

Consider the general case of a harmonic displacement lagging the force by an angle ϕ, as in a forced vibration. By letting these two quantities be expressed by the equations

$$F = F_0 \sin \omega t, \qquad x = X \sin(\omega t - \phi) \qquad (3.7\text{-}1)$$

the work done in a cycle of motion is

$$U = \int F \, dx = \int F \frac{dx}{dt} \, dt$$

$$= \omega F_0 X \int_0^{2\pi/\omega} \sin \omega t \cos(\omega t - \phi) \, dt$$

$$= \pi F_0 X \sin \phi \qquad (3.7\text{-}2)$$

It is evident from this equation that, for a given amplitude, the maximum dissipation of energy takes place when $\phi = 90°$.

Of particular interest is the energy dissipated in forced harmonic vibration at resonance. We recall that in forced harmonic vibration the resonant condition $\omega/\omega_n = 1$ leads to $\phi = 90°$ and $X = F_0/2\zeta k$. Substituting these into Eq. (3.7-2), the work done per cycle at resonance is

$$U = 2\zeta\pi k X^2 = \pi c\omega_n X^2 \quad (3.7\text{-}3)$$

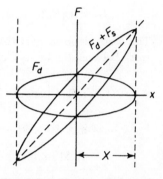

The energy dissipated per cycle by the damping force can be represented graphically as follows. Letting the displacement be given by Eq. (3.7-1), the velocity is

$$\dot{x} = \omega X \cos(\omega t - \phi)$$

$$= \pm\omega\sqrt{X^2 - x^2} \quad (3.7\text{-}4)$$

and the viscous damping force is represented by the ellipse

$$F_d = c\dot{x} = \pm c\omega\sqrt{X^2 - x^2} \quad (3.7\text{-}5)$$

Fig. 3.7-1. Energy dissipated by viscous damping.

The work done per cycle is then the area enclosed by the ellipse shown in Fig. 3.7-1.

If we consider the damping to be associated with the spring, it would be appropriate to plot the sum of the spring and damping forces. In this case, the equation for the total force would be

$$F_s + F_d = kx \pm c\omega\sqrt{X^2 - x^2} \quad (3.7\text{-}6)$$

which is also shown in Fig. 3.7-1.

EXAMPLE 3.7-1. Determine the expression for the power developed by a force $F = F_0 \sin(\omega t + \phi)$ acting on a displacement $x = X_0 \sin \omega t$.

Solution. Power is the rate of doing work which is the product of the force and velocity.

$$P = F\frac{dx}{dt} = (\omega X_0 F_0)\sin(\omega t + \phi)\cos \omega t$$

$$= (\omega X_0 F_0)[\cos\phi \cdot \sin\omega t \cos\omega t + \sin\phi \cdot \cos^2\omega t]$$

$$= \tfrac{1}{2}\omega X_0 F_0[\sin\phi + \sin(2\omega t + \phi)]$$

The first term is a constant, representing the steady flow of work per unit time. The second term is a sine wave of twice the frequency that represents the fluctuating component of power, the average value of which is zero over any interval of time that is a multiple of the period.

EXAMPLE 3.7-2. A force $F = 10 \sin \pi t$ lb. acts on a displacement of $x = 2 \sin (\pi t - \pi/6)$. Determine (a) the work done during the first 6 sec.; (b) the work done during the first $\frac{1}{2}$ sec.

Solution. The force in this case leads the displacement by 30 deg. Since the period of F and X is 2 sec., three complete cycles take place in the specified time in (a), and the work done is

$$U = 3(\pi F_0 X_0 \sin \phi) = 3\pi \times 10 \times 2 \times \sin 30° = 94.2 \text{ in. lb.}$$

The work done in part (b) is determined by integrating the expression for work between the limits 0 and $\frac{1}{2}$ sec.

$$U = \omega F_0 X_0 \left[\cos 30° \int_0^{1/2} \sin \pi t \cos \pi t \, dt + \sin 30° \int_0^{1/2} \sin^2 \pi t \, dt \right]$$

$$= \pi \times 10 \times 2 \left[-\frac{0.866}{4\pi} \cos 2\pi t + 0.50\left(\frac{t}{2} - \frac{\sin 2\pi t}{4\pi}\right) \right]_0^{1/2}$$

$$= 16.51 \text{ in. lb.}$$

3.8 Equivalent Viscous Damping

In an actual oscillatory system, many different types of damping forces may be present. The differential equation including such damping forces would, in general, be beyond the possibility of solution. In fact, only under the assumption of viscous damping is the solution harmonic and manageable in an elementary manner.

As illustrated by the amplitude-frequency curves for forced vibration, the importance of damping is mainly in the limiting of the amplitude of response at resonance. In the case of viscous damping, the resonant amplitude at $\omega = \omega_n$ is

$$X = \frac{F_0}{c\omega_n} \tag{3.8-1}$$

For all other types of damping, the resonant amplitude can be estimated with good accuracy by using in the above equation the equivalent viscous damping c_{eq} evaluated on the basis of harmonic motion and equal energy dissipation. Since the energy dissipated per cycle by viscous damping at frequency ω is $\pi c \omega X^2$, the equivalent viscous damping is determined from the equation

$$\pi c_{eq} \omega X^2 = U \tag{3.8-2}$$

where U must be evaluated for the particular type of damping force.

EXAMPLE 3.8-1. Determine the equivalent viscous damping coefficient for a damping force proportional to the square of the velocity, and find its resonant amplitude.

Solution. Let the damping force be expressed by the equation

$$F_d = \pm a\dot{x}^2$$

where the negative sign must be used when $\dot{x}$ is positive, and vice versa. Assuming harmonic motion with time measured from the position of extreme negative displacement,

$$x = -X \cos \omega t$$

the energy dissipated per cycle is

$$U = 2\int_{-X}^{X} a\dot{x}^2 \, dx = -2a\omega^2 X^3 \int_0^\pi \sin^3 \omega t \, d(\omega t)$$
$$= \tfrac{8}{3}a\omega^2 X^3$$

The equivalent viscous damping, from Eq. (3.8-2), is then

$$c_{eq} = \frac{8}{3\pi} a\omega X$$

The amplitude at resonance is found by letting $\omega = \omega_n$ and by substituting c_{eq} for c in Eq. (3.8-1):

$$X = \sqrt{\frac{3\pi F_0}{8a\omega_n^2}}$$

EXAMPLE 3.8-2. An oscillatory system forced to vibrate by an exciting force $F_0 \sin \omega t$ is known to be acted upon by several different forms of damping. Develop the equation for the equivalent damping and indicate the procedure for determining the amplitude at resonance.

Solution. Let U_1, U_2, U_3, etc., be the energy dissipated per cycle by the various damping forces. Equating the total energy dissipated to that of equivalent viscous damping

$$\pi c_{eq} \omega X^2 = U_1 + U_2 + U_3 + \cdots$$

The equivalent viscous damping coefficient is found to be

$$c_{eq} = \frac{\sum U}{\pi \omega X^2}$$

To determine the amplitude, it is necessary to obtain expressions for U_1, U_2, U_3, etc., which will contain X raised to various powers. Substituting c_{eq}

into the expression

$$X = \frac{F_0}{c_{eq}\omega}$$

the equation with $\omega = \omega_n$ is solved for X.

3.9 Structural Damping

When structural material such as steel or aluminum is cyclicly stressed, energy is dissipated within the material. Experiments by several investigators[*] indicate that, for most materials, the energy dissipated per cycle of stress is independent of the frequency and proportional to the square of the strain amplitude. This implies that the shape of the hysteresis curve for the material remains unchanged with amplitude, and that the enclosed area is independent of the strain rate.

Assuming a linear law, the energy dissipated by structural damping will be proportional to the square of the amplitude of vibration:

$$U = \alpha X^2 \tag{3.9-1}$$

where α is independent of the frequency. The equivalent viscous damping can then be found from Eq. (3.8-2):

$$\pi c_{eq} \omega X^2 = \alpha X^2$$

or

$$c_{eq} = \frac{\alpha}{\pi \omega} \tag{3.9-2}$$

The structural damping force must then be of the form

$$F_d = c_{eq}\dot{x} = \frac{\alpha}{\pi\omega}\dot{x} \tag{3.9-3}$$

and the differential equation of motion may be written as

$$m\ddot{x} + \frac{\alpha}{\pi\omega}\dot{x} + kx = F_0 \sin \omega t \tag{3.9-4}$$

The steady-state amplitude then becomes

$$X = \frac{F_0}{\sqrt{(k - m\omega^2)^2 + \left(\dfrac{\alpha}{\pi\omega}\omega\right)^2}} \tag{3.9-5}$$

[*] A. L. Kimball, "Vibration Damping, Including the Case of Solid Damping," *Trans. ASME*, APM 51-21, 1929.

which at resonance has the value

$$X_{\text{res}} = \frac{\pi}{\alpha} F \qquad (3.9\text{-}6)$$

Let us reexamine the equation for $F_d = (\alpha/\pi\omega)\dot{x}$. Since in harmonic motion $\dot{x} = \omega x$, the equation for the damping force becomes $F_d = (\alpha/\pi)x$, which is proportional to the displacement. It is convenient then to express this in terms of the spring force kx multiplied by a nondimensional factor γ and called the structural damping factor. Then

$$\gamma k = \frac{\alpha}{\pi}$$

and from Eq. (3.9-6),

$$\frac{kX_{\text{res}}}{F} = \frac{1}{\gamma} \qquad (3.9\text{-}7)$$

With this substitution in Eq. 3.9-5, X can be expressed in the nondimensional form

$$\frac{kX}{F_0} = \frac{1}{\sqrt{\left[1 - \left(\frac{\omega}{\omega_n}\right)^2\right]^2 + \gamma^2}} \qquad (3.9\text{-}8)$$

which can be used in calculating the amplitude of a structurally damped system in the region of resonance. Compared to the equation for the viscously damped system which is

$$\frac{kX}{F_0} = \frac{1}{\sqrt{\left[1 - \left(\frac{\omega}{\omega_n}\right)^2\right]^2 + \left[2\zeta\left(\frac{\omega}{\omega_n}\right)\right]^2}}$$

with resonant value of $kX_{\text{res}}/F_0 = 1/2\zeta$, it is seen that $\gamma = 2\zeta$, or the structural damping factor as defined is twice the viscous damping factor.

3.10 Sharpness of Resonance

In forced vibration there is a quantity Q related to damping which is a measure of the sharpness of resonance. To determine its equation, we will assume viscous damping, and start with Eq. (3.1-8).

When $\omega/\omega_n = 1$, the resonant amplitude $X/X_0 = 1/2\zeta$, as shown in Fig. 3.10-1. We now seek the two frequencies on either side of resonance (often referred to as sidebands), where X/X_0 is $\frac{1}{2}\sqrt{2}\,(1/2\zeta)$. These points are also referred to as the half-power points.

Substituting into Eq. (3.1-8), we obtain

$$\frac{1}{2}\left(\frac{1}{2\zeta}\right)^2 = \frac{1}{\left[1 - \left(\frac{\omega}{\omega_n}\right)^2\right]^2 + \left[\frac{2\zeta\omega}{\omega_n}\right]^2} \tag{3.10-1}$$

which yields the equation

$$\left(\frac{\omega}{\omega_n}\right)^4 - 2(1 - 2\zeta^2)\left(\frac{\omega}{\omega_n}\right)^2 + (1 - 8\zeta^2) = 0$$

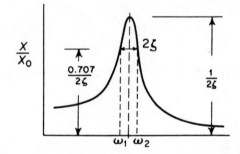

Fig. 3.10-1. Width of resonance curve at half-power points.

Solving, we obtain

$$\left(\frac{\omega}{\omega_n}\right)^2 = (1 - 2\zeta^2) \pm 2\zeta\sqrt{1 + \zeta^2} \tag{3.10-2}$$

If we now assume that $\zeta \ll 1$, and neglect higher-order terms, we arrive at the result

$$\left(\frac{\omega}{\omega_n}\right)^2 = 1 \pm 2\zeta \tag{3.10-3}$$

Letting the two frequencies corresponding to the roots of this equation be ω_2 and ω_1, we obtain

$$4\zeta = \frac{\omega_2^2 - \omega_1^2}{\omega_n^2} = 2\left(\frac{\omega_2 - \omega_1}{\omega_n}\right) \tag{3.10-4}$$

The quantity Q is then defined as

$$Q = \frac{\omega_n}{\omega_2 - \omega_1} = \frac{f_n}{f_2 - f_1} = \frac{1}{2\zeta} \tag{3.10-5}$$

Here again equivalent damping can be used to define Q for systems with other forms of damping. Thus, for solid damping, Q is equal to

$$Q = \frac{f_n}{f_2 - f_1} = \frac{1}{\gamma} \tag{3.10-6}$$

3.11 Vibration Measuring Instruments

The basic element of many vibration-measuring instruments is the seismic unit of Fig. 3.11-1. The motion to be measured is established from the relative motion between the seismic mass and the base. Depending on the frequency range utilized, displacement, velocity, or acceleration is indicated.

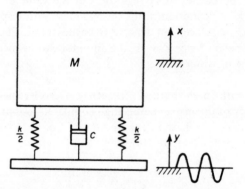

Fig. 3.11-1. Basic elements of vibration measuring instruments.

The equation of motion for a base excited system was studied in Sec. 3.4. We again let y be the motion of the base and x that of the mass M, in which case we can write

$$M\ddot{x} = -k(x - y) - c(\dot{x} - \dot{y}) \tag{3.11-1}$$

Since the quantity recorded by the instrument is the relative motion between the seismic mass and the case, we let

$$z = (x - y)$$

and rewrite Eq. (3.11-1) as

$$M\ddot{z} + c\dot{z} + kz = -M\ddot{y} \tag{3.11-2}$$

If the base motion is sinusoidal, $y = Y \sin \omega t$ which reduces to

$$M\ddot{z} + c\dot{z} + kz = M\omega^2 Y \sin \omega t \tag{3.11-3}$$

which is identical in form to Eq. (3.3-2) or to (3.1-1) with x replaced by z and F_0 by $M\omega^2 Y$. The steady-state solution $z = Z \sin(\omega t - \phi)$ can then be written down immediately by inspection:

$$Z = \frac{M\omega^2 Y}{\sqrt{(k - M\omega^2)^2 + (c\omega)^2}} = \frac{(\omega/\omega_n)^2 Y}{\sqrt{[1 - (\omega/\omega_n)^2]^2 + [2\zeta(\omega/\omega_n)]^2}} \qquad (3.11\text{-}4)$$

$$\phi = \tan^{-1}\frac{c\omega}{k - M\omega^2} = \tan^{-1}\frac{2\zeta(\omega/\omega_n)}{1 - (\omega/\omega_n)^2} \qquad (3.11\text{-}5)$$

and all of the conclusions of Sec. 3.3 apply.

Instrument with low natural frequency (Vibrometer). When the natural frequency of the instrument is low compared to that of the vibration to be measured, Z/Y approaches unity regardless of the value of ζ, as seen from Fig. 3.3-2. The relative motion of the seismic mass with respect to the frame is then equal to the displacement which is to be measured. Such an instrument is called a vibrometer. Since $Z = Y$, the amplitude to be measured is limited by the size of the instrument.

Instrument with high natural frequency (Accelerometer). When the natural frequency of the instrument is high compared to that of the vibration to be measured, $\omega/\omega_n \ll 1$, the denominator of Eq. (3.11-4) is nearly unity, resulting in the approximate relationship

$$Z = \frac{\omega^2 Y}{\omega_n^2} = \frac{\text{acceleration}}{\omega_n^2} \qquad (3.11\text{-}6)$$

Z is hence proportional to the acceleration, and such instruments are called accelerometers.

For an undamped accelerometer, the frequency range is quite limited because the denominator $1 - (\omega/\omega_n)^2$ drops off rapidly from unity as ω increases. However, with damping in the range $\zeta = 0.65$ to 0.70, the reduction in the term $1 - (\omega/\omega_n)^2$ is compensated by the additional term $2\zeta(\omega/\omega_n)$ to increase greatly the useful frequency range of the instrument.

Figure 3.11-2 shows the factor

$$\frac{1}{\sqrt{[1 - (\omega/\omega_n)^2]^2 + [2\zeta(\omega/\omega_n)]^2}}$$

for various damping plotted on a magnified scale. Most accelerometers utilize damping near $\zeta = 0.70$. Some instruments are damped with silicone oil, whereas others are damped electrically by a short-circuited coil in a magnetic field.

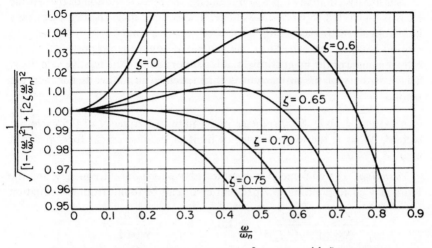

Fig. 3.11-2. Accelerometer error versus frequency with ζ as parameter.

Phase distortion. To reproduce a complex wave without a change in its shape, the phase of all harmonic components must be shifted equally along the time axis. This can be accomplished if the phase angle ϕ increases linearly with frequency. For example, if $\phi = \dfrac{\pi}{2} \dfrac{\omega}{\omega_n}$, which is nearly satisfied when $\zeta = 0.70$, a complex wave of frequency ω_1 and ω_2 given by the equation,

$$\ddot{y} = -\omega_1^2 Y_1 \sin \omega_1 t - \omega_2^2 Y_2 \sin \omega_2 t$$

will be indicated by the accelerometer as follows:

$$Z = \left(\frac{\omega_1}{\omega_n}\right)^2 Y_1 \sin \omega_1 \left(t - \frac{\pi}{2\omega_n}\right) + \left(\frac{\omega_2}{\omega_n}\right)^2 Y_2 \sin \omega_2 \left(t - \frac{\pi}{2\omega_n}\right)$$

Since both harmonic components are shifted along the time axis by the same amount $\pi/2\omega_n$, the original wave shape of acceleration is retained without phase distortion.

For the vibrometer ω/ω_n is large and ϕ is approximately 180 deg. for all harmonics. Thus no phase distortion is encountered. However, damping factor of approximately 0.7 is generally utilized to minimize the transient response of the instrument. In both the vibrometer and the accelerometer the above value of damping factor will extend the frequency range of the instrument.

Sensitivity. In many types of intruments the relative motion of the seismic mass is converted to electrical voltage. The seismic mass in such cases may be a magnet with the voltage generated in a stationary coil by the relative motion between the two. Since the generated voltage is proportional to the rate of change of the magnetic field, it is proportional to the relative velocity $\dot{z}$. The sensitivity of the instrument is thus specified in millivolts per inch per second.

EXAMPLE 3.11-1. A vibrometer has a natural frequency of 0.5 cycle per second. What is the lower frequency limit of the instrument for 2 per cent error if $\zeta = 0$?

Solution. The vibrometer is operated above its natural frequency, and the frequency corresponding to 2 per cent error can be found from the equation

$$\frac{Z}{X} = \frac{(f/f_n)^2}{1 - (f/f_n)^2} = -1.02, \qquad \frac{f}{f_n} = \sqrt{\frac{1.02}{0.02}} = 7.13$$

$$f = 7.13 \times 0.5 = 3.58 \text{ c.p.s.}$$

The error of the instrument will thus be less than 2 per cent for frequencies above 3.58 c.p.s.

EXAMPLE 3.11-2. A certain body is vibrating according to the equation

$$x = 0.10 \sin 4\pi t + 0.05 \sin 8\pi t$$

the units being the inch and the second. Show that the record of vibration z taken by a seismic instrument whose natural frequency is 3 vibrations per second, and damping factor $\zeta = 0.50$, is given by the equation

$$z = 0.0513 \sin (4\pi t - 50°) + 0.0575 \sin (8\pi t - 120°)$$

Solution. The frequencies of the two components of the given vibration are 2 and 4 c.p.s. The amplitude and phase recorded by the instrument are computed from Eqs. (3.11-4) and (3.11-5) as follows:

First Component:

$$Z_1 = \frac{(\tfrac{2}{3})^2 \times 0.10}{\sqrt{[1 - (\tfrac{2}{3})^2]^2 + [2 \times 0.50 \times \tfrac{2}{3}]^2}} = 0.0513$$

$$\phi_1 = \tan^{-1} \frac{2 \times 0.50 \times \tfrac{2}{3}}{1 - (\tfrac{2}{3})^2} = \tan^{-1}(1.2) = 50°$$

Second Component:

$$Z_2 = \frac{(\tfrac{4}{3})^2 \times 0.05}{\sqrt{[1 - (\tfrac{4}{3})^2]^2 + [2 \times 0.50 \times \tfrac{4}{3}]^2}} = 0.0575$$

$$\phi_2 = \tan^{-1} \frac{2 \times 0.50 \times \tfrac{4}{3}}{1 - (\tfrac{4}{3})^2} = \tan^{-1}(-1.71) = 120°$$

3.12 Whirling of Rotating Shafts

Rotating shafts tend to bow out and whirl at certain speeds called critical speed, whirling speed, or whipping speed. These phenomena result from various causes such as the mass unbalance of the rotating system, hysteresis damping in the shaft, gyroscopic forces, oil friction in the journal bearings,

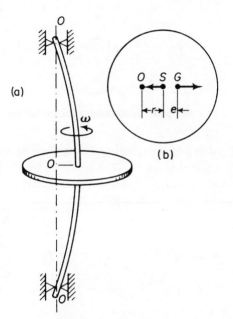

Fig. 3.12-1. Whirling of shaft due to unbalance.

unsymmetrical stiffness of the shaft or bearings, etc. Whirling is an interesting but rather subtle topic which cannot be treated lightly. The whirling of the shaft may take place in the same direction as that of the rotation of the shaft, or it is possible for the whirl to be in the opposite sense, and the whirling speed may or may not be equal to the rotation speed. It is possible, however, to consider here a simple case of synchronous whirl in connection with an idealized system shown in Fig. 3.12-1.

The idealized system consists of a single disk of mass m symmetrically located on a shaft supported by two bearings. The center of mass G of the disk is at a radial distance e from the geometric center S of the disk. The center line of the bearings intersects the plane of the disk at O and the shaft center is deflected by $OS = r$.

Neglecting the effect of gravity and friction, the disk is under the action of only two forces, namely, the restoring force of the shaft directed from S to O, and the centrifugal force acting in the outward direction through G. It is evident, then, that for these two forces to be in equilibrium, they must be collinear, equal in magnitude, and opposite in direction, thereby requiring points O, S, and G to lie along a straight line.

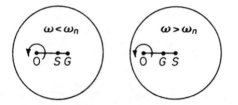

Fig. 3.12-2. Possible phase relationship for undamped system.

The lateral deflection r of the center of the disk can be determined simply by equating the two forces involved. The restoring force of the shaft is kr, where k is the lateral stiffness of the shaft at the disk, and the centrifugal force is equal to $m\omega^2(r + e)$. Equating the two,

$$kr = m\omega^2(r + e) \tag{3.12-1}$$

and solving for r, we obtain

$$r = \frac{m\omega^2 e}{k - m\omega^2} = \frac{(\omega/\omega_n)^2 e}{1 - (\omega/\omega_n)^2} \tag{3.12-2}$$

where $\omega_n = \sqrt{k/m}$, the natural frequency of lateral vibration of the shaft and disk at zero speed.

Equation (3.12-2) indicates that the critical speed of the shaft is equal to the natural frequency of lateral vibration ω_n of the shaft and disk. r is positive below the critical speed and negative for speeds greater than ω_n, which means that for ω less than ω_n the system rotates with the heavy side G outside S, whereas for ω greater than ω_n the light side, or the side opposite G, is outside S. Both conditions are shown in Fig. 3.12-2. For very high speeds, $\omega \gg \omega_n$, the amplitude r becomes equal to $-e$, and the points O and G coincide; that is, the disk rotates about its center of gravity G.

Friction included. The simple analysis of the previous section can be shown to be a special case of a more general analysis where friction is considerea. Forces such as air friction opposing the whirl can be resolved into a force F and a moment about the shaft center, the latter being overcome by the driving torque of the shaft.

For simplicity, the damping force F acting at S can be assumed to be of viscous nature and hence proportional to the tangential velocity $r\omega$. The expression for F may then be written as

$$F = cr\omega \qquad (3.12\text{-}3)$$

where c is the coefficient of viscous friction.

The presence of the frictional force F now enables the line e to lead the line r by an angle ϕ as shown in Fig. 3.12-3. The forces indicated in this diagram presuppose that the points O, S, and G remain fixed relative to each other, the configuration rotating together about O with

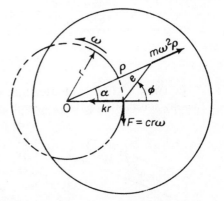

Fig. 3.12-3. Vector relationship for whirling shaft, including friction.

speed ω. From the previous analysis, this arbitrary assumption appears to be reasonable; it must, however, be substantiated by the condition of equilibrium, which will now be investigated.

Summing forces along r and perpendicular to r, we obtain the following equations:

$$-kr + m\omega^2\rho \cos \alpha = 0$$
$$-cr\omega + m\omega^2\rho \sin \alpha = 0 \qquad (3.12\text{-}4)$$

From the geometry of the figure, α and ϕ are related as follows:

$$\rho \sin \alpha = e \sin \phi$$
$$\rho \cos \alpha = r + e \cos \phi \qquad (3.12\text{-}5)$$

Substituting these relations, the equilibrium equations become

$$-kr + m\omega^2(r + e \cos \phi) = 0$$
$$-cr\omega + m\omega^2 e \sin \phi = 0 \qquad (3.12\text{-}6)$$

Solving simultaneously for ϕ and r (see Fig. 3.12-4),

$$\tan \phi = \frac{c\omega}{k - m\omega^2} = \frac{2\zeta(\omega/\omega_n)}{1 - (\omega/\omega_n)^2} \qquad (3.12\text{-}7)$$

$$r = \frac{m\omega^2 e \cos \phi}{k - m\omega^2} = \frac{m\omega^2 e}{\sqrt{(k - m\omega^2)^2 + (c\omega)^2}}$$

$$= \frac{e(\omega/\omega_n)^2}{\sqrt{[1 - (\omega/\omega_n)^2]^2 + [2\zeta(\omega/\omega_n)]^2}} \qquad (3.12\text{-}8)$$

Equation (3.12-7) indicates that ϕ is a constant at any given speed ω, thereby substantiating the assumption that the points O, S, and G remain fixed relative to each other and whirl about O with speed ω. It also indicates that when ω is very small compared to ω_n, $\phi \cong 0$, and the disk rotates with G outside S. When $\omega = \omega_n$, the eccentricity line e leads the lateral deflection line r by 90 deg. For speeds greater than the critical, ϕ is greater than 90 deg., approaching 180 deg. for very high speeds; that is, the light side remains outside S. Also for very high speeds, the term $m\omega^2$ becomes predominant, and Eq. (3.12-8) shows that r approaches e. The point O then coincides with point G, and the system rotates about its center of gravity. These relations are illustrated in Fig. 3.12-5.

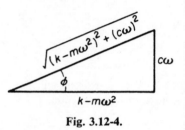

Fig. 3.12-4.

Both Eqs. (3.12-7) and (3.12-8) derived in this section are identical to those of Sec. 3.3. This result is to be expected, since in both cases the exciting force is due to the unbalance and equal to $me\omega^2$.

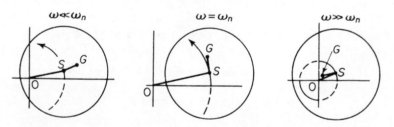

Fig. 3.12-5. Phase relationship of whirling shaft with viscous friction.

EXAMPLE 3.12-1. The rotor of a turbosupercharger weighing 20 lb. is keyed to the center of a 1-in. diameter steel shaft 16 in. between bearings. Determine (a) the critical speed, (b) the amplitude of vibration of rotor at a speed of 3200 r.p.m. if the eccentricity is 0.000620 in., and (c) the vibratory force transmitted to the bearings at this speed. (Assume the shaft to be simply supported.)

Solution. The stiffness of the shaft at mid-span is

$$k = \frac{48EI}{l^3} = \frac{48 \times 30 \times 10^6 \times \pi \times 1^4}{16^3 \times 64} = 17,230 \text{ lb./in.}$$

The effective weight of the shaft is $0.486w$ at the mid-span (see Chapter 1, Prob. 31), and the following weight must be added to the weight of the rotor:

$$0.486w = 0.486 \times \frac{\pi}{4} \times 16 \text{ in.} \times 0.283 = 1.7 \text{ lb.}$$

(a) The critical speed corresponds to the natural frequency of lateral vibration of the system, which is

$$f_n = \frac{60}{2\pi} \sqrt{\frac{k}{m}} = \frac{60}{2\pi} \sqrt{\frac{17,230 \times 386}{21.7}} = 5300 \text{ r.p.m.}$$

(b) The amplitude of vibration is determined from Eq. (3.12-2) as

$$r = \frac{(\omega/\omega_n)^2 e}{1 - (\omega/\omega_n)^2} = \frac{(32/53)^2 \times 0.000620}{1 - (32/53)^2} = 0.000356 \text{ in.}$$

(c) The force transmitted to the two bearings is

$$F = m(r + e)\omega^2 = \frac{21.7}{386}(0.000356 + 0.000620)\left(\frac{3200 \times 2\pi}{60}\right)^2$$

$$= 6.20 \text{ lb.}$$

EXAMPLE 3.12-2. A turbine operating above the critical speed must run through the dangerous speed each time it is started or stopped. Assuming the critical speed ω_n to be reached with zero amplitude, determine the equation for the amplitude of the shaft center.

Solution. For steady-state conditions it was possible to determine the amplitude and phase of a rotating disk without writing the differential equation. For a transient condition, such as the one encountered in this problem, it is necessary to start with the differential equations, summing forces in the x and y directions. Referring to Fig. 3-12-6, they are,

$$m \frac{d^2}{dt^2}(x + e \cos \omega_n t) = -kx$$

$$m \frac{d^2}{dt^2}(y + e \sin \omega_n t) = -ky$$

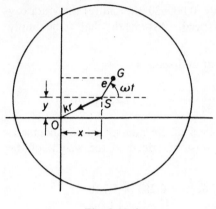

Fig. 3.12-6.

These equations can be rewritten as

$$\ddot{x} + \omega_n^2 x \doteq e\omega_n^2 \cos \omega_n t$$

$$\ddot{y} + \omega_n^2 y = e\omega_n^2 \sin \omega_n t$$

It should be noted here that the solution of the form

$$A \sin \omega_n t + B \cos \omega_n t$$

will satisfy only the homogeneous equation where the right side of the differential equation is zero. We must therefore try functions of the form $t \sin \omega_n t$ and $t \cos \omega_n t$ in addition to the solution of the homogeneous equation. The solution for initial displacement equal to zero then becomes

$$x = \frac{e\omega_n t}{2} \sin \omega_n t + \frac{\dot{x}(0)}{\omega_n} \sin \omega_n t$$

$$y = \frac{e}{2}(\sin \omega_n t - \omega_n t \cos \omega_n t) + \frac{\dot{y}(0)}{\omega_n} \sin \omega_n t$$

where $\dot{x}(0)$ and $\dot{y}(0)$ are the components of the initial velocity. If the initial velocity is also zero, then the displacement of the center of the shaft becomes

$$r = \frac{e\omega_n t}{2} \sqrt{1 - \frac{2 \cos \omega_n t \sin \omega_n t}{\omega_n t} + \left(\frac{\sin \omega_n t}{\omega_n t}\right)^2}$$

which for large time approaches the value

$$r = \frac{e\omega_n t}{2}$$

3.13 Rigid Shaft Supported by Flexible Bearings

Consider a rigid wheel and shaft supported in bearings with unequal flexibility in the x and y directions as shown in Fig. 3.13-1. Again assuming the mass unbalance to be designated by the eccentricity e, the equations of motion become

$$m\ddot{x} + k_x x = me\omega^2 \cos \omega t$$

$$m\ddot{y} + k_y y = me\omega^2 \sin \omega t$$

(3.13-1)

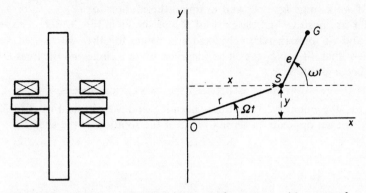

Fig. 3.13-1. Whirling of rigid disk—shaft system with unequal bearing flexibility.

which has the steady-state solution

$$x = \frac{me\omega^2}{k_x - m\omega^2} \cos \omega t = X \cos \omega t$$

$$y = \frac{me\omega^2}{k_y - m\omega^2} \sin \omega t = Y \sin \omega t$$

(3.13-2)

By squaring and adding, the simple equation of the ellipse is obtained

$$\frac{x^2}{X^2} + \frac{y^2}{Y^2} = 1$$

(3.13.3)

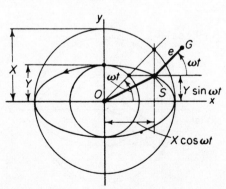

Fig. 3.13-2. Whirl with unequal bearing flexibility—ω less than both critical speeds.

If we assume $k_x < k_y$ and examine the motion for $\omega < \sqrt{k_x/m}$, both X and Y are positive and the path of S is as shown in Fig. 3.13-2. The vectors OS and SG remain relatively fixed and rotate together with speed ω. It is evident that if $k_x = k_y$ the ellipse degenerates to a circle and the lines OS and SG become collinear.

If the rotation speed is between the two critical speeds, $\sqrt{k_x/m} < \omega < \sqrt{k_y/m}$, the quantity X becomes negative and we obtain Fig. 3.13-3. The whirl is then opposite in sense to that of the rotation of the shaft.

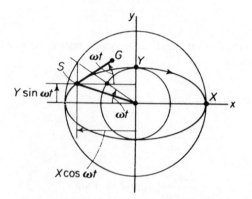

Fig. 3.13-3. ω between the two critical speeds.

When ω is greater than both the critical speeds, both X and Y are negative and the shaft whirls again in the same direction as that of the rotation. The distance OG, however, will tend to be less than OS, as shown in Fig. 3.13-4.

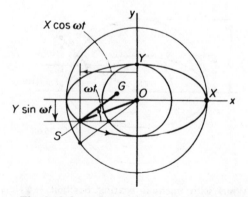

Fig. 3.13-4. ω greater than both critical speeds.

3.14 Static and Dynamic Balance

The unbalance of a single disk can be detected by allowing the disk to rotate on its axle between two parallel knife-edges, as shown in Fig. 3.14-1. The disk will rotate and come to rest with the heavy side on the bottom.

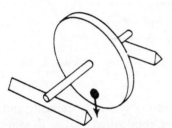

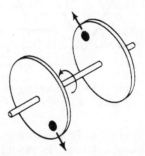

Fig. 3.14-1. System with static
unbalance.

Fig. 3.14-2. System with
dynamic unbalance.

This type of unbalance is called *static unbalance*, since it can be detected by static means.

In general, the mass of a rotor is distributed along the shaft such as in a motor armature or an automobile-engine crankshaft. A test similar to the one above may indicate that such parts are in static balance, but the system may show a considerable unbalance when rotated.

As an illustration, consider a shaft with two disks, as shown in Fig. 3.14-2. If the two unbalance weights are equal and 180 deg. apart, the system will be statically balanced about the axis of the shaft. However, when the system is rotated, each unbalanced disk would set up a rotating centrifugal force tending to rock the shaft on its bearings. Since this type of unbalance results only from rotation we refer to it as *dynamic unbalance*.

Figure 3.14-3 shows a general case where the system is both statically and dynamically unbalanced. It will now be shown that the unbalanced forces P and Q can always be eliminated by the addition of two correction weights in any two parallel planes of rotation.

Consider first the unbalance force P, which can be replaced by two parallel forces $P\,a/l$ and $P\,b/l$. In a similar manner Q can be replaced by two parallel forces $Q\,c/l$ and $Q\,d/l$. The two forces in each plane can then be combined into a single resultant force that can be balanced by a single correction weight as shown. The two correction weights C_1 and C_2 introduced in the two parallel

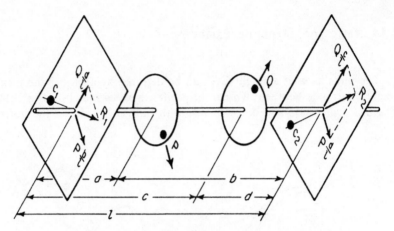

Fig. 3.14-3. Balancing of rotating bodies requires correction in two planes.

planes completely balanced P and Q, and the system is statically and dynamically balanced. It should be further emphasized that a dynamically balanced system is also statically balanced. The converse, however, is not always true; a statically balanced system may be dynamically unbalanced.

EXAMPLE 3.14-1. A rotor 4 in. long has an unbalance of 3 oz. in. in a plane 1 in. from the left end, and 2 oz. in. in the middle. plane. Its angular position is 90 deg. in the clockwise direction from the first unbalance when viewed from the left end. Determine the corrections in the two end planes, giving magnitude and angular positions.

Solution. The 3-oz. in. unbalance is equivalent to $2\frac{1}{4}$ oz. in. at the left end and $\frac{3}{4}$ oz. in. at the right end, as shown in Fig. 3.14-4. The 2 oz. in. at the middle is obviously equal to 1 oz. in. at the ends.

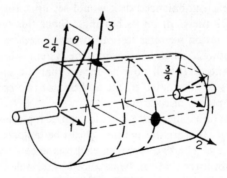

Fig. 3.14-4.

Combining the two unbalances at each end, the corrections are:

Left end:

$$C_1 = \sqrt{1^2 + (2.25)^2} = 2.47 \text{ oz. in. to be removed}$$

$$\theta_1 = \tan^{-1}\frac{1}{2.25} = 24°0' \text{ clockwise from plane of first unbalance}$$

Right end:

$$C_2 = \sqrt{(\tfrac{3}{4})^2 + 1^2} = 1.25 \text{ oz. in. to be removed}$$

$$\theta_2 = \tan^{-1}\frac{1}{(\tfrac{3}{4})} = 53° \text{ clockwise from plane of first unbalance}$$

3.15 Balancing of Rotating Machines

The amount and position of the unbalance in any rotating machine are usually unknown, and the proper corrections to be made in the two correction planes must be established by tests. E. L. Thearle* has outlined such a procedure for balancing rotating machines in their operating condition.

Disk. A thin disk may be balanced dynamically as follows. Run the disk at any speed which will result in a measurable amplitude. The high side a, shown in Fig. 3.15-1, is then marked and its amplitude oa which is the effect of the heavy spot is laid off to scale. Due to damping, the phase of the vector oa will lag the unknown position of the heavy spot.

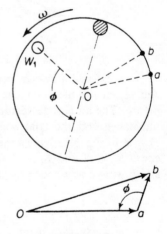

Next, a trial weight W_t is placed in any position and, with the disk running at the same speed as before, the new high side b is marked. The new amplitude ob now represents the effect of both the original unbalance and the added trial weight W_t. The vector difference $ab = ob - oa$ is then the effect of W_t alone.

If W_t is rotated counterclockwise by an angle ϕ, its displacement vector ab will be

Fig. 3.15-1.

parallel and opposite to oa, and if it is increased in the ratio oa/ab to equal the original unbalance, it will balance the disk.

* E. L. Thearle, "Dynamic Balancing of Rotating Machinery in the Field." *Trans. ASME*, APM 56–19, Vol. 56, 1934, p. 745–53.

Long rotor. The procedure here is an extension of the method discussed for the disk.

Referring to the two ends as near and far ends, as shown in Fig. 3.15-2, the rotor is first run "as is" and the amplitude vectors N and F relative to any reference mark on the end of the rotor are measured. N and F represent the effect of the actual unbalance in the rotor.

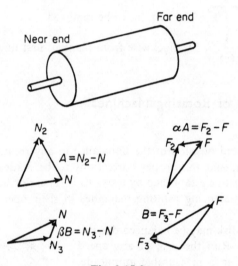

Fig. 3.15-2.

We next add a known trial weight W_{tN} at any position on the near end and repeat the measurement of amplitude and phase at the same speed as before. The amplitude vectors N_2 and F_2 now are the effects of the original unbalance and the trial weight W_{tN}. The vector difference $N_2 - N$ and $F_2 - F$ of Fig. 3.15-2 must then be the effect of W_{tN} alone on the near and far ends. $N_2 - N$ and $F_2 - F$ will vary directly with W_{tN} as it changes in size and position, and hence we can write

$$F_2 - F = \alpha(N_2 - N) = \alpha A \tag{3.15-1}$$

where α is a vector operator depending on the property of the machine and its mounts.

Removing W_{tN}, we next add a known trial weight W_{tF} on the far end and repeat the previous measurements and vector calculations shown in Fig. 3.15-2.

If we let the magnitude and direction of the required balance weights at the near and far ends be W_{bN} and W_{bF}, they can be obtained by changing the

size and position of the trial weights W_{tN} and W_{tF}. We can therefore write
the relationships

$$W_{bN} = \theta W_{tN}, \qquad W_{bF} = \phi W_{tF} \qquad (3.15\text{-}2)$$

where θ and ϕ are vector operators. The operator θ will have the same effect
on A and αA as it has on W_{tN}, and hence for complete balance of the rotor
we have

$$\theta A + \phi(\beta B) = -N, \qquad \phi B + \theta(\alpha A) = -F \qquad (3.15\text{-}3)$$

Solving for θ and ϕ,

$$\theta = \frac{\beta F - N}{(1 - \alpha\beta)A}, \qquad \phi = \frac{\alpha N - F}{(1 - \alpha\beta)B} \qquad (3.15\text{-}4)$$

where N and F are the amplitude vectors of the original vibrations. α and β
are obtained by the division $\alpha = \alpha A/A$ and $\beta = \beta B/B$. With θ and ϕ known,
the balance weights W_{bN} and W_{bF} are found from Eq. (3.15-2).

PROBLEMS

1. A machine part weighing 4.3 lb. vibrates in a viscous medium. Determine the
damping coefficient when a harmonic exciting force of 5.5 lb. results in a
resonant amplitude of 0.50 in. with a period of 0.20 sec.

2. If the system of Prob. 1 is excited by a harmonic force of frequency 4 c.p.s.,
what will be the percentage increase in the amplitude of forced vibration when
the dashpot is removed?

3. A weight attached to a spring of stiffness 3.0 lb./in. has a viscous damping
device. When the weight was displaced and released, the period of vibration
was found to be 1.80 sec., and the ratio of consecutive amplitudes was 4.2 to
1.0. Determine the amplitude and phase when a force $F = 2 \cos 3t$ acts on
the system.

4. Show that for the system of Fig. 3.1-1 the peak amplitude occurs at a frequency
ratio given by the expression,

$$(\omega/\omega_n)_p = \sqrt{1 - 2\zeta^2}$$

5. A spring-mass system is excited by a force $F_0 \sin \omega t$. At resonance the amplitude
was measured to be 0.58 in. At 0.80 resonant frequency, the amplitude was
measured to be 0.46 in. Determine the damping factor c of the system. (*Hint:*
Assume damping term is negligible at 0.80 resonance.)

6. A weight of 12 lb. suspended by a spring of stiffness 6 lb./in. is forced to
vibrate by a harmonic force of 2.0 lb. Assuming viscous damping of $c = 0.430$
lb./in. per sec., find (a) the resonant frequency, (b) the amplitude at resonance,
(c) the phase angle at resonance, (d) the frequency corresponding to the peak

amplitude, (e) the peak amplitude, and (f) the phase angle corresponding to peak amplitude.

7. A torsional pendulum with viscous damping has the following moment of inertia and stiffness:

$$J = 18 \text{ lb. in. sec.}^2, \qquad K = 10^6 \text{ in. lb./rad.}$$

When the pendulum was subjected to a harmonic torque of 80 in. lb., the resonant amplitude was found to be 0.00917 deg. Determine (a) the natural frequency, (b) the damping factor, and (c) the coefficient of viscous damping in torsion.

8. Determine the solution for the displacement of an undamped spring-mass system when a force $F_0 \sin (\omega t + \varphi)$ acts on the mass with initial displacement and velocity equal to zero. Discuss special cases for $\varphi = 0$ and $\varphi = \pi/2$.

9. For a spring-mass system with damping, determine the complex amplitude of the velocity vector. Determine the mechanical impedance, defined as the ratio of force to the velocity.

10. For the system shown in Fig. 3-1, set up the equation of motion and solve for the steady-state amplitude and phase angle by the method of complex algebra.

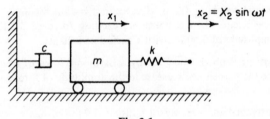

Fig. 3-1.

11. Set up the equation of motion for the system of Fig. 3-2 and solve for the steady-state amplitude and phase angle by the method of complex algebra.

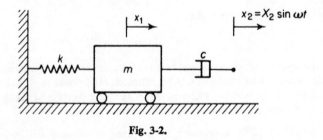

Fig. 3-2.

12. A counterrotating eccentric weight exciter, shown in Fig. 3.3-3, is used to determine the vibrational characteristics of a structure weighing 400 lb. At a speed

of 900 r.p.m. a stroboscope showed the eccentric weights to be at the top at the instant the structure was moving upward through its static equilibrium position, and the corresponding amplitude was 0.85 in. If the unbalance of each wheel of the exciter is 4 lb. in., determine (a) the natural frequency of the structure, (b) the damping factor of the structure, (c) the amplitude at 1200 r.p.m., and (d) the angular position of the eccentrics at the instant the structure is moving upward through its equilibrium position.

13. A refrigerator unit weighing 65 lb. is to be supported by three springs of stiffness k lb./in. each. If the unit operates at 580 r.p.m., what should be the value of the spring constant k if only 10 per cent of the shaking force of the unit is to be transmitted to the supporting structure?

14. An aircraft instrument board, including instruments, weighs 40 lb. and is supported by four rubber mounts rated at $\frac{3}{32}$ in. per 12 lb. What percentage of the engine vibration is transmitted to the instruments at an engine speed of (a) 2200 r.p.m.; (b) 1500 r.p.m.?

15. Discuss the significance of mounting a machine on a large concrete block for isolation. If the spring stiffness is also increased so that the natural frequency is unchanged, what effect does the large mass have on (a) the transmissibility, (b) the amplitude of vibration?

16. An industrial machine weighing 1000 lb. is supported on springs with a statical deflection of 0.20 in. If the machine has a rotating unbalance of 20 lb. in., determine (a) the force transmitted to the floor at 1200 r.p.m., (b) the dynamical amplitude at this speed. (Assume damping to be negligible.)

   If the machine is mounted on a large concrete block weighing 2500 lb. and the stiffness of the springs or pads under the block is increased so that the statical deflection is still 0.20 in., what will be the dynamical amplitude?

17. A machine weighing 150 lb. is mounted on springs of stiffness $k = 6000$ lb./in. with an assumed damping factor of $\zeta = 0.20$. A piston within the machine weighing 4 lb. has a reciprocating motion with a stroke of 3 in. and a speed of 3000 c.p.m. Assuming the motion of the piston to be harmonic, determine (a) the amplitude of motion of the machine, (b) its phase angle with respect to the exciting force, (c) the transmissibility and the force transmitted to the foundation, and (d) the phase angle of the transmitted force with respect to the exciting force.

18. A single-cylinder reciprocating engine supported on springs has the following data:
   total weight = 900 lb.
   equivalent weight of reciprocating part = 35 lb.
   stroke = 8.0 in. (assumed harmonic)
   statical deflection of springs due to weight of engine = 2.0 in.
   ratio of consecutive amplitudes in free vibration = 1.00 to 0.42
   Determine (a) the amplitude of engine vibration at 250 r.p.m., (b) the transmissibility, and (c) the force transmitted to the base at this speed.

19. An aircraft radio weighing 24 lb. is to be isolated from engine vibrations ranging in frequencies from 1600 to 2200 c.p.m. If rubber isolaters are used, what statical deflection must it have for 85 per cent isolation?

20. An optical interferometer weighing 2 tons is to be isolated from the vibrations of a nearby railroad. To obtain satisfactory isolation the natural frequency of the system must be 1.0 c.p.s. Design air springs to support the interferometer at three points with air pressure of 70 lb./sq. in.

21. Figure 3-3 represents a simplified diagram of a spring-supported vehicle traveling over a rough road. Determine the equation for the amplitude of W as a function of the speed and determine the most unfavorable speed.

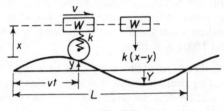

Fig. 3-3.

22. The springs of an automobile trailer are compressed 4 in. under its weight. Find the critical speed when the trailer is traveling over a road with a profile approximated by a sine wave of amplitude 3.0 in. and wave length of 48 ft. What will be the amplitude of vibration at 40 m.p.h.? (Neglect damping.)

23. Figure 3-4(a) represents an automobile front wheel—tire, springs and shock absorber—which is idealized to the vibrational system of Fig. 3-4(b). Set up the differential equation of the system, assuming the lower end of spring k_0

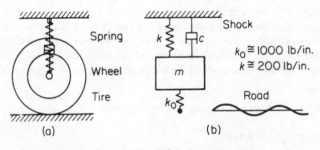

Fig. 3-4.

to be undergoing a harmonic displacement of $x_2 = X_2 \sin \omega t$. To design the shock absorber properly one must know the critical damping of the system. Obtain an expression for the critical damping coefficient of the system in terms of the quantities given.

24. Figure 3-5 shows a simplified diagram of the driving wheel of a locomotive. The spring k_1 between the wheel and the body has a stiffness of 20,000 lb./in., and the railroad track has a vertical stiffness k_2 of 120,000 lb./in. If the wheel assembly including the counterweight weighs 10,000 lb., and its diameter is 6.0 ft., determine the forward speed of the locomotive at which resonant forced vibration would occur. (Assume that the wheel remains in contact with the track.)

If the wheel is overbalanced by 500 lb. ft., determine the amplitude of vertical vibration of the wheel when running at 80 mph.

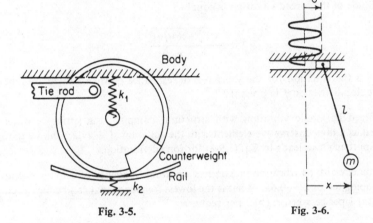

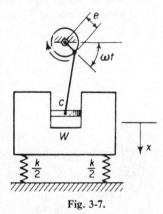

Fig. 3-5.   Fig. 3-6.

25. The point of suspension of a simple pendulum is given a harmonic motion $x_0 = X_0 \sin \omega t$ along a horizontal line, as shown in Fig. 3-6. Write the differential equation of motion for small amplitude of oscillation, using the coordinates shown. Determine the solution for x/x_0 and show that when $\omega = \sqrt{2}\, \omega_n$ the node is found at the mid-point of l. Show that in general the distance h from the mass to the node is given by the relation $h = l(\omega_n/\omega)^2$ where $\omega_n = \sqrt{g/l}$.

26. Assuming viscous damping between the piston and the block, write the differential equation for the motion x of the system shown in, Fig. 3-7. Identify the equation with one of the systems discussed in this chapter and write the solution for the amplitude of the block by comparison.

27. Express the ellipse for the energy dissipated by viscous damping in terms of the damping factor $\zeta = c/c_c$ and determine the semi-major and semi-minor axes.

Fig. 3-7.

28. Show that under structural damping, the plot of the damping force versus displacement is an ellipse with semi-major axis equal to X and semi-minor axis equal to $\gamma k X$.

29. Determine the equation for the logarithmic decrement for a system with structural damping.

30. A system with Coulomb damping f is in forced vibration under harmonic excitation $F_0 \sin \omega t$. For small to moderate damping, the motion is nearly harmonic and we may assume $x = X \sin(\omega t - \phi)$. The energy dissipated per cycle is then $4fX$. Determine the equivalent viscous damping and show that the amplitude of the forced vibration is equal to

$$X = \frac{\sqrt{F_0^2 - (4f/\pi)^2}}{k - m\omega^2}$$

31. What is the equation for the sharpness of resonance Q when a small amount of Coulomb friction f is present?

32. In forced harmonic vibration with structural damping, the spring is often viewed as a nonconservative element with the equation $(1 + i\gamma)kx$. Show that this approach also leads to Eq. (3.9-8) for forced vibration.

33. A commercial-type vibration pickup has a natural frequency of 4.75 c.p.s. and a damping factor $\zeta = 0.65$. What is the lowest frequency that can be measured with (a) 1 per cent error, (b) 2 per cent error?

34. An undamped vibration pickup having a natural frequency of 1 c.p.s. is used to measure a harmonic vibration of 4 c.p.s. If the amplitude indicated by the pickup (relative amplitude between pickup mass and frame) is 0.052 in., what is the correct amplitude?

35. Derive the equation for the relative amplitude Z of the end of the cantilever reed shown in Fig. 3-8 with respect to the base, when the base is subject to harmonic motion $Y_0 \sin \omega t$ in a direction perpendicular to the reed.

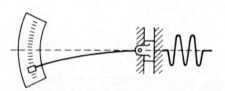

Fig. 3-8.

Fig. 3-9.

36. The shaft of a torsiograph, shown in Fig. 3-9, undergoes harmonic torsional oscillation $\theta_0 \sin \omega t$. Determine the expression for the relative amplitude of the outer wheel with respect to (a) the shaft, (b) a fixed reference.

37. Plot x and z of Example 3.11-2. Discuss the limitations of this instrument in reproducing the vibratory motion.

38. Discuss the requirements of a seismic instrument from the standpoint of limiting phase distortion of complex waves.

39. A solid disk of weight 10 lb. is keyed to the center of a $\frac{1}{2}$-in. steel shaft 2 ft. between bearings. Determine the lowest critical speed. (Assume shaft to be simply supported at the bearings.)

40. A circular disk rotating about its geometric axis has two holes A and B drilled through it. The diameter and position of the holes are $d_A = 1.0$ in., $r_A = 3.0$ in., $\theta_A = 0$ deg.; $d_B = \frac{1}{2}$ in., $r_B = 2.0$ in., $\theta_B = 90$ deg. Determine the diameter and position of a third hole at 1 in. radius that will balance the disk.

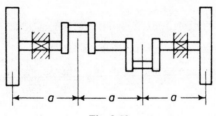

Fig. 3-10.

41. The crank arm and pin of the two-cylinder crankshaft shown in Fig. 3-10 is equivalent to an eccentric weight of w lb. at a radius of r in. Determine the counterweights necessary at the two flywheels if they are also placed at a radial distance of r in.

42. Show that an unbalanced rotor can be balanced by adding weights in two planes, both of which are on one side of the rotor.

43. The rotor of a turbine weighing 30 lb. is supported at the midspan of a shaft with bearings 16 in. apart, as shown in Fig. 3-11. The rotor is known to have an unbalance of 4 oz. in. Determine the forces exerted on the bearings at a speed of 6000 r.p.m. if the diameter of the steel shaft is 1.0 in. Compare this result with that of the same rotor mounted on a steel shaft of diameter $\frac{3}{4}$ in. (Assume the shaft to be simply supported at the bearings.)

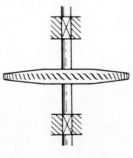

Fig. 3-11.

44. Using the results of Example 3.1-1, show that if damping is small the amplitude of lateral vibration of a shaft at the critical speed build up according to the equation

$$r = \frac{e}{2\zeta}(1 - e^{-\zeta \omega_n t})$$

where e is the eccentricity.

45. For turbines operating above the critical speed, stops are provided to limit the amplitude as it runs through the critical speed. In the turbine of Prob. 43, if the clearance between the 1-in. shaft and the stops is 0.02 in., and the eccentricity is $\frac{1}{120}$ in., determine the time required for the shaft to hit the stops, assuming that the critical speed is reached with zero amplitude.

46. The thin disk of Fig. 3.15-1 is rotated counterclockwise at 900 r.p.m. The following data were obtained by using a portable dynamic balancing equipment. (All angles were measured counterclockwise from an arbitrary index point 0.) amplitude and angular position due to initial unbalance

$$= 0.006 \text{ in. at } 300°$$

trial weight $W_t = 5$ grams at 75°
amplitude and angular position with trial weight added

$$= 0.010 \text{ in. at } 45°$$

Determine the magnitude and location of the weight required to balance the disk.

4

Transient and Nonperiodic Vibration

When a dynamical system is excited by a suddenly applied nonperiodic excitation such as a pulse, the response to such excitations are called transients since steady-state oscillations are generally not produced. Such oscillations take place at the natural frequency of the system where the amplitude will vary, depending on the type of excitation.

We first study the response of a spring mass system to an impulse excitation because this case is important in the understanding of the more general problem of transients.

4.1 Impulsive Excitation

We frequently encounter a force of very large magnitude which acts for a very short time, but with a time integral which is finite. Such forces are called impulsive and their time integral is designated by the symbol $\hat{F}$ (lb-sec) and defined by the equation

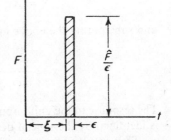

Fig. 4.1-1. Impulsive force.

$$\hat{F} = \int_t^{t+\epsilon} F \, dt \qquad (4.1\text{-}1)$$

Figure 4.1-1 shows an impulsive force of magnitude $\hat{F}/\epsilon$ with a time duration of ϵ. As ϵ approaches zero, such forces tend to become infinite; however, the impulse defined by its time integral is $\hat{F}$. When $\hat{F}$ is equal to unity, such force in the limiting case $\epsilon \to 0$ is called either the *unit impulse* or the *delta function*, and is identified by the symbol $\delta(t - \xi)$ where the latter function has the following properties:

$$\delta(t + \xi) = 0 \qquad \text{for all } t \neq \xi$$

$$\int_0^\infty \delta(t - \xi) \, d\xi = 1.0 \tag{4.1-2}$$

$$\int_0^\infty f(\xi)\delta(t - \xi) \, d\xi = f(\xi)$$

Since $F \, dt = m \, dv$ the impulse $\hat{F}$ acting on the mass will result in a sudden change in its velocity without an appreciable change in its displacement. Thus a spring-mass system excited by an impulse $\hat{F}$ responds in free vibration with initial conditions $x(0) = 0$ and $\dot{x}(0) = v_0 = \hat{F}/m$. Under free-vibration we have found that the undamped spring mass system with initial conditions behaves according to [see Eq. (1.2-5)]:

$$x = \frac{v_0}{\omega_n} \sin \omega_n t + x_0 \cos \omega_n t$$

and hence by letting $x_0 = 0$ and $v_0 = \hat{F}/m$ the response to the impulse $\hat{F}$ is

$$x = \frac{\hat{F}}{m\omega_n} \sin \omega_n t \tag{4.1-3}$$

where $\omega_n = \sqrt{k/m}$.

When damping is present we can start with the free vibration equation

$$x = Xe^{-\zeta\omega_n t} \sin (\sqrt{1 - \zeta^2} \, \omega_n t - \phi)$$

and substituting the above initial conditions we would arrive at the equation

$$x = \frac{\hat{F}}{m\omega_n\sqrt{1 - \zeta^2}} e^{-\zeta\omega_n t} \sin \sqrt{1 - \zeta^2} \, \omega_n t \tag{4.1-4}$$

The response to the unit impulse is of importance to our later discussions, and is identified by the special designation $g(t)$.

Thus in either the undamped or damped case the equation for the impulsive response can be expressed in the form

$$x = \hat{F}g(t) \tag{4.1-5}$$

Figure 4.1-2 shows Eq. (4.1-4) plotted in nondimensional form with ζ as parameter.

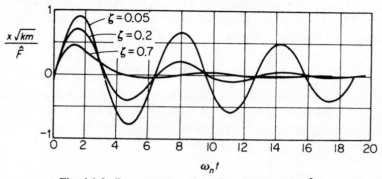

Fig. 4.1-2. Response to a delta function impulse $\hat{F}$.

4.2 Arbitrary Excitation

Having the response $g(t)$ to a unit impulse excitation, it is possible to establish the equation for the response of the system excited by an arbitrary force $f(t)$. For this development, we consider the arbitrary pulse to be a series of impulses as shown in Fig. 4.2-1. If we examine one of the impulses (shown cross-hatched) which starts at time ξ, its magnitude is $\hat{F} = f(\xi)\,\Delta\xi$ and its contribution to the response at the time t is found by replacing the time with the elapsed time $(t - \xi)$.

$$f(\xi)\,\Delta\xi g(t - \xi) \tag{4.2-1}$$

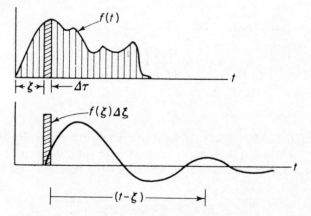

Fig. 4.2-1. Arbitrary pulse as a series of impulses.

Since the system we are considering is linear, the principle of superposition holds. Thus by summing all such contributions the response to the arbitrary excitation $f(t)$ is represented by the integral

$$x(t) = \int_0^t f(\xi)g(t - \xi)\, d\xi \tag{4.2-2}$$

The above integral is called the Duhamel integral or is sometimes referred to as the superposition integral. Note that $f(\xi) = 0$ for ξ greater than the pulse time, so that the upper limit of the integral remains equal to the pulse time for response beyond the pulse time.

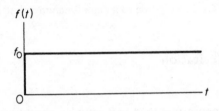

Fig. 4.3-1. Step function excitation.

4.3 Response to Step Excitation

Let us determine the response to the step excitation shown in Fig. 4.3-1, $f(t) = F_0$. Considering the undamped system, we have

$$g(t) = \frac{1}{m\omega_n} \sin \omega_n t \tag{4.3-1}$$

and by substituting into Eq. (4.2-2) we obtain the response as

$$x(t) = \frac{F_0}{m\omega_n} \int_0^t \sin \omega_n(t - \xi)\, d\xi$$

$$= \frac{F_0}{k}(1 - \cos \omega_n t) \tag{4.3-2}$$

We find then that the peak response to the step excitation is equal to twice the value F_0/k, the displacement if the load were applied slowly to the system.

For a damped system the procedure can be repeated with

$$g(t) = \frac{e^{-\zeta\omega_n t}}{m\omega_n\sqrt{1 - \zeta^2}} \sin \sqrt{1 - \zeta^2}\, \omega_n t \tag{4.3-3}$$

or, alternatively, we can simply consider the differential equation

$$\ddot{x} + 2\zeta\omega_n\dot{x} + \omega_n^2 x = \frac{F_0}{m} \qquad (4.3\text{-}4)$$

whose solution is the sum of the solutions to the homogeneous equation and

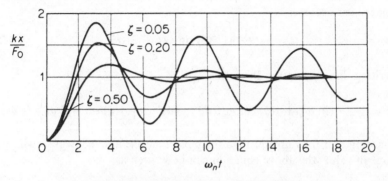

Fig. 4.3-2. Response to a unit step function.

that of the particular solution which for this case is $F_0/m\omega_n^2$. Thus the equation

$$x(t) = Xe^{-\zeta\omega_n t} \sin\left(\sqrt{1 - \zeta^2}\,\omega_n t - \phi\right) + \frac{F_0}{m\omega_n^2} \qquad (4.3\text{-}5)$$

fitted to the initial conditions of $x(0) = \dot{x}(0) = 0$ will result in the solution which is given as

$$x = \frac{F_0}{k}\left[1 - \frac{e^{-\zeta\omega_n t}}{\sqrt{1 - \zeta^2}} \cos\left(\sqrt{1 - \zeta^2}\,\omega_n t - \psi\right)\right] \qquad (4.3\text{-}6)$$

where

$$\tan\psi = \frac{\zeta}{\sqrt{1 - \zeta^2}}$$

Figure 4.3-2 shows a plot of xk/F_0 versus $\omega_n t$ with ζ as parameter, and it is evident that the peak response is less than $2F_0/k$ when damping is present.

4.4 Base Excitation

Often the support of the dynamical system is subjected to a sudden movement specified by its displacement, velocity or acceleration. As shown in Sec. 3.11, the equation of motion can be expressed in terms of the relative displacement $z = x - y$ as follows:

$$\ddot{z} + 2\zeta\omega_n\dot{z} + \omega_n^2 z = -\ddot{y} \qquad (4.4-1)$$

and hence all of the results for the force-excited system apply to the base-excited system for z when the term F_0/m is replaced by $-\ddot{y}$ or the negative of the base acceleration.

For an undamped system initially at rest, the solution for the relative displacement becomes

$$z = -\frac{1}{\omega_n} \int_0^t \ddot{y}(\xi) \sin \omega(t - \xi) \, d\xi \qquad (4.4-2)$$

Rewriting this equation in the form

$$z = -\frac{1}{\omega_n}\left\{\sin \omega_n t \int_0^t \ddot{y}(\xi) \cos \omega_n\xi \, d\xi - \cos \omega_n t \int_0^t \ddot{y}(\xi) \sin \omega_n\xi \, d\xi\right\} \quad (4.4-3)$$

and noting that for t greater than the pulse time the above integrals will not change in value, the above equation can be written as

$$z = -\frac{1}{\omega_n} A \sin (\omega_n t - \phi) \quad (4.4-4)$$

where we have made the substitution

$$A \sin \phi = \int_0^{t_p} \ddot{y}(\xi) \sin \omega_n\xi \, d\xi$$
$$A \cos \phi = \int_0^{t_p} \ddot{y}(\xi) \cos \omega_n\xi \, d\xi \qquad (4.4-5)$$

where t_p = pulse time. Thus the motion of the system after the passage of the pulse is harmonic, and if the base returns to its original position at $t = t_p$, $z = x$ for $t > t_p$.

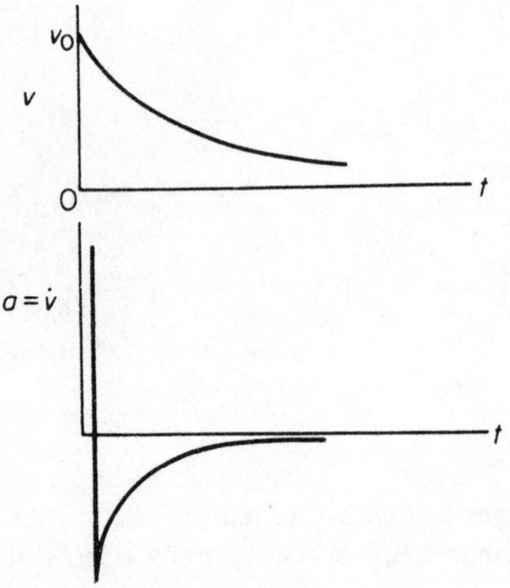

EXAMPLE 4.4-1. Consider an undamped spring-mass system where the motion of the base is specified by a velocity pulse of the form

$$\dot{y}(t) = v_0 e^{-t/t_0}$$

Fig. 4.4-1.

which is shown in Fig. 4.4-1 together with its time rate of change $a = \dot{v}$.

Solution. The acceleration of the base is then equal to

$$\ddot{y}(t) = v_0\delta(t) - \frac{v_0}{t_0} e^{-t/t_0}$$

where $\delta(t)$ is the delta function at $t = 0$. The delta function in this equation is easily justified when we recognize the fact that

$$\int_0^t v_0 \, \delta(t) \, dt = v_0$$

and that $\delta(t) = 0$ for $t > 0$.
Substitution of $\ddot{y}(t)$ into Eq. (4.4-2), yields

$$z(t) = -\frac{v_0}{\omega_n} \int_0^t \{\delta(\xi) - \frac{1}{t_0} e^{-\xi/t_0}\} \sin \omega_n(t - \xi) \, d\xi$$

$$= \frac{v_0 t_0}{1 + (\omega_n t_0)^2} \{e^{-t/t_0} - \omega_n t_0 \sin \omega_n t - \cos \omega_n t\}$$

4.5 Shock Spectrum

A shock is a disturbance or excitation pulse of displacement, velocity, acceleration, or force, whose duration is short compared to the characteristic period of the system. The application of this shock pulse to a single-degree-of-freedom system (which for convenience we will refer to as an oscillator), results in a time response of the oscillator. The maximum value of the time response, for a given shock pulse, depends on the natural frequency and damping of the oscillator. The plot of the maximum response of the oscillator against the natural frequency of the oscillator is the shock spectrum of the disturbance.

As an example, for a base excitation of $\ddot{y}(t)$ the relative displacement z is given by Eq. (4.4-2) as

$$z = -\frac{1}{\omega} \int_0^t \ddot{y}(\xi) \sin \omega(t - \xi) \, d\xi$$

Thus its maximum value

$$z_{\text{max}} = \left[-\frac{1}{\omega} \int_0^t \ddot{y}(\xi) \sin \omega(t - \xi) \, d\xi \right]_{\text{max}} \tag{4.5-1}$$

plotted against ωt_1, where t_1 is some characteristic pulse time, would represent a shock spectrum.

EXAMPLE 4.5-1. Determine the shock spectrum of a step function as a function of the damping.

Solution. The response of a damped oscillator to a step function of magnitude F_0 was found in Sec. 4.3 and given as Eq. (4.3-6), which is rewritten as

$$\frac{xk}{F_0} = 1 - \frac{e^{-\zeta\omega_n t}}{\sqrt{1 - \zeta^2}} \cos\left(\sqrt{1 - \zeta^2}\,\omega_n t - \psi\right)$$

$$\tan\psi = \frac{\zeta}{\sqrt{1 - \zeta^2}}$$

By differentiating and setting the velocity equal to zero, we find the time t_p corresponding to the peak response.

$$\frac{\dot{x}k}{F_0} = \omega_n e^{-\zeta\omega_n t}\left[\sin\left(\sqrt{1 - \zeta^2}\,\omega_n t - \psi\right) + \frac{\zeta}{\sqrt{1 - \zeta^2}}\cos\left(\sqrt{1 - \zeta^2}\,\omega_n t - \psi\right)\right]$$

$$= \frac{\omega_n e^{-\zeta\omega_n t}}{\sqrt{1 - \zeta^2}}\left[\cos\psi\sin\left(\sqrt{1 - \zeta^2}\,\omega_n t - \psi\right) + \sin\psi\cos\left(\sqrt{1 - \zeta^2}\,\omega_n t - \psi\right)\right]$$

$$= \frac{\omega_n e^{-\zeta\omega_n t}}{\sqrt{1 - \zeta^2}}\sin\sqrt{1 - \zeta^2}\,\omega_n t$$

Thus the time corresponding to the peak response is found from

$$\sin\sqrt{1 - \zeta^2}\,\omega_n t_p = 0,$$

or

$$\omega_n t_p = \frac{\pi}{\sqrt{1 - \zeta^2}}$$

Substituting this value into the displacement equation, the peak response becomes

$$\left(\frac{xk}{F_0}\right)_{max} = 1 - \frac{e^{-\pi\zeta/\sqrt{1-\zeta^2}}}{\sqrt{1 - \zeta^2}}\cos\left(\pi - \tan^{-1}\frac{\zeta}{\sqrt{1 - \zeta^2}}\right)$$

$$= 1 + e^{-\pi\zeta/\sqrt{1-\zeta^2}}$$

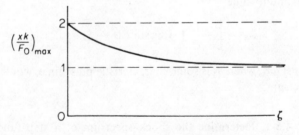

Fig. 4.5-1. Shock spectrum of a step function as a function of ζ.

We find in this case that the peak response is independent of the natural frequency ω_n of the oscillator and depends only on the damping ζ. A plot of this equation appears in Fig. 4.5-1.

EXAMPLE 4.5-2. Determine the undamped shock spectrum for a step function with a rise time t, shown in Fig. 4.5-2.

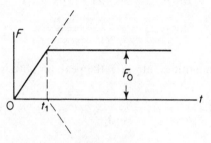

Fig. 4.5-2.

Solution. The input can be considered to be the sum of two ramp functions $F_0(t/t_1)$, the second of which is negative and delayed by the time t_1. For the first ramp function the terms of Duhamel's equation are:

$$f(t) = F_0(t/t_1)$$

$$g(t) = \frac{1}{m\omega_n} \sin \omega_n t = \frac{\omega_n}{k} \sin \omega_n t$$

and the response becomes

$$x(t) = \frac{\omega_n}{k} \int_0^t \frac{F_0\xi}{t_1} \sin \omega_n(t - \xi)\, d\xi$$

$$= \frac{F_0}{k}\left(\frac{t}{t_1} - \frac{\sin \omega_n t}{\omega_n t_1}\right), \qquad t < t_1$$

For the second ramp function starting at t_1, the solution can be written down by inspection of the above equation as

$$x(t) = -\frac{F_0}{k}\left[\frac{(t - t_1)}{t_1} - \frac{\sin \omega_n(t - t_1)}{\omega_n t_1}\right]$$

By superimposing these two equations the response for $t > t_1$ becomes

$$x(t) = \frac{F_0}{k}\left[1 - \frac{\sin \omega_n t}{\omega_n t_1} + \frac{1}{\omega_n t_1} \sin \omega_n(t - t_1)\right] \qquad t > t_1$$

Differentiating and equating to zero, the peak time is obtained as

$$\tan \omega_n t_p = \frac{1 - \cos \omega_n t_1}{\sin \omega_n t_1}$$

Since $\omega_n t_p$ must be greater than π, we also obtain

$$\sin \omega_n t_p = -\sqrt{\tfrac{1}{2}(1 - \cos \omega_n t_1)}$$

$$\cos \omega_n t_p = \frac{-\sin \omega_n t_1}{\sqrt{2(1 - \cos \omega_n t_1)}}$$

Substituting these quantities into $x(t)$, the peak amplitude is found as

$$\left(\frac{xk}{F_0}\right)_{\max} = 1 + \frac{1}{\omega_n t_1}\sqrt{2(1 - \cos 2\omega_n t_1)}$$

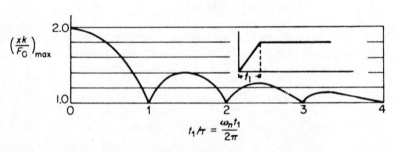

Fig. 4.5-3. Shock spectrum for step-ramp function.

Letting $\tau = 2\pi/\omega_n$ be the period of the oscillator, the above equation is plotted against t_1/τ in Fig. 4.5-3.

EXAMPLE 4.5-3. Determine the shock spectrum for the base velocity input, $\dot{y}(t) = v_0 e^{-t/t_0}$ of Example 4.4-1.

Solution. The relative displacement $z(t)$ was found in Example 4.4-1 to be

$$z(t) = \frac{v_0 t_0}{1 + (\omega_n t_0)^2} \times (e^{-t/t_0} - \omega_n t_0 \sin \omega_n t - \cos \omega_n t)$$

To determine the peak value z_p, the usual procedure is to differentiate the equation with respect to time t, set it equal to zero, and substitute this time back into the equation for $z(t)$. It is evident that for this problem this results in a transcendental equation which must be solved by plotting. To avoid this numerical task, we will consider a different approach as follows.

For a very stiff system, which corresponds to large ω_n, the peak response will certainly occur at small t, and we would obtain for the time varying part of the equation the peak value

$$(1 - \omega_n t_0 - 1) \cong -\omega_n t_0$$

Thus for large ω_n the peak value will be nearly equal to

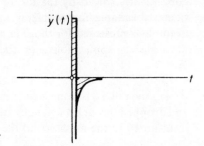

$$|z_p| \cong \frac{v_0 t_0}{1 + (\omega_n t_0)^2}\,(\omega_n t_0) \cong \frac{v_0 t_0}{\omega_n t_0}$$

so that $\dfrac{z_p}{v_0 t_0}$ plots against $\omega_n t_0$ as a rectangular hyperbola.

Fig. 4.5-4. Impulsive doublet.

For small ω_n, or a very soft spring, the duration of the input would be small compared to the period of the system. Hence the input would appear as an impulsive doublet shown in Fig. 4.5-4 with the equation $v_0 t_0 \delta'(t)$. The solution for $z(t)$ is then

$$z(t) = v_0 t_0 \cos \omega_n t$$

and its peak value is

$$|z_p| \cong v_0 t_0$$

With these extreme conditions evaluated, we can now fill in the shock spectrum which is shown in Fig. 4.5-5.

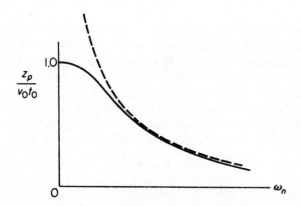

Fig. 4.5-5. Shock spectrum for the base velocity input $\dot{y}(t) = v_0 e^{-t/t_0}$.

4.6 Solution by Laplace Transforms

Differential equations with given initial or boundary conditions can be conveniently solved by the use of Laplace transforms. The method is particularly suitable for problems in transient vibrations, and the following example is presented for those familiar with the transform theory, mainly to illustrate its application. For the theory of Laplace transformation, see Ref. 7.

EXAMPLE 4.6-1. A mass m is packaged in a box, as shown in Fig. 4.6-1, and dropped through a height h. It is desired to determine the maximum force transmitted to the mass m and the required rattle space.

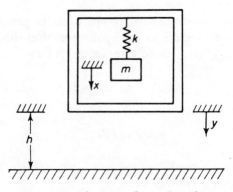

Fig. 4.6-1. Drop test of a packaged mass.

We make the following idealized assumptions: (1) The mass m is supported within the box by a linear spring of stiffness k lb/in. (2) The mass of the box is large compared to that of the enclosed mass m, so that the free fall of the box is not influenced by the relative motion of the mass m. (3) On striking the floor, the box remains in contact with the floor.

Letting x be the displacement of m relative to the box, measured downward from the static equilibrium position, and y the displacement of the box from the starting position, the general equation of motion of m is

$$m(\ddot{x} + \ddot{y}) = -kx \qquad (4.6\text{-}1)$$

With $\omega_n^2 = k/m$, this equation becomes

$$\ddot{x} + \omega_n^2 x = -\ddot{y} \qquad (4.6\text{-}2)$$

For initial conditions $x(0)$, $\dot{x}(0)$, $y(0)$, $\dot{y}(0)$, the Laplace transform of the above equation is

$$\bar{x}(s) = [x(0) + y(0)] \frac{s}{s^2 + \omega_n^2} + [\dot{x}(0) + \dot{y}(0)] \frac{1}{s^2 + \omega_n^2} - \frac{s^2 \bar{y}(s)}{s^2 + \omega_n^2} \quad (4.6\text{-}3)$$

and its inverse can be written as

$$x(t) = [x(0) + y(0)] \cos \omega_n t + \frac{1}{\omega_n} [\dot{x}(0) + \dot{y}(0)] \sin \omega_n t - \mathscr{L}^{-1} \frac{s^2 \bar{y}(s)}{s^2 + \omega_n^2}$$
$$(4.6\text{-}4)$$

We can now specialize this equation for the conditions of our problem. Of particular interest are the displacement $x(t_0)$ and the velocity $\dot{x}(t_0)$ of m at the time t_0 when the box strikes the floor.

The initial conditions for the free-fall interval are $x(0) = y(0) = \dot{x}(0) = \dot{y}(0) = 0$, and the motion of the box and its transform are

$$y(t) = \tfrac{1}{2} g t^2, \qquad \bar{y}(s) = \frac{g}{s^3} \quad (4.6\text{-}5)$$

The motion of m during free fall is then found from substituting Eq. (4.6-5) into (4.6-4) with the zero initial conditions, which gives

$$x(t) = -\mathscr{L}^{-1} \frac{g}{s(s^2 + \omega_n^2)} = -\frac{g}{\omega_n^2} (1 - \cos \omega_n t) \quad (4.6\text{-}6)$$

Since the time to fall through the height h is $t_0 = \sqrt{2h/g}$, we arrive at the quantities of interest as

$$x(t_0) = -\frac{g}{\omega^2} (1 - \cos \omega_n t_0)$$

$$\dot{x}(t_0) = -\frac{g}{\omega_n} \sin \omega_n t_0.$$

These quantities become the initial conditions for the second phase of the problem after impact of the box with the floor.

Redefining the time from the instant of impact, the initial conditions for the second phase of the problem are,

$$y(0) = 0, \qquad x(0) = -\frac{g}{\omega_n^2} (1 - \cos \omega_n t_0)$$

$$\dot{y}(0) = g t_0, \qquad \dot{x}(0) = -\frac{g}{\omega_n} \sin \omega_n t_0$$

From the general equation, Eq. (4.6-4), the equation for the displacement of

m after impact becomes

$$x(t) = -\frac{g}{\omega_n^2}(1 - \cos \omega t_0) \cos \omega_n t + \left(\frac{g t_0}{\omega_n} - \frac{g}{\omega_n^2} \sin \omega_n t_0\right) \sin \omega_n t$$

$$= \frac{g}{\omega_n^2}\sqrt{(1 - \cos \omega_n t_0)^2 + (\omega_n t_0 - \sin \omega_n t_0)^2} \sin (\omega_n t - \phi) \quad (4.6\text{-}7)$$

where

$$\tan \phi = \frac{(1 - \cos \omega_n t_0)}{(\omega_n t_0 - \sin \omega_n t_0)} \quad (4.6\text{-}8)$$

Thus the maximum amplitude attained by m is

$$X_1 = \frac{g}{\omega_n^2}\sqrt{(1 - \cos \omega_n t_0)^2 + (\omega_n t_0 - \sin \omega_n t_0)^2} \quad (4.6\text{-}9)$$

which occurs at the time $(\omega_n t_1 - \phi) = \pi/2$. The maximum force is simply kX_1.

For any drop height h, or drop time $t_0 = \sqrt{2h/g}$, X_1 has a maximum value of h for $\omega_n \to 0$. This can be shown by replacing $\sin \omega_n t_0$ and $\cos \omega_n t_0$ with

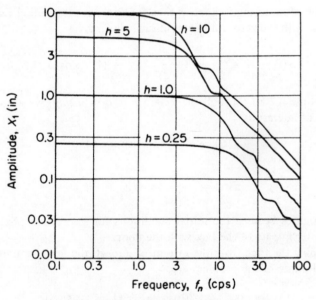

Fig. 4.6-2. Displacement spectra for drop test.

their series form and allowing $\omega_n t_0$ to approach zero. Figure 4.6-2 shows a displacement spectrum of X_1 versus frequency $f_n = (1/2\pi)\sqrt{k/m}$ for $h = 10$, 5, 1, and 0.25 in. Equation (4.6-9) however, indicates that $\omega_n^2 X_1/g = X_1/\delta_{st}$

is a function only of

$$\omega_n t_0 = \omega_n \sqrt{2h/g} = \sqrt{2h/\delta_{st}} \qquad (4.6\text{-}10)$$

so that the curves of Fig. 4.6-2 plot as a single nondimensional curve as shown in Fig. 4.6-3.

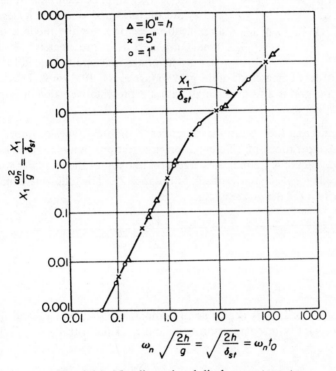

Fig. 4.6-3. Nondimensional displacement spectra.

4.7 The Analogue Computer

Our brief encounter with transients and shock spectra sufficiently indicates the algebraic difficulties confronted in even the very simple problems. Such problems are conveniently solved by the analogue computer which is ideally suited for the solution of ordinary differential equations. These computers are also capable of solving nonlinear problems for which analytic techniques are either nonexistent or too complex for practical use.

The basic element of the analogue computer is the high gain d.c. operational amplifier, shown symbolically in Fig. 4.7-1. It is not necessary, however, to know the details of its electronics. Such amplifiers are characterized by the equation

$$e_0 = -\mu e_g \qquad (4.7\text{-}1)$$

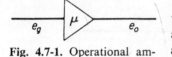

Fig. 4.7-1. Operational amplifier.

where μ is the amplification factor, and e_g and e_0 are the grid and output voltages. The amplification factor μ for the modern operational amplifiers is approximately 10^8. Since the output voltage is generally limited to ± 100 volts, the order of magnitude of the grid voltage e_g is $\pm 10^{-6}$ volts. The current drawn by the grid is also quite small with a representative value of approximately 10^{-7} amps.

By connecting the operational amplifier to different types of impedances the various operations of differentiation, integration, summing, etc., can be performed. Figure 4.7-2 indicates a general hookup of the amplifier with an input impedance Z_i and a feedback impedance Z_f. The following equations can be written for the above circuit.

$$e_i - e_g = i_i Z_i$$

$$e_g - e_0 = i_f Z_f$$

$$i_i = i_f + i_g$$

$$e_0 = -\mu e_g$$

In these equations, e_g is negligible in comparison to e_0 and e_i, and i_g is a negligible quantity in comparison to i_i and i_f. Thus with $e_g = i_g = 0$, the

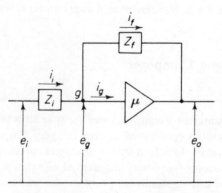

Fig. 4.7-2. Operational amplifier circuit.

above equations become,

$$e_i = i_i Z_i$$

$$-e_0 = i_f Z_f$$

$$i_i = i_f$$

From these, we obtain the output-input relationship

$$\frac{e_0}{e_i} = -\frac{Z_f}{Z_i} \tag{4.7-2}$$

which is fundamental to the analogue computer.

(a) *Sign Change.* The simplest operation is that of changing sign. By letting $Z_i = R_i$ and $Z_f = R_f$, Eq. (4.7-2) becomes

$$e_0 = -\frac{R_f}{R_i} e_i \tag{4.7-3}$$

so that if $R_f = R_i$ we simply get

$$e_0 = -e_i$$

The circuit for the sign change is shown in Fig. 4.7-3.

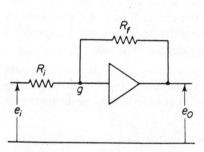

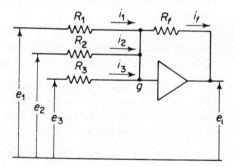

Fig. 4.7-3. Circuit for scale factor, R_f/R_i.   **Fig. 4.7-4.** Circuit for summation.

(b) *Summation.* If more than one input is connected to point g, as shown in Fig. 4.7-4, then i_f is the sum of the input currents:

$$i_1 + i_2 + i_3 = i_f$$

$$\frac{e_1}{R_1} + \frac{e_2}{R_2} + \frac{e_3}{R_3} = -\frac{e_0}{R_f}$$

and the output voltage is given by the sum

$$e_0 = -\sum_i \frac{R_f}{R_i} e_i \qquad (4.7\text{-}4)$$

If all the resistances are equal, this then leads to the sum of the input voltages.

$$e_0 = -\sum_i e_i$$

(c) *Integration.* If the feedback impedance is a capacitor C, as shown in Fig. 4.7-5, the circuit will perform the function of integration. With initial voltage $e(0)$ across the capacitor, the capacitor voltage at any time t is (remembering that $e_g \cong 0$):

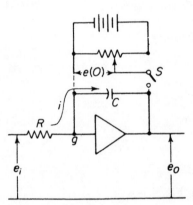

$$e_0 = -\frac{1}{C} \int_0^t i\, dt + e(0)$$

But $i = e_i/R$ so that the above equation becomes

$$e_0 = -\frac{1}{RC} \int_0^t e_i\, dt + e(0) \qquad (4.7\text{-}5)$$

With $R = 1$ megohm and $C = 1$ microfarad, $RC = 1$ second, and the computer time is directly in seconds.

Fig. 4.7-5. Integrating circuit with initial conditions.

The initial voltage $e(0)$ is obtained from the circuit shown in dotted lines by closing switch S prior to starting the computation. When the computation is initiated, the switch S is simultaneously opened by a relay.

(d) *Differentiation.* Differentiation is avoided in analogue computers because the unavoidable noise signal in the input is amplified by μ, thereby causing the amplifier to saturate. Instead, it is usually possible to rearrange the equations for integration.

(e) *Voltage Division.* The potentiometer, often used to obtain a fraction k times the input voltage

$$e_2 = ke_1$$

is designated symbolically in Fig. 4.7-6(a). The fractional setting k on potentiometers holds only when the output is open circuited. When a load resistance R_L is placed across the output, as in Fig. 4.7-6(b), the output voltage can be shown to be equal to

$$e_2 = ke_1 \left[\frac{1}{1 + \dfrac{R}{R_L} k(1 - k)} \right] \qquad (4.7\text{-}6)$$

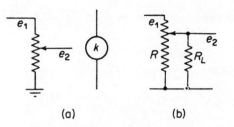

Fig. 4.7-6. Voltage division.

where R is the resistance of the potentiometer. It is evident that this equation approaches ke_1 when $R/R_L \rightarrow 0$.

(*f*) *Multiplication.* Multiplication is one of the most difficult operations for the analogue computer. In one method, the principle of the unloaded potentiometer is exploited by using a servo-drive together with a ganged potentiometer, as shown in Fig. 4.7-7. The first potentiometer is connected to ± 100 volts, whereas the second potentiometer is connected to $\pm e_2$, the voltage to be multiplied by e_1.

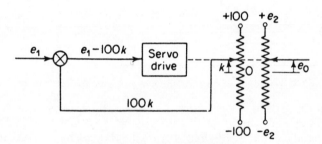

Fig. 4.7-7. Servo-multiplier.

The function of the servo drive is to position the slider of the ganged potentiometer to zero the error between the output of the first potentiometer and the input voltage e_1. Since the output of each potentiometer is proportional to k we have

$$e_1 = 100k \quad \text{and} \quad e_0 = ke_2$$

Eliminating k, we obtain the equation

$$e_0 = \frac{e_1 e_2}{100} \tag{4.7-7}$$

(*g*) *Computer Circuit for the Single-Degree-of-Freedom System.* The use of the analogue computer for the solution of the single-degree-of-freedom linear system is demonstrated by the circuit of Fig. 4.7-8. The equation represented is

$$m\ddot{x} + c\dot{x} + kx = F(t)$$

which is rearranged to

$$\ddot{x} = -2\zeta\omega_n\dot{x} - \omega_n^2 x + \frac{1}{m}F(t)$$

Assuming the input of the first amplifier to be $\ddot{x}$, its output is $-\dot{x}$, etc. The voltages in the three terminals ①, ②, and ③, which are equal to $\ddot{x}$ are those of the right side of the above equation. Note that the sign changes

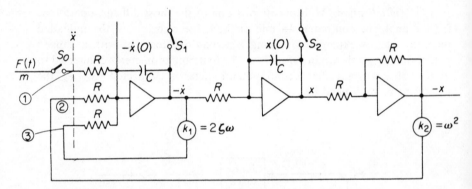

Fig. 4.7-8. Analogue circuit for the single degree of freedom system.

across each amplifier and that the potentiometer settings are for $2\zeta\omega_n$ and ω_n^2. The initial conditions $x(0)$ and $\dot{x}(0)$ are the voltages on the two capacitors at time $t = 0$. When the computer is set into operation by closing switch S_0, the switches S_1 and S_2 open simultaneously.

(h) *Scale Change*. If a linear transformation is made in the variables of the problem, the characteristics of the system remain unchanged, and we merely expand or contract the scale of the variables. In solving a problem on a computer, it is often necessary to make such a scale change because of the limitations of the computer. A change in the time scale may be necessary because the actual vibration may be too high in frequency for the computer and the recorder to follow. On the other hand, if the actual vibration is too low in frequency, the computer time may become too long, thereby introducing errors from drifts. Amplitude scaling is also necessary in order to operate the amplifiers within the limits of ±100 volts. For maximum accuracy the peak response should be close to the ±100-volt limits.

Assume that the equation for the actual system is

$$\ddot{x}(t) + 2\zeta\omega_n\dot{x}(t) + \omega_n^2 x(t) = \frac{1}{m} F(t) \qquad (4.7\text{-}8)$$

with the initial conditions

$$\dot{x}(0) \quad \text{and} \quad x(0)$$

If we wish to change the time scale of the problem by α, we let

$$\tau = \alpha t \qquad (4.7\text{-}9)$$

The derivatives are then related by the equations

$$\alpha \frac{d}{d\tau} = \frac{d}{dt}, \qquad \alpha^2 \frac{d^2}{d\tau^2} = \frac{d^2}{dt^2} \qquad (4.7\text{-}10)$$

and the original differential equation with its initial conditions becomes

$$\alpha^2 \frac{d^2 x(\tau)}{d\tau^2} + 2\zeta\omega_n\alpha \frac{dx(\tau)}{d\tau} + \omega_n^2 x(\tau) = \frac{1}{m} F(\tau)$$

$$\frac{dx(0)}{d\tau} = \frac{1}{\alpha} \frac{dx(0)}{dt}, \qquad x_r(0) = x_t(0) \qquad (4.7\text{-}11)$$

Dividing through by α^2, this equation takes the form

$$\frac{d^2 x(\tau)}{d\tau^2} + 2\zeta\left(\frac{\omega_n}{\alpha}\right)\frac{dx(\tau)}{d\tau} + \left(\frac{\omega_n}{\alpha}\right)^2 x(\tau) = \frac{1}{\alpha^2}\frac{F(\tau)}{m} \qquad (4.7\text{-}12)$$

which shows that the natural frequency of the system has been changed from ω_n to $\Omega = \omega_n/\alpha$. The damping factor ζ, however, has not been changed since the critical damping $c_{cr} = 2m\omega_n$ for the original equation has been changed to $c_{cr} = 2m\Omega$ for the new equation.

Theoretically, it is possible to devise a computer circuit to solve this new equation and interpret its results in terms of the original variables. However, there are problems concerning the orders of magnitudes of the voltages of the computer circuit which need further attention, and which can best be discussed in terms of the following example.

EXAMPLE. The equation for a certain mechanical system excited by a step load $f(t) = 2000$ lb. is given as

$$0.10\ddot{x} + 5\dot{x} + 4000x = 2000 \text{ lb.}$$

with initial conditions

$$\dot{x}(0) = -20 \text{ in./sec.} \quad \text{and} \quad x(0) = 0.25 \text{ in.}$$

Rewrite the equation for the computer and establish a workable circuit for its computation.

Solution. Writing the equation in the form

$$\ddot{x} + 50\dot{x} + 40{,}000x = 20{,}000$$

the natural frequency of the system is found to be

$$\omega_n = \sqrt{k/m} = \sqrt{40{,}000} = 200 \text{ rad./sec.}$$

This frequency is too high for the computer, so we arbitrarily choose $\alpha = 100$ to slow down the time scale by a factor of 100. The new equation with τ as the independent variable is then [see Eq. (4.7-12)]:

$$\frac{d^2x(\tau)}{d\tau^2} + 0.50 \frac{dx(\tau)}{d\tau} + 4.0x(\tau) = 2.0$$

with the initial conditions

$$\frac{dx(0)}{d\tau} = \frac{-20}{100} = -0.20, \qquad x_r(0) = 0.25$$

and we observe that the natural frequency has been reduced to $\Omega = 2$ rad./sec

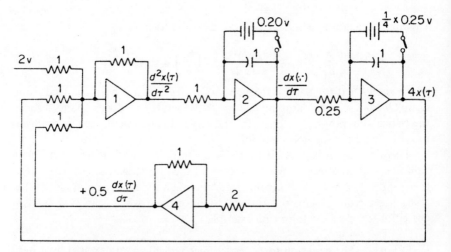

Fig. 4.7-9. Analogue circuit for single degree of freedom system.

If we ignore orders of magnitudes, the following circuit of Fig. 4.7-9 would satisfy these equations. However, for the computer to give reliable results, the output voltages of the amplifiers must not exceed ± 100 volts, nor be too small. For these considerations, it is necessary to make an estimate of the peak displacement, velocity, and acceleration to be encountered in the revised differential equation, and establish proper scale factors which will give peak output voltages of amplifiers near their allowable maximum of ± 100 volts.

From the revised equation, we recognize that the damping will eventually eliminate all oscillations and the final displacement attained will be equal to

$$x(\tau)_{\tau=\infty} = \frac{2.0}{4.0} = 0.5 \text{ in.}$$

We also note that without damping, the peak amplitude under a step function excitation is twice above value, or

$$\underset{\text{max}}{x(\tau)} = 1.0 \text{ in.}$$

The peak velocity and acceleration to be encountered can be estimated on the basis of harmonic motion to be

$$\left[\frac{dx(\tau)}{d\tau}\right]_{\text{max}} = \Omega \underset{\text{max}}{x(\tau)} = 2 \text{ in./sec.}$$

$$\left[\frac{d^2x(\tau)}{d\tau^2}\right]_{\text{max}} = \Omega^2 \underset{\text{max}}{x(\tau)} = 4 \text{ in./sec.}^2$$

If we now examine the circuit of Fig. 4.7-9 with the above maximum values, amplifiers 1, 2, and 3 are found to have peak output voltages of only 4, 2, and 4 volts, and no accuracy can be obtained from such a circuit.

To overcome this difficulty, we make a scale change so that with $\underset{\text{max}}{x(\tau)} =$ 1 in., the output will be more nearly equal to the allowable peak of 100 volts. To avoid exceeding the 100-volt limit, we will let this voltage be 80 volts. By multiplying the revised differential equation by 20, we get

$$20\frac{d^2x(\tau)}{d\tau^2} = -10\frac{dx(\tau)}{d\tau} - 80x(\tau) + 40$$

The circuit for this scaled-up equation might now take the form shown in Fig. 4.7-10, where the initial voltages are scaled according to the output of the amplifiers, i.e., with $dx(0)/d\tau = -0.20$ in./sec., and the output of amplifier 2 equal to $-10(dx(\tau)/d\tau)$, its value should be $-10(-0.20) = +2$ volts. Similarly, the initial voltage of amplifier 3 should be $80 \times 0.25 = 20$ volts.

We now have an output of 80 volts for amplifier 3, corresponding to the maximum expected amplitude of 1 in. However, since the maximum expected velocity is 2 in./sec., the peak output of amplifier 2 is only $10(dx/d\tau) = 20$ volts. It is preferable then to change the gain of amplifier 2 by a factor of 4 and reduce the gain of amplifier 3 accordingly, as shown in Fig. 4.7-11. The maximum expected value of the acceleration being 4 in./sec., the gain of amplifier 1 need not be changed. The gain of the amplifier is determined by the quantity RC which is unity for $R = 1$ megohm and $C = 1$ microfarad.

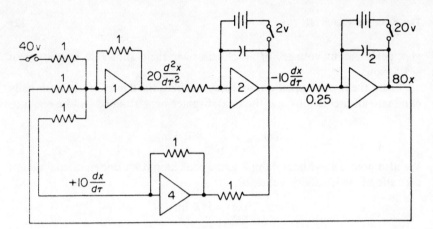

Fig. 4.7-10.

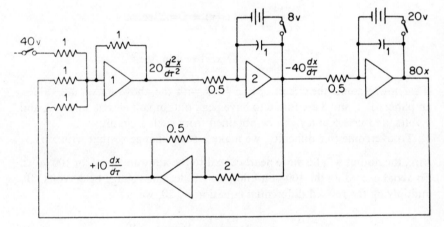

Fig. 4.7-11.

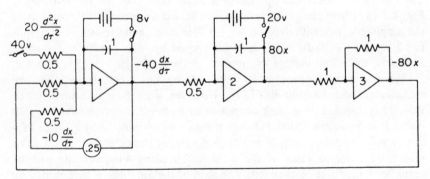

Fig. 4.7-12.

Note also that these changes require a further change in the initial voltage of amplifier 2 from 2 volts to 8 volts.

Finally, it should be apparent that these computer circuits are not unique, and the same equation can be solved by different circuits. Thus the equation for this problem can also be solved by the circuit of Fig. 4.7-12.

REFERENCES

1. Ayres, R. S., "Transient Response to Step and Pulse Functions," *Shock and Vibration Handbook*, Vol. 1, Ch. 8. (New York: McGraw-Hill Book Company, 1961).

2. Crede, C. E., *Vibration and Shock Isolation*. (New York: John Wiley & Sons, Inc., 1951).

3. Fung, Y. C., and M. V. Barton, "Some Characteristics and Uses of Shock Spectra," *J. Appl. Mech.*, **25**, No. 3, pp. 365–372 (Sept., 1958).

4. Hudson, D. E., and G. W. Housner, "Structural Vibrations Produced by Ground Motion," *Proc. of ASCE*, **81**, Paper No. 816 (Oct., 1955).

5. Mindlin, R. D., "Dynamics of Package Cushioning," *Bell System Technical Jour.*, **24**, pp. 353–461 (July, 1945).

6. Soroka, W. W., *Analog Methods in Computation and Simulation*. (New York: McGraw-Hill Book Company, 1954).

7. Thomson, W. T., *Laplace Transformation*, Second Ed. (Englewood Cliffs, N.J.: Prentice-Hall, Inc., 1960).

8. Thomson, W. T., "Single Degree of Freedom System," *Applied Mechanics Handbook*, Ch. 56. (New York: McGraw-Hill Book Company, 1962).

PROBLEMS

1. Show that the time t_p corresponding to the peak response for the impulsively excited spring-mass system is given by the equation:

$$\tan \sqrt{1 - \zeta^2}\, \omega_n t_p = \sqrt{1 - \zeta^2}/\zeta$$

2. Determine the peak displacement for the impulsively excited spring-mass system, and show that it can be expressed in the form

$$\frac{x_{\text{peak}}\sqrt{km}}{\hat{F}} = \exp\left(-\frac{\zeta}{\sqrt{1 - \zeta^2}}\tan^{-1}\frac{\sqrt{1 - \zeta^2}}{\zeta}\right)$$

Plot this result as a function of ζ.

3. Show that the time t_p corresponding to the peak response of the damped spring-mass system excited by a step force F_0 is $\omega_n t_p = \pi/\sqrt{1 - \zeta^2}$.

4. For the system of Prob. 3, show that the peak response is equal to

$$\left(\frac{xk}{F_0}\right)_{max} = 1 + \exp\left(-\frac{\zeta\pi}{\sqrt{1 - \zeta^2}}\right)$$

5. A rectangular pulse of height F_0 and duration t_0 is applied to an undamped spring-mass system. Considering the pulse to be the sum of two step pulses, as shown in Fig. 4-1, determine its response for $t > t_0$ by the superposition of the undamped solutions.

6. If an arbitrary force $f(t)$ is applied to an undamped oscillator which has initial conditions other than zero, show that the solution must be in the form

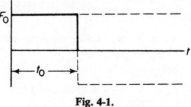

Fig. 4-1.

$$x(t) = x_0 \cos \omega_n t + \frac{v_0}{\omega_n} \sin \omega_n t + \frac{1}{m\omega_n}\int_0^t f(\xi) \sin \omega_n(t - \xi)\, d\xi$$

7. Show that the response to a unit step function, designated by $h(t)$, is related to the impulsive response $g(t)$ by the equation $g(t) = \dot{h}(t)$. '

8. Show that Duhamel's equation [Eq. (4.2-2)] can also be written in terms of $h(t)$ as

$$x(t) = f(0)h(t) + \int_0^t f'(\xi)h(t - \xi)\, d\xi$$

9. Show that the shock spectrum for the rectangular pulse of Fig. 4-1 is given by

$$\left(\frac{xk}{F_0}\right)_{max} = 2 \sin \frac{\pi t_0}{\tau}, \qquad \frac{t_0}{\tau} < 0.50$$

$$= 2 \qquad \frac{t_0}{\tau} > 0.50$$

and

$$\tau = \frac{2\pi}{\omega_n}$$

which is shown in Fig. 4-2.

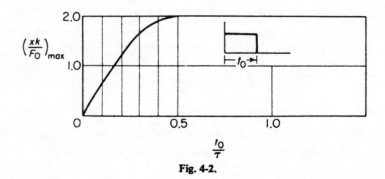

Fig. 4-2.

10. Figure 4-3 shows a sinusoidal pulse, which again can be considered to be the superposition of two sine waves. Show that its shock spectrum can be determined from the equations

$$\left(\frac{xk}{F_0}\right) = \frac{1}{(\tau/2t_1 - 2t_1/\tau)}\left(\sin\frac{2\pi t}{\tau} - \frac{2t_1}{\tau}\sin\frac{\pi t}{t_1}\right) \qquad t < t_1$$

$$\left(\frac{xk}{F_0}\right) = \frac{1}{(\tau/2t_1 - 2t_1/\tau)}\left[\left(\sin\frac{2\pi t}{\tau} - \frac{2t_1}{\tau}\sin\frac{\pi t}{t_1}\right)\right.$$
$$\left. + \left(\sin 2\pi\frac{t - t_1}{\tau} - \frac{2t_1}{\tau}\sin \pi\frac{t - t_1}{t_1}\right)\right], \qquad t > t_1$$

where $\tau = 2\pi/\omega$.

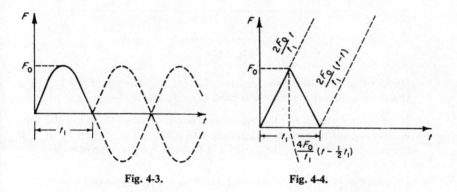

Fig. 4-3.   Fig. 4-4.

11. For the triangular pulse shown in Fig. 4-4, show that the response is

$$x = \frac{2F_0}{k}\left(\frac{t}{t_1} - \frac{\tau}{2\pi t_1}\sin 2\pi\frac{t}{\tau}\right), \qquad 0 < t < \tfrac{1}{2}t_1$$

$$x = \frac{2F_0}{k}\left\{1 - \frac{t}{t_1} + \frac{\tau}{2\pi t_1}\left[2\sin\frac{2\pi}{\tau}(t - \tfrac{1}{2}t_1) - \sin 2\pi\frac{t}{\tau}\right]\right\}, \qquad \tfrac{1}{2}t_1 < t < t_1$$

$$x = \frac{2F_0}{k}\left\{\frac{\tau}{2\pi t_1}\left[2\sin\frac{2\pi}{\tau}(t - \tfrac{1}{2}t_1) - \sin\frac{2\pi}{\tau}(t - t_1) - \sin 2\pi\frac{t}{\tau}\right]\right\}, \qquad t > t_1$$

12. Figure 4-5 shows the shock spectrum for the sine pulse of Prob. 10. Show that for small values of t_1/τ the peak response occurs in the region $t > t_1$. Show that when $t_1/\tau = \tfrac{1}{2}$ the peak response occurs at $t = t_1$.

13. For the triangular pulse of Prob. 11, show that when $t_1/\tau = \tfrac{1}{2}$, the peak response occurs at $t = t_1$, which can be established from the equation

$$2\cos\frac{2\pi t_1}{\tau}\left(\frac{t_p}{t_1} - 0.5\right) - \cos 2\pi\frac{t_1}{\tau}\left(\frac{t_p}{t_1} - 1\right) - \cos\frac{2\pi t_1}{\tau}\frac{t_p}{t_1} = 0$$

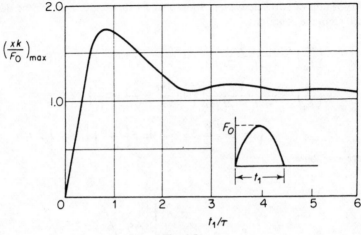

Fig. 4-5.

found by differentiating the equation for the displacement for $t > t_1$. The shock spectrum for the triangular pulse is shown in Fig. 4-6.

14. If the natural period τ of the oscillator is large compared to that of the pulse duration t_1, the maximum peak response will occur in the region $t > t_1$. For the undamped oscillator, the integrals in Duhamel's equation written as

$$x = \frac{\omega_n}{k} \left\{ \sin \omega_n t \int_0^t f(\xi) \cos \omega_n \xi \, d\xi - \cos \omega_n t \int_0^t f(\xi) \sin \omega_n \xi \, d\xi \right\}$$

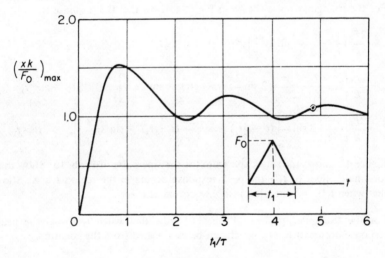

Fig. 4-6.

will then not change for $t > t_1$, since in this region $f(t) = 0$. Thus, by making the substitution

$$A \cos \phi = \omega_n \int_0^{t_1} f(\xi) \cos \omega_n \xi \, d\xi$$

$$A \sin \phi = \omega_n \int_0^{t_1} f(\xi) \sin \omega_n \xi \, d\xi$$

the response for $t > t_1$, is a simple harmonic motion with amplitude A. Discuss the nature of the shock spectrum for this case.

15. An undamped spring-mass system, m, k, is given a force excitation $F(t)$ shown in Fig. 4-7. Show that for $t < t_0$

$$\frac{kx(t)}{F_0} = \frac{1}{\omega_n t_0} (\omega_n t - \sin \omega_n t)$$

and for $t > t_0$.

$$\frac{kx(t)}{F_0} = \frac{1}{\omega_n t_0} [\sin \omega_n (t - t_0) - \sin \omega_n t] + \cos \omega_n (t - t_0)$$

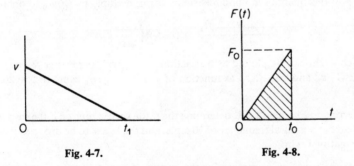

Fig. 4-7.   Fig. 4-8.

16. The base of an undamped spring-mass system, m, k, is given a velocity pulse shown in Fig. 4-8. Show that if the peak occurs at $t < t_1$, the shock spectrum is given by the equation

$$\frac{\omega z_{max}}{v_0} = \frac{1}{\omega_n t_1} - \frac{1}{\omega_n t_1 \sqrt{1 + (\omega_n t_1)^2}} - \frac{\omega_n t_1}{\sqrt{1 + (\omega_n t_1)^2}}$$

Plot this result.

17. In Prob. 16, if $t > t_1$, show that the solution is

$$\frac{\omega_n z}{v_0} = -\sin \omega_n t + \frac{1}{\omega_n t_1} [\cos \omega_n (t - t_1) - \cos \omega_n t]$$

18. The system of Fig. 4-9 has a Coulomb damper which exerts a constant friction force f. For a base excitation, show that the solution is

$$\frac{\omega_n z}{v_0} = \frac{1}{\omega_n t_1}\left(1 + \frac{f t_1}{m v_0}\right)(1 - \cos \omega_n t) - \sin \omega_n t$$

19. Show that the peak response for Prob. 18 is

$$\frac{\omega_n z_{max}}{v_0} = \frac{1}{\omega_n t_1}\left(1 + \frac{f t_1}{m v_0}\right)\left\{1 - \frac{\frac{1}{\omega_n t_1}\left(1 + \frac{f t_1}{m v_0}\right)}{\sqrt{1 + \left[\frac{1}{\omega_n t_1}\left(1 + \frac{f t_1}{m v_0}\right)\right]^2}}\right\}$$
$$- \frac{1}{\sqrt{1 + \left[\frac{1}{\omega_n t_1}\left(1 + \frac{f t_1}{m v_0}\right)\right]^2}}$$

By dividing by $\omega_n t_1$, the quantity $z_{max}/v_0 t_1$ can be plotted as a function of $\omega_n t_1$ with $f t_1/m v_0$ as parameter.

20. In Prob. 19 the maximum force transmitted to m is

$$F_{max} = f + |k z_{max}|$$

To plot this quantity in nondimensional form, multiply by $t_1/m v_0$ to obtain

$$\frac{F_{max} t_1}{m v_0} = \frac{f t_1}{m v_0} + (\omega_n t_1)^2\left(\frac{z_{max}}{v_0 t_1}\right)$$

which again can be plotted as a function of ωt_1 with parameter $f t_1/m v_0$. Plot $|\omega_n z_{max}/v_0|$ and $|z_{max}/v_0 t_1|$ as function of $\omega_n t_1$ for $f t_1/m v_0$ equal to 0, 0.20, and 1.0.

21. Referring to Example 4.6-1, determine the rattle space required if the suspended system has a natural frequency of 10 c.p.s. and the box is to be dropped through a height of 1.0 in.

22. Write the equation of motion for the system of Fig. 4-10 with base excitation $y(t)$. Draw the analogue computer circuit and show how the quantities $z = (x - y)$ and x can be measured.

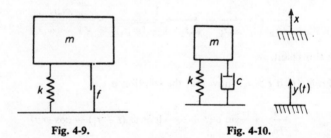

Fig. 4-9.   Fig. 4-10.

23. Set up an analogue computer circuit to solve the base-excited undamped system of Prob. 16. Verify the shock spectrum of Fig. 4.5-5 when the excitation is

$$\dot{y}(t) = 60e^{-t/10}, \qquad \dot{y}(t) = 60(1 - 5t)$$

The shock spectra for the above excitation are shown to scale in Fig. 4-11.

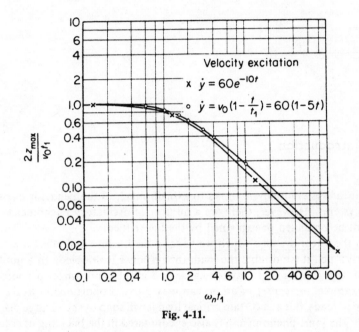

Fig. 4-11.

24. A spring-mass system has the equation

$$\ddot{x} + 2\dot{x} + 100x = 0$$

with initial conditions $x(0) = 1.0$ in. and $x(0) = 3$ in./sec. Slow down the computer equation by a factor 10, and determine the circuit diagram and scale factors for its efficient computation.

25. For a certain single-degree-of-freedom system the following values are given: $m = 0.122$ lb. sec.^{2}/in., $k = 6100$ lb./in., $c = 0.10c_c$. Choose a computer time which is 500 times greater than the actual time and write the equation for the computer for an arbitrary excitation $F(\tau)$. Develop the circuit diagram with appropriate scale.

5 | Nonlinear Vibrations

5.1 Introduction

Linear system analysis serves to explain much of the behavior of oscillatory systems. However, there are a number of oscillatory phenomena which cannot be predicted or explained by the linear theory.

In the linear systems which we have studied, cause and effect are related linearly; i.e., if we double the load, the response is doubled. In a nonlinear system this relationship between cause and effect is no longer proportional. For example, the center of an oil can may move proportionally to the force for small loads, but at a certain critical load it will snap over to a large displacement. The same phenomenon is also encountered in the buckling of columns, electrical oscillations of circuits containing inductance with an iron core, and vibration of mechanical systems with nonlinear restoring forces. For such systems, no linear relationship exists between cause and effect. Expressed in another way, the properties of the nonlinear system depend on the dependent variables. As examples, the stiffness of a nonlinear spring depends on the displacement; the period of oscillation of a pendulum for large amplitudes depends on the amplitude, etc.

For the single-degree-of-freedom nonlinear system, the general form of the equation of motion may be presented as

$$m\ddot{x} + f(\dot{x}, x, t) = 0 \tag{5.1-1}$$

Such equations are distinguished from linear equations in that the principle of superposition does not hold for their solution. For example, consider one of the simpler equations of the form

$$m\ddot{x} + kx^3 = 0 \tag{5.1-2}$$

130

and assume that $x_1 = \varphi_1(t)$ and $x_2 = \varphi_2(t)$ are solutions which satisfy the above equation independently, i.e.,

$$m\ddot{\varphi}_1 + k\varphi_1^3 = 0 \qquad (5.1\text{-}3)$$

$$m\ddot{\varphi}_2 + k\varphi_2^3 = 0 \qquad (5.1\text{-}4)$$

Thus by adding these two equations, the relationship

$$m(\ddot{\varphi}_1 + \ddot{\varphi}_2) + k(\varphi_1^3 + \varphi_2^3) = 0 \qquad (5.1\text{-}5)$$

must also be a correct equation.

If now we assume $x = \varphi_1 + \varphi_2$, and substitute it into Eq. (5.1-2), we would obtain

$$m(\ddot{\varphi}_1 + \ddot{\varphi}_2) + k(\varphi_1^3 + 3\varphi_1^2\varphi_2 + 3\varphi_1\varphi_2^2 + \varphi_2^3) = 0 \qquad (5.1\text{-}6)$$

Thus, due to the presence of the terms $3\varphi_1^2\varphi_2$ and $3\varphi_1\varphi_2^2$, the sum $\varphi_1 + \varphi_2$ cannot be a solution, indicating that the principle of superposition does not hold for the nonlinear system.

Unfortunately, analytical procedures for the treatment of nonlinear systems are difficult and require extensive mathematical study. Exact solutions which are known are relatively few, and a large part of the progress in the knowledge of nonlinear systems comes from approximate and graphical solutions, and from studies made on machine computers. Yet much of our new knowledge of the behavior of dynamical systems lies in the area of nonlinear systems, which is a fertile field for advanced research.

5.2 The Phase Plane[11]*

A differential equation is said to be *autonomous* if the properties of the system are not affected by the independent variable. For vibration problems, the system is autonomous if the independent variable t appears only as a differential in the equations of motion. It is then allowable to shift the time origin or change the time scale without influencing the behavior of the system.

For autonomous systems, the phase plane representation offers one useful approach for the study of the nonlinear system. Consider the free vibration of a single-degree-of-freedom system whose equation of motion is

$$\ddot{x} + f(\dot{x}, x) = 0 \qquad (5.2\text{-}1)$$

Since $dt = dx/\dot{x}$, the equation can be rewritten as

$$\dot{x}\frac{d\dot{x}}{dx} + f(\dot{x}, x) = 0 \qquad (5.2\text{-}2)$$

or

$$\frac{d\dot{x}}{dx} = \phi(\dot{x}, x) \qquad (5.2\text{-}3)$$

* Superscript refers to reference list at end of chapter.

Thus, if we use $\dot{x}$ and x as coordinates, there will be associated for any point $(\dot{x}, x)$ a slope $d\dot{x}/dx$ which can be represented by a short line segment. With a sufficient number of points, the entire plane can be filled by such short line segments as shown in Fig. 5.2-1, which graphically presents a family of solutions to the differential equation.

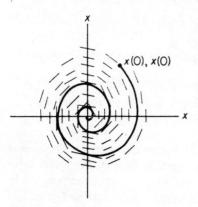

Fig. 5.2-1. Slope segments representing differential equation.

Given the initial conditions $\dot{x}(0)$ and $x(0)$, we simply proceed tangentially to these slopes with a curve called the trajectory. The trajectory drawn in this manner represents the solution for the specified initial conditions displayed in a phase plane. Because of the large amount of labor involved in plotting slopes, it is not a practical approach when only one solution is desired. Its value lies mainly in the study of the behavior of systems in the neighborhood of singular points which have bearing on the question of stability.

When the linear or nonlinear system is conservative, the trajectories in the phase plane have certain geometrical properties associated with the potential energy of the system. Consider the conservative system defined by the equation

$$\ddot{x} + f(x) = 0 \qquad (5.2\text{-}4)$$

and rewrite it in the form

$$\frac{d\dot{x}}{dx} = -\frac{f(x)}{\dot{x}} \qquad (5.2\text{-}5)$$

We define a singular point as one where both $f(x)$ and $\dot{x}$ become zero so that the slope $d\dot{x}/dx$ is indeterminate; i.e., $d\dot{x}/dx = 0/0$. Such singular points represent points of equilibrium in that, with $f(x) = 0$ and $\dot{x} = 0$, the acceleration $\ddot{x}$, or force, must also be zero. In the phase plane, such singular points must all lie along the x-axis.

If $\dot{x} = 0$ and $f(x) \neq 0$, the slope $d\dot{x}/dx$ becomes infinite. Thus, for a conservative system, the trajectories of the phase plane must all cross the x-axis at right angles.

Letting $U(x)$ be the potential energy, the equation for the conservative system can be written as

$$\tfrac{1}{2}\dot{x}^2 + U(x) = E = \text{constant} \qquad (5.2\text{-}6)$$

where E is the total energy. Solving for $\dot{x}$,

$$\dot{x} = \pm\sqrt{2E - U(x)} \qquad (5.2\text{-}7)$$

which indicates that the phase-plane trajectories must be symmetrical about the x-axis. Furthermore, for a conservative system the potential energy is related to $f(x)$ by the equation

$$U(x) = -\int_0^x f(x)\,dx \qquad (5.2\text{-}8)$$

or

$$\frac{dU}{dx} = -f(x) = \ddot{x} \qquad (5.2\text{-}9)$$

Thus the points for which $dU/dx = 0$ are points of equilibrium, and both $\ddot{x}$ and $\dot{x}$ are zero as shown in Fig. 5.2-2. The singular points indicated by the

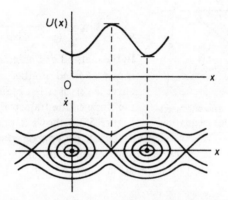

Fig. 5.2-2. Potential energy and phase-plane trajectories.

minimum of $U(x)$ are stable equilibrium positions, whereas the saddle points indicated by the maximum of $U(x)$ are positions of unstable equilibrium. If a motion is initiated inside the saddle-point trajectories, its motion is described by a closed trajectory associated with a period. Outside the saddle point trajectories, the trajectories may not be closed, in which case no period can be associated with such motion. The period for the periodic trajectories can be found from Eq. (5.2-7) to be

$$\tau = \oint \frac{dx}{\sqrt{2E - U(x)}} \qquad (5.2\text{-}10)$$

EXAMPLE 5.2-1. One of the interesting nonlinear equations which has been studied extensively is the Van der Pol equation:

$$\ddot{x} - \mu\dot{x}(1 - x^2) + x = 0$$

The equation somewhat resembles that of free vibration of a spring mass system with viscous damping; however, the damping term of this equation is nonlinear in that it depends on the amplitude x. When x is small, the damping is negative and the amplitude will grow until x exceeds unity. When x is greater than unity, the damping is positive and the amplitude will diminish with time. If the system is initiated with $x(0)$ and $\dot{x}(0)$, it will increase or decrease in amplitude, depending on whether x is small or large, and it will finally reach a stable amplitude known as the *limit cycle*.

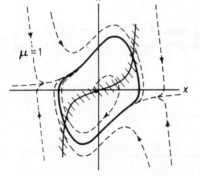

Fig. 5.2-3. Phase-plane representation of Van der Pol's equation.

Rewriting Van der Pol's equation in the form

$$\dot{x}\frac{d\dot{x}}{dx} - \mu\dot{x}(1 - x^2) + x = 0$$

and solving for $\dot{x}$, we obtain

$$\dot{x} = \frac{x}{(\mu - d\dot{x}/dx) - \mu x^2}$$

In the method of *isoclines*, the slope $d\dot{x}/dx$ is given a fixed number α, in which case a curve $\dot{x}$ versus x is obtained along which the slope of the trajectories is a constant. Figure 5.2-3 shows a phase plane representation of Van der Pol's equation for $\mu = 1.0$. The solid closed curve, approached by the inner and outer trajectories, represents the limit cycle. Also shown is one of the isocline curves for a given slope.

5.3 The Delta Method[9]

The delta method is a graphical procedure which applies to the solution of equations of the form

$$\ddot{x} + f(\dot{x}, x, t) = 0 \tag{5.3-1}$$

where $f(\dot{x}, x, t)$ must be continuous and single-valued. The equation is first rewritten by adding and subtracting a term $\omega_n^2 x$

$$\ddot{x} + [f(\dot{x}, x, t) - \omega_n^2 x] + \omega_n^2 x = 0 \tag{5.3-2}$$

and letting

$$\delta = \frac{1}{\omega_n^2}[f(\dot{x}, x, t) - \omega_n^2 x], \qquad v = \frac{\dot{x}}{\omega_n} \tag{5.3-3}$$

Since $\ddot{x} = \dot{x}(d\dot{x}/dx)$, Eq. (5.3-2) with the substitution of Eq. (5.3-3), becomes

$$v\frac{dv}{dx} + x + \delta = 0 \qquad (5.3\text{-}4)$$

or

$$\frac{dv}{dx} = \frac{x + \delta}{-v} \qquad (5.3\text{-}5)$$

This equation is now similar in form to Eq. (5.2-5), and a trajectory in the v, x plane can be found.

We recognize that in the substitution of $v = \dot{x}/\omega_n$ we have nondimensionalized the time to $\tau = \omega_n t$; thus δ is a function of:

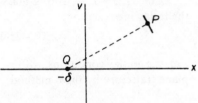

$$\delta = \delta(v, x, \tau) \qquad (5.3\text{-}6)$$

Starting with initial conditions $v(0)$, $x(0)$ at $\tau = 0$, the slope dv/dx at

$$P[v(0), x(0)]$$

can be represented by a line segment as before, as shown in Fig. 5.3-1.

Fig. 5.3-1. Delta method of phase-plane construction.

Over a short interval of time $\Delta\tau$ we can assume δ to be a constant, in which case Eq. (5.3-5) can be integrated as:

$$-\int v\, dv = \int (x + \delta)\, dx$$

or

$$(x + \delta)^2 + v^2 = \rho^2 \qquad (5.3\text{-}7)$$

The above equation is that of a circle of radius ρ with its center at $(v, x) = (0, -\delta)$. Thus the line segment at P, representing the slope of the trajectory, is tangent to the circle and perpendicular to the radial line QP, as shown in Fig. 5.3-1. The length of the circular arc over which the solution is valid depends on the variation of δ which must be held relatively constant, so that judgment must be exercised as to the time interval to be used.

The relationship between the length of arc and the angular rotation $d\theta$ of the line QP is

$$d\theta = \frac{ds}{\rho} = \frac{\sqrt{1 + (dv/dx)^2}\, dx}{\sqrt{(x + \delta)^2 + v^2}}$$

$$= \frac{\sqrt{1 + [(x + \delta)/-v]^2}\, dx}{\sqrt{1 + [(x + \delta)/-v]^2}\, v} = \frac{dx}{v} \qquad (5.3\text{-}8)$$

Substituting $v = \dot{x}/\omega_n = dx/\omega_n \, dt$ into the above equation, we arrive at the result

$$d\theta = \omega_n \, dt = d\tau \qquad (5.3\text{-}9)$$

With the coordinates chosen as in Fig. 5.3-1, $d\theta$ is clockwise for increasing time.

In the actual construction of the phase trajectory, the increment $\Delta\theta = \Delta\tau$ to be used must be small, and hence is difficult to measure with any accuracy. For this reason, it is preferable to use average values of v or x in establishing an average δ. The procedure can be best described by an example.

EXAMPLE 5.3-1. A spring-mass system with cubic hardening spring has the following equation:

$$\ddot{x} + 25(1 + 0.2x^2)x = 0$$

For the initial conditions, $x(0) = 2.40$ and $\dot{x}(0) = 0$, determine the phase-plane trajectory by the δ method.

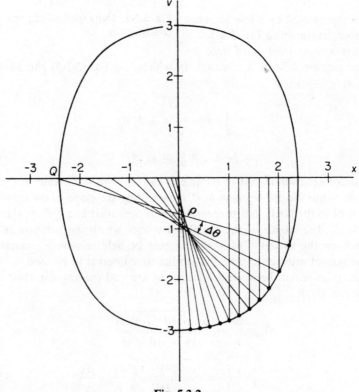

Fig. 5.3-2.

Solution. Comparison with Eq. (5.3-2) indicates that $\omega_n = 5$, $v = \dot{x}/5$, and $\delta = 0.20x^3$. When δ is a complicated function of x or v, it is preferable to first plot δ versus x or v; however, in this problem it is not necessary.

Assuming an increment $\Delta x = -0.20$, the average value of x for the first interval is 2.30, and the average δ is $\delta_{av} = 0.20(2.3)^3 = 2.44$. The point Q is then located as ($v = 0$, $x = -2.44$). The radius of curvature from Eq. (5.3-7) is $\rho = 4.84$, and the first point is located by the intersection of the arc with center at Q and the line $x = x(0) + \Delta x = 2.20$, as shown in Fig. 5.3-2. The procedure can now be repeated with the newly established point. The calculation of ρ is no longer necessary if we choose ρ from the new center to the former point on the trajectory.

The system investigated is conservative and the conclusions of Sec. 5.2 apply to this problem. The trajectory is symmetric about the x-axis and intersects it at right angles. Because of the nonlinear term, the trajectory deviates from the circular curve of the linear system.

5.4 Analytical Methods

A general method for the exact solution of nonlinear differential equations is as yet unknown, and most analytical methods which have been developed yield approximate solutions. Also, available techniques vary greatly with the type of nonlinear equation.

In mechanical vibrations the nonlinear equations encountered are often of the form

$$\ddot{x} + \omega_n^2 x + \mu f(x, \dot{x}, t) = 0 \tag{5.4-1}$$

where $f(x, \dot{x}, t)$ may represent forcing terms as well as the nonlinear term. Such equations are called quasi-linear if μ is small. If $\mu = 0$, the solution of this equation is periodic with frequency ω_n. Poincaré pointed out that, in the vicinity of these solutions for $\mu = 0$, there may exist periodic solutions for very small values of μ. Efforts in the search for such solutions have led to a number of analytical techniques yielding approximate solutions. In this section we cite a few of these methods, as well as a direct method of integration.

(*a*) *Direct Integration.* In Sec. 5.2, the equation

$$\ddot{x} + f(x) = 0$$

was put in the form

$$v \, dv = -f(x) \, dx$$

where $\dot{x} = v$. If $x = X$ when $v = 0$, its integral is

$$|v| = \left| \frac{dx}{dt} \right| = \sqrt{\int_x^X f(\xi) \, d\xi} \tag{5.4-2}$$

and its second integral yields

$$t - t_0 = \int_0^x \frac{d\eta}{\sqrt{\int_\eta^X f(\xi)\,d\xi}} \qquad (5.4\text{-}3)$$

where t_0 is the time corresponding to $x = 0$. The above equation then expresses time as a function of the displacement, and its inverse is the displacement-time relationship.

It is evident from the phase plane v versus x of Fig. 5.2-2 that if the motion is periodic with period τ, the time corresponding to the motion from $x = 0$ ($t = t_0$) to $x = X$ ($v = 0$) represents a quarter-period, and hence

$$\tau = 4\int_0^X \frac{d\eta}{\sqrt{\int_\eta^X f(\xi)\,d\xi}} \qquad (5.4\text{-}4)$$

Thus the period becomes a function of the amplitude X.

If we introduce the variable $\theta = \omega_n t$, and recognize that $\omega_n \tau = 2\pi$, we would obtain in place of Eq. (5.4-4) the equation

$$\frac{1}{\omega_n} = \frac{2}{\pi}\int_0^X \frac{d\eta}{\sqrt{\int_\eta^X f(\xi)\,d\xi}} \qquad (5.4\text{-}5)$$

which is the form given by Rauscher.[12]

EXAMPLE 5.4-1. For a simple pendulum, the equation of motion is

$$\ddot{\theta} + \omega_n^2 \sin\theta = 0$$

where $\omega_n = \sqrt{g/l}$. Hence the integral corresponding to $\int_\eta^X f(\xi)\,d\xi$ is

$$\omega_n^2 \int_\eta^X \sin\xi\,d\xi = \omega_n^2(\cos\eta - \cos X)$$

Substituting into Eq. (5.4-3),

$$t - t_0 = \frac{1}{\omega_n}\int_0^\theta \frac{d\eta}{\sqrt{\cos\eta - \cos X}}$$

This equation can be put into the form of an elliptical integral of the first kind by the substitution

$$\cos\eta = 1 - 2k^2 \sin^2\zeta$$

$$k = \sin\frac{X}{2}$$

The result is

$$t - t_0 = \frac{1}{\omega_n} \int_0^\theta \frac{d\zeta}{\sqrt{1 - k^2 \sin^2 \zeta}}$$

and the period is given by

$$\tau = \frac{4}{\omega_n} \int_0^{\pi/2} \frac{d\zeta}{\sqrt{1 - k^2 \sin^2 \zeta}} = 4\sqrt{\frac{l}{g}} \phi(k)$$

which is a function of the amplitude.

(b) *Free Oscillation.* Consider the free oscillation of a mass on a non-linear spring which is defined by the equation

$$\ddot{x} + \omega_n^2 x + \mu x^3 = 0 \tag{5.4-6}$$

In the *perturbation method*[1] we seek a solution in the form of an infinite series of the nonlinear parameter μ, as follows:

$$x = x_0(t) + \mu x_1(t) + \mu^2 x_2(t) + \cdots \tag{5.4-7}$$

Furthermore, our experience with the simple pendulum indicates that the fundamental frequency of oscillation will vary with the amplitude A, so that we should also assume

$$\omega = \omega_n + \mu \omega_1(A) + \mu^2 \omega_2(A) + \cdots \tag{5.4-8}$$

However, since only ω_n^2 appears in the differential equation, it is convenient to reformulate the above equation as

$$\omega^2 = \omega_n^2 + \mu \alpha_1(A) + \mu^2 \alpha_2(A) + \cdots \tag{5.4-9}$$

We will consider only the first two terms of Eqs. (5.4-7) and (5.4-9), which will adequately illustrate the procedure, and substituting these into Eq. (5.4-6), we obtain,

$$\ddot{x}_0 + \mu \ddot{x}_1 + (\omega^2 - \mu \alpha_1)(x_0 + \mu x_1) + \mu(x_0^3 + 3\mu x_0^2 x_1 + \cdots) = 0 \tag{5.4-10}$$

Since the nonlinear parameter μ could have been chosen arbitrarily, the coefficients of the various powers of μ must be equated to zero. This leads to a system of equations which can be solved recursively.

$$\ddot{x}_0 + \omega^2 x_0 = 0$$
$$\ddot{x}_1 + \omega^2 x_1 = \alpha_1 x_0 - x_0^3 \tag{5.4-11}$$

.

.

.

The solution to the first equation, subject to the initial conditions, $x(0) = A$, $\dot{x}(0) = 0$, is

$$x_0 = A \cos \omega t \tag{5.4-12}$$

which is called the generating solution. Substituting this into the right side of the second equation in (5.4-11), we obtain

$$\ddot{x}_1 + \omega^2 x_1 = \alpha_1 A \cos \omega t - A^3 \cos^3 \omega t$$

$$= (\alpha_1 - \tfrac{3}{4}A^2)A \cos \omega t - \tfrac{1}{4}A^3 \cos 3\omega t \qquad (5.4\text{-}13)$$

where $\cos^3 \omega t = \tfrac{3}{4} \cos \omega t + \tfrac{1}{4} \cos 3\omega t$ has been used.

The general solution of Eq. (5.4-13) is

$$x_1 = C_1 \cos \omega t + C_2 \sin \omega t + (\alpha_1 - \tfrac{3}{4}A^2)A \frac{\omega t}{2} \cos \omega t + \frac{A^3}{32\omega^2} \cos 3\omega t$$

$$(5.4\text{-}14)$$

Since the initial conditions have been satisfied by x_0, we let $x_1(0) = 0$, and $\dot{x}_1(0) = 0$. Then $C_2 = 0$, $C_1 = -A^3/32\omega^2$, and

$$x_1 = (\alpha_1 - \tfrac{3}{4}A^2)A \frac{\omega t}{2} \cos \omega t - \frac{A^3}{32\omega^2}(\cos \omega t - \cos 3\omega t) \qquad (5.4\text{-}15)$$

We note here a *secular term*, $t \cos \omega t$, which arose because the frequency of the excitation $\cos \omega t$ coincided with ω on the left side of Eq. (5.4-13); (i.e., we have a condition of resonance). Such terms violate the initial assumption that x_1 be a small correction to the linear solution x_0, and hence we must impose the condition

$$(\alpha_1 - \tfrac{3}{4}A^2) = 0 \qquad (5.4\text{-}16)$$

We thus obtain for the solution, including the first-order correction

$$x = A \cos \omega t - \mu \frac{A^3}{32\omega^2}(\cos \omega t - \cos 3\omega t) \qquad (5.4\text{-}17)$$

with

$$\omega^2 = \omega_n^2 + \tfrac{3}{4}\mu A^2 \qquad (5.4\text{-}18)$$

The procedure can now be repeated for higher-order corrections, where the secular terms can be removed in each step.

(c) *Forced Oscillation, Duffing's Equation.* G. Duffing[7] made an exhaustive study of the equation

$$\ddot{x} + \omega_n^2 x \pm \mu x^3 = \frac{F}{m} \begin{Bmatrix} \sin \omega t \\ \cos \omega t \end{Bmatrix} \qquad (5.4\text{-}19)$$

which represents a mass on a cubic spring, excited harmonically. The $\pm$ sign establishes whether the spring is a hardening or a softening spring.

In seeking a steady-state solution, the frequency of the oscillations must be the same as that of the excitation plus harmonics (see Eq. (5.4-31)). Generally the region of interest lies in the neighborhood of $\omega_n^2 = k/m$ so that if the perturbation method is used, the frequency of oscillation ω can be expressed in terms of ω_n by a relationship of the form

$$\omega^2 = \omega_n^2 + \mu\alpha_1 + \mu^2\alpha_2 + \cdots \qquad (5.4\text{-}20)$$

In dealing with oscillation near ω_n, we should recognize that such oscillations are largely free oscillations because the force required at resonance approaches zero. Thus, if a perturbation solution is sought near resonance, it would be appropriate to associate μ with the forcing function and look for the solution of the equation

$$\ddot{x} + \omega_n^2 x \pm \mu x^3 = \mu F_0 \begin{Bmatrix} \sin \omega t \\ \cos \omega t \end{Bmatrix} \qquad (5.4\text{-}21)$$

Assuming again that x can be represented in a power series of the non-linear parameter μ,

$$x = x_0 + \mu x_1 + \mu^2 x_2 + \cdots \qquad (5.4\text{-}22)$$

we will investigate the equation

$$\ddot{x} + \omega_n^2 x - \mu x^3 = \mu F_0 \cos \omega t \qquad (5.4\text{-}23)$$

corresponding to the softening spring. Limiting the series of Eqs. (5.4-20) and (5.4-22) to two terms, and substituting into Eq. (5.4-23) we obtain

$$\ddot{x}_0 + \mu\ddot{x}_1 + (\omega^2 - \mu\alpha_1)(x_0 + \mu x_1)$$
$$- \mu(x_0^3 + 3\mu x_0^2 x_1 + \cdots) = \mu F_0 \cos \omega t \qquad (5.4\text{-}24)$$

The recurring equations from the coefficients of μ^0 and μ^1 are then

$$\ddot{x}_0 + \omega^2 x_0 = 0 \qquad (5.4\text{-}25)$$

$$\ddot{x}_1 + \omega^2 x_1 = \alpha_1 x_0 + x_0^3 + F_0 \cos \omega t$$

The generating solution based on initial conditions $x_0(0) = A$ and $\dot{x}_0(0) = 0$ is

$$x_0 = A \cos \omega t \qquad (5.4\text{-}26)$$

Substituting this into the second equation for x_1, leads to

$$\ddot{x}_1 + \omega^2 x_1 = \left(\alpha_1 + \tfrac{3}{4}A^2 + \frac{F_0}{A}\right)A \cos \omega t + \tfrac{1}{4}A^3 \cos 3\omega t \qquad (5.4\text{-}27)$$

Again we must impose the condition

$$\left(\alpha_1 + \tfrac{3}{4}A^2 + \frac{F_0}{A}\right) = 0 \qquad (5.4\text{-}28)$$

to suppress the secular term in the solution for x_1. Thus the solution of the equation

$$\ddot{x}_1 + \omega^2 x_1 = \tfrac{1}{4}A^3 \cos 3\omega t \qquad (5.4\text{-}29)$$

for the initial conditions $x_1(0) = \dot{x}_1(0) = 0$ is

$$x_1 = C_1 \cos \omega t + C_2 \sin \omega t + C_3 \cos 3\omega t$$

$$= \frac{A^3}{32\omega^2} (\cos \omega t - \cos 3\omega t) \qquad (5.4\text{-}30)$$

Substituting into Eq. (5.4-22), the first-order solution to the forced vibration problem in the neighborhood of resonance is

$$x = A \cos \omega t + \frac{\mu A^3}{32\omega^2} (\cos \omega t - \cos 3\omega t) \qquad (5.4\text{-}31)$$

(d) *The Jump Phenomenon.* In problems of this type, it is found that the amplitude A undergoes a sudden discontinuous jump near resonance. We can investigate this phenomenon by replacing α_1 in Eq. (5.4-28) with

$$(\omega^2 - \omega_n^2)(1/\mu)$$

from Eq. (5.4-20). Thus we have the amplitude-frequency relationship

$$(\omega^2 - \omega_n^2)\frac{1}{\mu} + \tfrac{3}{4}A^2 + \frac{F_0}{A} = 0$$

which can be rearranged to

$$\tfrac{3}{4}\mu \frac{A^3}{\omega_n^2} = \left(1 - \frac{\omega^2}{\omega_n^2}\right)A - \frac{\mu F_0}{\omega_n^2} \qquad (5.4\text{-}32)$$

The solution of this amplitude equation can be visualized from Fig. 5.4-1. In this diagram, the cubic curve represents the left side of the equation, and the straight line with the intercept $-\mu F_0/\omega_n^2$ represents the right side, its slope being dependent on ω/ω_n. For $\omega/\omega_n < 1$, the curves intersect at

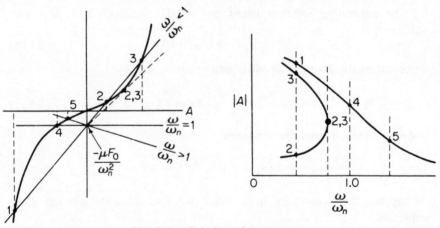

Fig. 5.4-1. Solution of Eq. 5.4-31.

three points 1, 2, 3, which are shown in the amplitude-frequency plot. As ω/ω_n increases toward unity, the points 2 and 3 approach each other, after which only one value of the amplitude will satisfy Eq. (5.4-32). When $\omega/\omega_n = 1$, or when $\omega/\omega_n > 1$, these points are 4 or 5.

The jump phenomenon can be described as follows. With increasing frequency of excitation, the amplitude gradually increases until point a in Fig. 5.4-2 is reached. It then suddenly jumps to a larger value indicated by the point b, and diminishes along the curve to its right. In decreasing the frequency from some noint c, the amplitude continues to increase beyond b to point d, and suddenly drops to a smaller value at e. A stability analysis shows the middle branch to be unstable. Thus

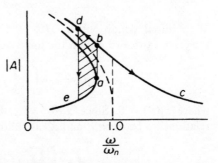

Fig. 5.4-2. The jump phenomenon for the softening spring.

there is a jump phenomenon, and the shaded region in the amplitude-frequency plot is unstable, its width being dependent on a number of factors, such as the amount of damping present, the rate of change of the exciting frequency, etc. If, instead of the softening spring, a hardening spring had been chosen, the same type of analysis would be applicable, and the result would be a curve of the type shown in Fig. 5.4-3.

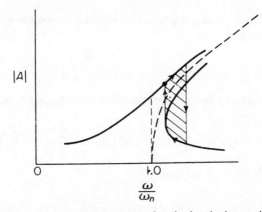

Fig. 5.4-3. The jump phenomenon for the hardening spring.

(e) *Method of Iteration.*[2] In the method of iteration, an assumed solution $x_0(t)$ is substituted into the differential equation to obtain a solution of improved accuracy. The method converges for small nonlinearities and the procedure is repeated until the desired accuracy is obtained.

For example, the equation

$$\ddot{x} + f(x, t) = 0$$

can be written in terms of the assumed solution as

$$\ddot{x} = -f[x_0(t), t]$$

With initial conditions of $x(0) = A$ and $\dot{x}(0) = 0$, this equation can be integrated twice to obtain the improved solution $x_1(t)$ as follows:

$$x_1(t) = -\int_{t'} \left\{ \int_0^\eta f[x_0(\xi), \xi] \, d\xi \right\} d\eta \tag{5.4-33}$$

where $x_1(t') = 0$. Since $x_1(0) = A$, the above equation also leads to the relationship between t' and A, as follows:

$$A = \int_0^{t'} \left\{ \int_0^\eta f[x_0(\xi), \xi] \, d\xi \right\} d\eta \tag{5.4-34}$$

The procedure is now repeated with $x_1(t)$ to obtain

$$x_2(t) = \int_t^{t'} \left\{ \int_0^\eta f[x_1(\xi), \xi] \, d\xi \right\} d\eta \tag{5.4-33'}$$

and again the initial conditions result in the equation

$$A = \int_0^{t'} \left\{ \int_0^\eta f[x_1(\xi), \xi] \, d\xi \right\} d\eta \tag{5.4-34'}$$

EXAMPLE 5.4-2. Using the iteration method, solve for the period of the linear equation

$$\ddot{x} + \omega_n^2 x = 0$$

with the initial conditions $x(0) = A$ and $\dot{x}(0) = 0$.

Solution. Assume for the first solution $x = 1$, and substitute into Eq. (5.4-32).

$$x_1(t) = -\int_{t'}^t \left\{ \omega_n^2 \int_0^\eta 1 \, d\xi \right\} d\eta = \frac{\omega_n^2 t'^2}{2} \left(1 - \frac{t^2}{t'^2} \right)$$

When $t = 0$, $x_1(0) = A$, so that

$$x_1(t) = A \left(1 - \frac{t^2}{t'^2} \right)$$

We now repeat the procedure with $x_1(t)$ to obtain

$$x_2(t) = \int_t^{t'} \left\{ \omega_n^2 A \int_0^\eta \left(1 - \frac{\xi^2}{t'^2} \right) d\xi \right\} d\eta = \omega_n^2 A \left\{ \frac{5}{12} t'^2 - \left(\frac{t^2}{2} - \frac{t^4}{12 t'^2} \right) \right\}$$

Again letting $x_2(0) = A$, we obtain

$$t' = \frac{1}{\omega_n}\sqrt{\frac{12}{5}} = \frac{1.55}{\omega_n} = 1.55\left(\frac{\tau}{2\pi}\right) = \frac{\tau}{4.05}$$

Thus we find that after two iterations the value of t' is found to be nearly equal to the exact value of $\tau/4$.

5.5 Stability of Oscillation (Self-Excited Oscillations)

When the exciting force is a function of the displacement, velocity, or acceleration, the oscillation is called self-excited. Shimmy of automobile wheels, flutter of airplane wings, and the electrical oscillations of an electronic oscillator are some examples of self-excited oscillations.

Self-excited oscillations may be linear or nonlinear, depending on whether the equations of motion are linear or nonlinear. If the motion of the system tends to increase the energy of the system, the amplitude will increase and the system is considered to be unstable.

Consider a single-degree-of-freedom system excited by a force proportional to the velocity. The equation of motion then becomes,

$$m\ddot{x} + c\dot{x} + kx = F_0\dot{x} \tag{5.5-1}$$

which is linear. Rearranging the above equation to

$$\ddot{x} + \left(\frac{c - F_0}{m}\right)\dot{x} + \frac{k}{m}x = 0 \tag{5.5-2}$$

we can recognize the possibility of negative damping for $F_0 > c$.

Letting $x = e^{st}$, the characteristic equation for the system becomes

$$s^2 + \left(\frac{c - F_0}{m}\right)s + \frac{k}{m} = 0$$

$$s = -\left(\frac{c - F_0}{2m}\right) \pm i\sqrt{\frac{k}{m} - \left(\frac{c - F_0}{2m}\right)^2} \tag{5.5-3}$$

Thus, when $F_0 > c$, we obtain a diverging oscillation of frequency:

$$\omega = \sqrt{\frac{k}{m} - \left(\frac{c - F_0}{2m}\right)^2}, \quad \text{provided} \quad \frac{k}{m} > \left(\frac{c - F_0}{2m}\right)^2 \tag{5.5-4}$$

A diverging non-oscillatory motion results when

$$\frac{k}{m} < \left(\frac{c - F_0}{2m}\right)^2 \tag{5.5-5}$$

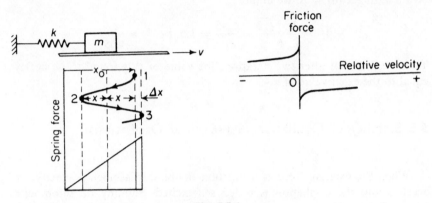

Fig. 5.5-1.

EXAMPLE 5.5-1. The coefficient of kinetic friction μ_k is generally less than the coefficient of static friction μ_s, this difference increasing somewhat with the velocity. Thus if the belt of Fig. 5.5-1 is started, the mass will move with the belt until the spring force is balanced by the static friction.

$$kx_0 = \mu_s mg \qquad (a)$$

At this point the mass will start to move back to the left, and the forces will again be balanced on the basis of kinetic friction when

$$k(x_0 - x) = \mu_{kl} mg$$

From these two equations, the amplitude of oscillation is found to be

$$x = x_0 - \mu_{kl}\frac{mg}{k} = \frac{(\mu_s - \mu_{kl})g}{\omega_n^2} \qquad (b)$$

While the mass is moving to the left, the relative velocity between it and the belt is greater than when it is moving to the right, thus μ_{kl} is less than μ_{kr} where the subscripts l and r refer to left and right. It is evident then that the work done by the friction force while moving to the right is greater than that while moving to the left, so that more energy is put into the spring-mass system than taken out. This then represents one type of self-excited oscillation and the amplitude will continue to increase.

The work done by the spring from 2 to 3 is

$$-\tfrac{1}{2}k[(x_0 + \Delta x) + (x_0 - 2x)](2x + \Delta x)$$

The work done by friction from 2 to 3 is

$$\mu_{kr} mg(2x + \Delta x)$$

Equating the net work done between 2 and 3 to the change in kinetic energy which is zero,

$$-\tfrac{1}{2}k(2x_0 - 2x + \Delta x) + \mu_{kr}mg = 0 \qquad (c)$$

Substituting (a) and (b) into (c), the increase in amplitude per cycle of oscillation is found to be

$$\Delta x = \frac{2g(\mu_{kr} - \mu_{kl})}{\omega_n^2} \qquad (d)$$

Mathieu Equation. Consider the nonlinear equation

$$\ddot{x} + \omega_n^2 x + \mu x^3 = F \cos \omega t \qquad (5.5\text{-}6)$$

and assume a perturbation solution

$$x = x_1(t) + \xi(t) \qquad (5.5\text{-}7)$$

Substituting Eq. (5.5-7) into (5.5-6), we obtain the following two equations

$$\ddot{x}_1 + \omega_n^2 x_1 + \mu x_1^3 = F \cos \omega t \qquad (5.5\text{-}8)$$

$$\ddot{\xi} + (\omega_n^2 + 3x_1^2)\xi = 0 \qquad (5.5\text{-}9)$$

If μ is assumed to be small, we can let

$$x_1 \cong A \sin \omega t \qquad (5.5\text{-}10)$$

and substitute it into Eq. (5.5-9), which becomes

$$\ddot{\xi} + \left[\left(\omega_n^2 + \frac{3}{2}A^2\right) - \frac{3}{2}A^2 \cos 2\omega t\right]\xi = 0 \qquad (5.5\text{-}11)$$

This equation is of the form

$$\ddot{\xi} + (a - 2b \cos 2\omega t)\xi = 0 \qquad (5.5\text{-}12)$$

which is known as the Mathieu equation.[5] The stable and unstable regions of the Mathieu equation depend on the parameters a and b, and are shown in Fig. 5.5-2. Since the Mathieu equation in this case represents a

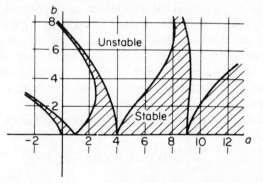

Fig. 5.5-2. Stable region of Mathieu equation indicated by shaded area.

perturbation of the periodic solution associated with the nonlinear equation, the stability is here examined in terms of the stability of the Mathieu equation.

5.6 Shock Spectra of Nonlinear Systems

For a designer, the time history of the response is of little interest. He is generally interested in the maximum peak response of the system. For a linear single-degree-of-freedom system, the maximum peak response for a specified excitation $F_0 f(t)$ is a function of the following

$$x_{max} = F_0 \Phi[k, m, c] \qquad (5.6\text{-}1)$$

which may be rearranged in the nondimensional form

$$\frac{x_{max}}{\delta} = \psi[\omega t_e, \zeta] \qquad (5.6\text{-}2)$$

where

$$\delta = \frac{F}{k}, \qquad \zeta = \frac{c}{2\sqrt{km}}$$

$$\omega_n = \sqrt{\frac{k}{m}}, \qquad t_e = \begin{array}{l}\text{characteristic time of the}\\\text{excitation; i.e., pulse time.}\end{array}$$

For the nonlinear single-degree-of-freedom system, there is an additional parameter of nonlinearity which is a function of the displacement, and the maximum peak response is no longer a linear function of the magnitude of the exciting force.

Quite often, the space clearance and the maximum allowable stress will limit the maximum peak displacement. Fung and Barton[8] suggested that the maximum displacement be eliminated as a parameter by a definite choice of its value, and an examination be made of the linear and the nonlinear systems for the force required to produce the same maximum response in the two systems. The ratio of the forces, called the *load ratio* then becomes a significant design parameter for the specified peak response, which is to be used together with the linear shock response curves to establish the safe nonlinear load.

Let the equations of motion for the nonlinear and linear systems be

$$\frac{d^2 z}{d\tau^2} + \Omega^2[z + \mu\phi(z)] = F_{nl} f(\tau) \qquad (5.6\text{-}3)$$

$$\frac{d^2 z'}{d\tau^2} + \Omega^2 z' = F_l f(\tau) \qquad (5.6\text{-}4)$$

where

$$z = \frac{x}{x_{\max}} \qquad \tau = \frac{t}{T}$$

$$\omega_n = \sqrt{\frac{k}{m}} \qquad \Omega = \omega_n T$$

Then for $x = x_{\max}$, $z = 1$ and the load ratio becomes

$$L = \frac{F_{nl}}{F_l} \qquad (5.6\text{-}5)$$

EXAMPLE 5.6-1. Determine the load ratio of a nonlinear system excited by a suddenly applied step function, F.

Solution. With no damping we can equate the work done on m to the change in its kinetic energy. Since the system starts at rest and has zero velocity at maximum displacement, the change in its kinetic energy is zero. The work done on the nonlinear system is

$$F_{nl} \int_0^1 1 \, dz - \Omega^2 \int_0^1 [z + \mu\phi(z)] \, dz = 0$$

or

$$F_{nl} = \tfrac{1}{2}\Omega^2 + \mu\Omega^2 \int_0^1 \phi(z) \, dz \qquad (5.6\text{-}6)$$

For the linear system, the work done is

$$F_l \int_0^1 1 \, dz' - \Omega^2 \int_0^1 z' dz' = 0$$

and

$$F_l = \tfrac{1}{2}\Omega^2 \qquad (5.6\text{-}7)$$

The load ratio is then established as

$$L = \frac{F_{nl}}{F_l} = 1 + 2\mu \int_0^1 \phi(z) \, dz \qquad (5.6\text{-}8)$$

If $\int_0^1 \phi(z) \, dz > 0$, then for the nonlinear system to have the same peak amplitude as the linear system, a larger step loading is required.

EXAMPLE 6.5-2. If the load is impulsive, determine the load ratio.

Solution. When the impulse $\hat{F}$ is applied to m, the mass acquires a velocity $\dot{x}_0 = \hat{F}/m$, after which the system undergoes free vibration. The work done on the mass of the nonlinear system is equal to the change in the kinetic energy, which is $-\frac{1}{2}m\dot{x}_0^2 = -\frac{1}{2}m(\hat{F}/m)^2$.

$$-\Omega^2\int_0^1 z\,dz - \mu\Omega^2\int_0^1 \phi(z)\,dz = -\frac{1}{2}\left(\frac{\hat{F}_{nl}}{m}\right)^2 \qquad (5.6\text{-}9)$$

The work done on the linear system is

$$-\Omega^2\int_0^1 z'\,dz' = -\frac{1}{2}\left(\frac{\hat{F}_l}{m}\right)^2 \qquad (5.6\text{-}10)$$

Taking the ratio of the two force impulses, we arrive at the result

$$\frac{\hat{F}_{nl}}{\hat{F}_l} = \left[1 + 2\mu\int_0^1 \phi(z)\,dz\right]^{1/2} \qquad (5.6\text{-}11)$$

Bilinear System. It is sometimes possible to replace the stiffness curve of a nonlinear spring by two straight lines. Such systems are then called bilinear, and although the equations of motion are linear in the two regions,

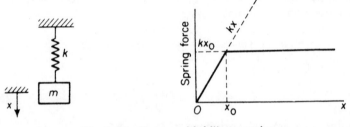

Fig. 5.6-1. System with bilinear spring.

the system must be considered as nonlinear because the response is not linearly proportional to the excitation.

Figure 5.6-1 shows a bilinear system with a constant spring force beyond the yield point deflection x_0. Since the system is conservative the work and energy method may be applied between the initial state and the state corresponding to the maximum peak response.

If initiated into motion by an impulse $\hat{F}$, the system acquires a velocity $v_0 = \hat{F}/m$, and thereafter undergoes free vibration. The work and energy method then results in the equation

$$\tfrac{1}{2}mv_0^2 = \tfrac{1}{2}kx_0^2 + kx_0(x_{\max} - x_0) \qquad (5.6\text{-}12)$$

With $v_0 = \hat{F}/m$ and $\omega_n^2 = k/m$, the equation can be rewritten in the following nondimensional form:

$$\left(\frac{\omega_n\hat{F}}{kx_0}\right)^2 = 2\left(\frac{x_{\max}}{x_0}\right) - 1 \qquad (5.6\text{-}13)$$

which plots as a parabola shown in Fig. 5.6-2.

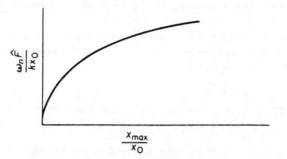

Fig. 5.6-2. Plot of Eq. 5.6-13.

The stress-strain relationship of many materials approaches the bilinear stiffness curve of Fig. 5.6-1. If the maximum peak displacement is made to correspond to the ultimate breaking strain, and x_0 to the elastic limit, the impulse corresponding to the elastic limit is found from Eq. (5.6-13) by letting $x_{max}/x_0 = 1$, or

$$\left(\frac{\omega_n \hat{F}_{e.l.}}{kx_0}\right)^2 = 1 \tag{5.6-14}$$

The load ratio of the impulses to attain ultimate strain to that required to attain elastic limit is then

$$\left(\frac{\omega_n \hat{F}_{ult}}{kx_0}\right)^2 \left(\frac{kx_0}{\omega_n \hat{F}_{e.l.}}\right)^2 = 2\left(\frac{x_{ult}}{x_{e.l.}}\right) - 1 \tag{5.6-15}$$

so that Eq. (5.6-13) is also the load ratio of the bilinear system for impulsive loads. B. Cummings[4] points out that this ratio is quite large for certain engineering materials which are identified on Fig. 5.6-3.

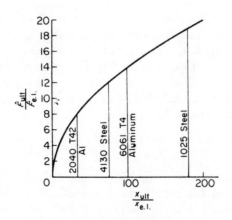

Fig. 5.6-3. Plot of Eq. 5.6-15.

When damping is present, the problem of the bilinear system can be solved by writing the equations of motion for each interval $0 < x \leq x_0$ and $x_0 \leq x$, and matching the displacement and velocity at $x = x_0$. Results of such a study have been presented in various nondimensional and phase-plane plots by W. T. Thomson.[14]

5.7 Analogue Computer Circuits for Nonlinear Systems

Many nonlinear systems can be studied by the use of the electronic analogue computer. Presented in this section are some of the circuit diagrams

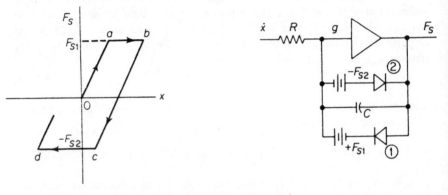

Fig. 5.7-1. Bilinear-hysteresis.

associated with nonlinear systems with a brief discussion as to their working principles.

(a) System with Hysteresis Damping. Figure 5.7-1 shows a typical variation for the spring force leading to hysteresis damping, together with an integrator circuit proposed by T. K. Caughey[3] which limits the output voltage by means of diodes. The circuit works in the following manner. Assuming the voltage across the capacitor C to be initially zero, we apply a positive velocity $\dot{x}$ to its input. The circuit being that of an integrator, the output voltage begins to build up according to the equation

$$F_s = \frac{1}{RC} \int_0^t \dot{x}\, dt = kx$$

Noting that the potential of the grid g is essentially zero, this voltage appears across C. During this time diode ② cannot conduct and hence circuit ② appears as open. Diode ① can conduct only when its voltage from the output side exceeds the bias voltage $+F_{s1}$ which is set to the level indicated by a in the force displacement curve. When diode ① conducts, the voltage across C is limited to $+F_{s1}$ until $\dot{x}$ becomes negative at b, at which time diode ① becomes nonconducting and the circuit appears as if only C is present across the amplifier. When the output voltage reaches some negative value set by $-F_{s2}$ at c, the diode ② conducts and limits the negative voltage on C until again the velocity $\dot{x}$ becomes positive at d.

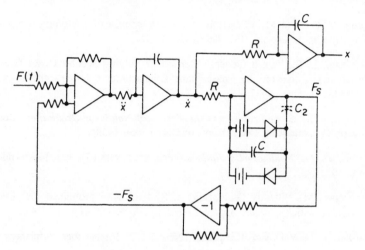

Fig. 5.7-2. Circuit for Eq. 5.7-1.

Figure 5.7-2 shows how the limited integrator is incorporated into the circuit to solve the equation

$$\frac{d^2x}{d\tau^2} + F_s = F(t) \tag{5.7-1}$$

By inserting an additional capacitor C_2 (shown dotted), it is possible to give the line ab and cd of the stiffness curve a slope other than zero.

REFERENCES

1. Bellman, R., *Perturbation Techniques in Mathematics, Physics and Engineering.* (New York: Holt, Rinehart & Winston, Inc., 1964).

2. Brock, J. E., "An Iterative Numerical Method for Nonlinear Vibrations," *Jour. Appl'd. Mech.*, pp. 1–11 (March, 1951).

3. Caughey, T. K., "Sinusoidal Excitation of a System with Bilinear Hysteresis," *Jour. Appl'd. Mech.*, pp. 640–643 (Dec., 1960).

4. Cummings, B., "Shock Response of an Elastic-Plastic System," *Jour. Appl'd. Mech.*, pp. 541–542 (Sept., 1964).

5. Cunningham, W. J., *Introduction to Nonlinear Analysis*. (New York: McGraw-Hill Book Company, 1958).

6. Davis, H. T., *Introduction to Nonlinear Differential and Integral Equations*. (Washington, D.C.: Govt. Printing Office, 1956).

7. Duffing, G., *Erwungene Schwingungen bei veranderlicher Eigenfrequenz*. (Braunschweig: F. Vieweg u. Sohn, 1918).

8. Fung, Y. C., and M. V. Barton, "Shock Response of a Nonlinear System," *Jour. Appl'd. Mech.*, pp. 465–476 (Sept., 1962).

9. Jacobsen, L. S., "On a General Method of Solving Second Order Ordinary Differential Equations by Phase Plane Displacements," *Jour. Appl'd. Mech.*, pp. 543–553 (Dec., 1952).

10. Malkin, I. G., *Some Problems in the Theory of Nonlinear Oscillations*, Books I and II. (Washington, D.C.: Dept. of Commerce, 1959).

11. Minorsky, N., *Nonlinear Oscillations*. (Princeton, N.J.: D. Van Nostrand Co., Inc., 1962).

12. Rauscher, M., "Steady Oscillations of Systems with Nonlinear and Unsymmetrical Elasticity," *Jour. Appl'd. Mech.*, pp. A169–177 (Dec., 1934).

13. Stoker, J. J., *Nonlinear Vibrations*. (New York: Interscience Publishers, Inc., 1950).

14. Thomson, W. T., "Shock Spectra of a Nonlinear System," *Jour. Appl'd. Mech.*, pp. 528–534 (Sept., 1960).

PROBLEMS

1. Plot the phase-plane trajectory and the isoclines for an undamped linear spring-mass system.

2. Plot the $U(x)$ versus x curve for Prob. 1, and determine the period from the equation

$$\tau = 4 \int_0^A \frac{dx}{\sqrt{2E - U(x)}}$$

3. A spring-mass linear system with viscous damping is displaced by x_0 and released with zero velocity. Draw its phase-plane trajectory. What do the isoclines look like?

4. If the system of Prob. 3 is initiated by an impulse from its equilibrium position, draw its phase-plane trajectory.

5. Expressing the equation $v = dx/dt$ in the integral form $t - t_0 = \int_{x_0}^{x} dx/v$, how would we avoid the difficulties encountered at $v = 0$?

6. Consider the equation $\ddot{x} + \omega_n^2 x + \mu x^3 = 0$ and replace $\ddot{x}$ with $v(dv/dx)$. Equation (5.2-6) then becomes

$$v^2 + \omega_0^2 x^2 + \tfrac{1}{2}\mu x^4 = 2E$$

with $v = 0$ when $x = A$. Show that the period is available from

$$\tau = 4 \int_0^A \frac{dx}{\sqrt{2E - U(x)}}$$

and discuss its evaluation.

7. Plot the isoclines for Van der Pol's equation with $\mu = 1.0$ and $d\dot{x}/dx = 0, -1$, and $+1$.

8. The equation for the free oscillation of a nonlinear system is given as

$$m\ddot{x} + c\dot{x} + kx + \mu x^3 = 0$$

Express this equation in the form $d\dot{x}/dx = \phi(\dot{x}, x)$.

or

$$\frac{dv}{dx} = \phi(v, x) \qquad v = \frac{\dot{x}}{\omega_n}$$

9. The equation of motion for a spring-mass system with constant Coulomb damping can be written as

$$\ddot{x} + \omega_n^2 x + C \operatorname{sgn}(\dot{x}) = 0$$

where $\operatorname{sgn}(\dot{x})$ signifies either a $+$ or $-$ sign equal to that of the sign of $\dot{x}$. Express this equation in the form $d\dot{x}/dx = \phi(\dot{x}, x)$.

10. Plot the phase-plane trajectory for the system of Prob. 9, with initial conditions $x(0) = A$, and $\dot{x}(0) = 0$. Also obtain the displacement-time curve from the phase-plane, and show that the amplitude varies linearly with time.

11. A mass is attached to the midpoint of a string of length $2l$ stretched to a tension T, as shown in Fig. 5-1. Determine its nonlinear equation for its lateral displacement, and determine its solution by the phase-plane delta method when the motion is initiated with $x(0) = A$ and $\dot{x}(0) = 0$.

12. Develop the phase-plane delta trajectory for the system of Fig. 5-2.

13. Look up $\phi(k) = \int_0^{\pi/2} \frac{d\zeta}{\sqrt{1 - k^2 \sin^2 \zeta}}$ in a table of elliptic functions and plot the period of a simple pendulum as a function of the amplitude $X = \theta_{max}$. (See Example 5.4-1.)

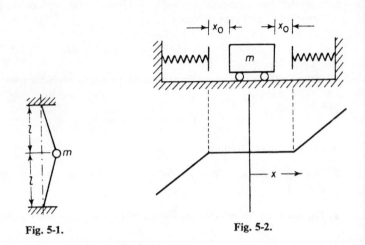

Fig. 5-1.   Fig. 5-2.

14. For large angles, express the equation of motion of a simple pendulum in a form suitable for the phase-plane delta method. Plot the free motion trajectory for the initial conditions $\dot{x}(0) = 0$ and $x(0) = 60$ deg.

15. Apply the perturbation method to the simple pendulum of Prob. 14 with $\sin \theta$ replaced by $\theta - \frac{1}{6}\theta^3$. Use only the first two terms of Eqs. (5.4-7) and (5.4-9).

16. Determine from the perturbation method the equation for the period of the simple pendulum as a function of the amplitude.

17. Apply the perturbation method to Duffing's equation with a hardening spring, and develop the amplitude-frequency curve.

18. Complete the solution for $x(t)$ in Example 5.4-2 and show that the iteration procedure approaches the result:

$$x(t) = A \cos \omega_n t = 1 - \frac{(\omega_n t)^2}{2} + \frac{(\omega_n t)^4}{4!} - \cdots$$

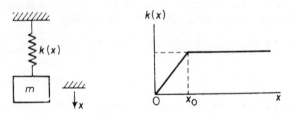

Fig. 5-3.

19. Apply the iteration procedure to the equation

$$\ddot{x} + \omega_n^2 x + \mu x^3 = 0$$

with the initial conditions $x(0) = A$, $\dot{x}(0) = 0$.

20. A constant force F_0 is suddenly applied to the bilinear system shown in Fig. 5-3. Construct the phase-plane delta trajectory with initial conditions $v(0) = 0$, and $x(0) = 2x_0$.

21. Solve Prob. 20 by the method of direct integration.

22. The cord of a simple pendulum shown in Fig. 5-4 is wrapped around a fixed cylinder of radius R such that its length is l when in the vertical position. Determine the differential equation of motion.

23. By multiplying Eq. (b) of Example 5.5-1 by $\dot{x}$ it can be written in the form

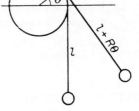

$$\frac{d}{dt}(\tfrac{1}{2}m\dot{x}^2 + \tfrac{1}{2}kx^2) = (F_0 - c)\dot{x}^2$$

Discuss the stability of the system from energy considerations when $(F_0 - c)$ is positive and negative.

24. Show that a divergent nonoscillatory motion takes place in systems of single degree of freedom if the exciting force is proportional to the displacement or acceleration.

Fig. 5-4.

25. The supporting end of a simple pendulum is given a motion as shown in Fig. 5-5. Show that the equation of motion is

$$\ddot{\theta} + \left(\frac{g}{l} - \omega^2 y_0 \cos 2\omega t\right) \sin \theta = 0$$

26. For a given value of g/l, determine the frequencies of the excitation for which the simple pendulum with a stiff arm l will be stable in the inverted position.

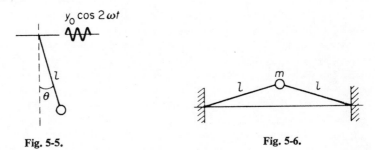

Fig. 5-5.

Fig. 5-6.

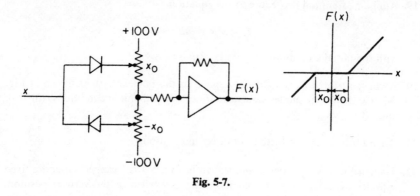

Fig. 5-7.

27. Determine the perturbation solution for the system of Fig. 5-6 leading to a Mathieu equation. Use initial conditions $\dot{x}(0) = 0$, $x(0) = A$.

28. Figure 5-7 shows a circuit which simulates a dead zone in the spring stiffness. Complete the analogue circuit to solve the problem shown in Fig. 5-2.

6 | Systems with Two Degrees of Freedom

6.1 Introduction

The degrees of freedom of a system are equal to the number of independent coordinates necessary to describe the motion of the system. A system with n degrees of freedom will have n natural frequencies which characterize the behavior of the system.

As in the single-degree-of-freedom system, we can describe the behavior of multi-degrees-of-freedom systems under *free vibration* and *forced vibration*. We will find again that the free vibrations take place at the natural frequencies of the system and that, in general, the motion will consist of several simultaneous oscillations at the various natural frequencies of the system. However, under certain specified conditions, all the coordinates will undergo harmonic motion corresponding to one of the natural frequencies of the system. When such motion takes place in every part of the system, the motion is called the principal mode of vibration or the normal-mode vibration. The number of such normal-mode vibrations will correspond to the degrees of freedom of the system so that an n-degrees-of-freedom system will have n natural frequencies characterized by their normal modes of vibration. Although normal-mode vibrations represent a very restricted type of motion, they are extremely important in that the more general types of motion can be represented by the superposition of normal-mode vibrations.

For systems of two degrees of freedom, the algebraic operations are fairly simple and the matrix notation is not essential. However, as preparation for systems with multidegrees of freedom, we will occasionally introduce the matrix notation as an alternative way of expressing the equations of motion.

6.2 Normal Mode Vibrations

All of the fundamental concepts of the multi-degrees-of-freedom system can be described in terms of a two-degrees-of-freedom system without becoming burdened with the algebraic difficulties of the higher multi-degrees-of-freedom systems. For this reason we will introduce many of the concepts of multi-degrees-of-freedom systems in terms of the two-degrees-of-freedom system, and later discuss additional numerical techniques necessary tor the multi-degrees-of-freedom systems.

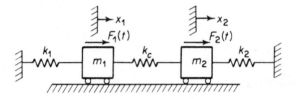

Fig. 6.2-1. Two degrees of freedom spring-mass system.

Any system which requires two independent coordinates to describe its motion is a two-degrees-of-freedom system. We could describe many such systems, but their equations of motion will all reduce to a common form which can be treated in a similar manner.

Figure 6.2-1 shows a two-degrees-of-freedom system which can be described by two linear and independent coordinates x_1 and x_2, measured from the equilibrium positions of the masses. From the free-body diagrams of the two masses the equations of motion are found as

$$m_1\ddot{x}_1 = -k_1 x_1 + k_c(x_2 - x_1) + F_1(t)$$
$$m_2\ddot{x}_2 = -k_2 x_2 - k_c(x_2 - x_1) + F_2(t)$$

(6.2-1)

On rearranging, these equations become

$$m_1\ddot{x}_1 + (k_1 + k_c)x_1 - k_c x_2 = F_1(t)$$
$$m_2\ddot{x}_2 + (k_2 + k_c)x_2 - k_c x_1 = F_2(t)$$

(6.2-2)

These are a set of linear differential equations of second order, and it is evident that the coupling between the two coordinates is due to k_c.

Before proceeding with the solution of the above equations we will express them in a matrix notation as follows:

$$\begin{bmatrix} m_1 & 0 \\ 0 & m_2 \end{bmatrix} \begin{Bmatrix} \ddot{x}_1 \\ \ddot{x}_2 \end{Bmatrix} + \begin{bmatrix} (k_1 + k_c) & -k_c \\ -k_c & (k_2 + k_c) \end{bmatrix} \begin{Bmatrix} x_1 \\ x_2 \end{Bmatrix} = \begin{Bmatrix} F_1(t) \\ F_2(t) \end{Bmatrix}$$

(6.2-3)

and state that it represents a special case of the general equation

$$\begin{bmatrix} m_{11} & m_{12} \\ m_{21} & m_{22} \end{bmatrix} \begin{Bmatrix} \ddot{x}_1 \\ \ddot{x}_2 \end{Bmatrix} + \begin{bmatrix} k_{11} & k_{12} \\ k_{21} & k_{22} \end{bmatrix} \begin{Bmatrix} x_1 \\ x_2 \end{Bmatrix} = \begin{Bmatrix} F_1(t) \\ F_2(t) \end{Bmatrix} \qquad (6.2\text{-}4)$$

which applies to all two-degrees-of-freedom systems. The matrix

$$\begin{bmatrix} m_{11} & m_{12} \\ m_{21} & m_{22} \end{bmatrix}$$

is called the mass matrix whereas the matrix

$$\begin{bmatrix} k_{11} & k_{12} \\ k_{21} & k. \end{bmatrix}$$

is called the stiffness matrix. The matrix notation here simply implies what the original equation states; i.e., by multiplying the elements of each row by the corresponding elements of each column, the original equation is obtained.* Our present interest here is in pointing out the general form of the equations of motion for the two-degrees-of-freedom system, and we will state nothing further about the matrix equation at this time, and return to Eq. (6.2-2).

To investigate the free vibrations of the system shown in Fig. 6.2-1 we let $F_1(t) = F_2(t) = 0$ and rewrite the equations:

$$m_1 \ddot{x}_1 + (k_1 + k_c)x_1 - k_c x_2 = 0$$
$$m_2 \ddot{x}_2 + (k_2 + k_c)x_2 - k_c x_1 = 0 \qquad (6.2\text{-}5)$$

We will first look for the normal modes of the system; i.e., we assume the motion of every point in the system to be harmonic:

$$x_1 = A_1 \sin \omega t$$
$$x_2 = A_2 \sin \omega t \qquad (6.2\text{-}6)$$

and examine the conditions under which such motion is possible.

Substituting these assumed solutions into the differential equations the resulting algebraic equations of the amplitudes are

$$(k_1 + k_c - m_1\omega^2)A_1 - k_c A_2 = 0$$
$$-k_c A_1 + (k_2 + k_c - m_2\omega^2)A_2 = 0 \qquad (6.2\text{-}7)$$

These equations are satisfied for any A_1 and A_2 only if the following determinant is zero:

$$\begin{vmatrix} (k_1 + k_c - m_1\omega^2) & -k_c \\ -k_c & (k_2 + k_c - m_2\omega^2) \end{vmatrix} = 0 \qquad (6.2\text{-}8)$$

* See Appendix A.

We call this determinant the *characteristic equation* of the system and its solution leads to the natural frequencies of the system. Multiplying out the determinant, we obtain the frequency equation

$$\omega^4 - \left[\frac{k_1 + k_c}{m_1} + \frac{k_2 + k_c}{m_2}\right]\omega^2 + \frac{k_1 k_2 + (k_1 + k_2)k_c}{m_1 m_2} = 0 \quad (6.2\text{-}9)$$

which is a quadratic in ω^2 leading to ω_1^2 and ω_2^2 when the k_s' and m_s' are known. Although each of these lead to frequencies $= \pm\omega$, we discard the negative signs as being of no physical significance, and arrive at two natural frequencies ω_1 and ω_2.

Furthermore, the amplitude ratio describing the normal modes of the system are found from Eq. 6.2-7 to be

$$\frac{A_1}{A_2} = \frac{k_2 + k_c - m_2\omega^2}{k_c} \quad (6.2\text{-}10)$$

where in place of ω^2 we substitute either ω_1^2 or ω_2^2. Since only the ratio of the amplitudes has significance in describing the normal modes, it is generally the practice to assign one of the amplitudes to be unity, in which case the modes are said to be normalized to unity.

Summarizing, we have found two normal mode vibrations for the system; the first one at the natural frequency ω_1, with an amplitude ratio of

$$\left(\frac{A_1}{A_2}\right)^{(1)} = \frac{k_1 + k_c - m_2\omega_1^2}{k_c} \quad (6.2\text{-}11)$$

and a second one at the natural frequency ω_2 with an amplitude ratio of

$$\left(\frac{A_1}{A_2}\right)^{(2)} = \frac{k_1 + k_c - m_2\omega_2^2}{k_c} \quad (6.2\text{-}12)$$

The superscript here is used to identify the particular mode.

EXAMPLE 6.2-1. Determine the normal modes of the system of Fig. 6.2-2 when the springs and masses are equal.

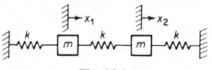

Fig. 6.2-2.

Solution. The frequency equation from Eq. (6.2-9) is

$$\omega^4 - \frac{4k}{m}\omega^2 + \frac{3k^2}{m^2} = 0$$

which results in the two roots

$$\omega_{1,2}^2 = \frac{2k}{m} \pm \sqrt{4(k/m)^2 - 3(k/m)^2}$$

$$= \frac{k}{m}[2 \pm 1]$$

The two natural frequencies are therefore

$$\omega_1 = \sqrt{k/m}$$

$$\omega_2 = \sqrt{3k/m}$$

and the corresponding amplitude ratios from Eqs. (6.2-11) and (6.2-12) are

$$\left(\frac{A_1}{A_2}\right)^{(1)} = \frac{2k - m\left(\dfrac{k}{m}\right)}{k} = 1$$

$$\left(\frac{A_1}{A_2}\right)^{(2)} = \frac{2k - m\left(\dfrac{3k}{m}\right)}{k} = -1$$

In the first mode at frequency $\omega_1 = \sqrt{k/m}$, the two masses appear to move as a single mass of $2m$ working against the two outer springs of combined stiffness $2k$; i.e., the center spring is not deflected.

In the second mode at frequency $\omega_2 = \sqrt{3k/m}$, the two masses move in opposition and there is a node at the center of the middle spring. Each half then behaves as a single-degree-of-freedom system. The behavior of the system in the two normal modes can be. interpreted in terms of the systems shown in Fig. 6.2-3(a) and (b).

EXAMPLE 6.2-2. Determine the normal modes and natural frequencies of the system of Fig. 6.2-1 when the relative values of the masses and springs are given as follows:

$$m_2 = 2m_1, \qquad k_2 = k_1, \qquad k_c = ck_1$$

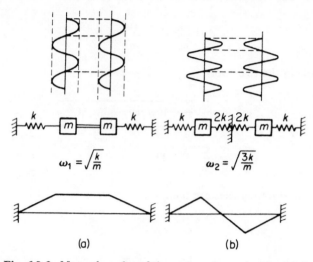

Fig. 6.2-3. Normal modes of the system shown in Fig. 6.2-2.

Solution. It is convenient to let $k_1/m_1 = \omega_{11}^2$ in which case Eq. 6.2-9 reduces to

$$\omega^4 - \tfrac{3}{2}\omega_{11}^2(1 + c)\omega^2 + \tfrac{1}{2}\omega_{11}^4(1 + 2c) = 0$$

The two normal-mode frequencies in terms of ω_{11}^2 then become

$$\frac{\omega_{1,2}^2}{\omega_{11}^2} = \tfrac{3}{4}(1 + c) \pm \sqrt{\tfrac{9}{16}(1 + c)^2 - \tfrac{1}{2}(1 + 2c)}$$

On varying the stiffness of the coupling spring from zero to ∞, the following numerical values for $(\omega_1/\omega_{11})^2$ and $(\omega_2/\omega_{11})^2$ are found and plotted in Fig. 6.2-4.

Normal-mode frequencies as function of c		
c	$(\omega_1/\omega_{11})^2$	$(\omega_2/\omega_{11})^2$
0	0.50	1.0
0.5	0.611	1.641
1.0	0.634	2.366
2.0	0.650	3.850
4.0	0.660	6.840
10.0	0.666	15.83
100.0	0.666	150.8
∞	0.666	∞

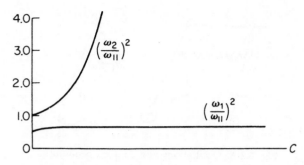

Fig. 6.2-4. The two natural frequencies of the system of Fig. 6.2-1, as function of coupling c.

The mode shapes can now be found for any c from Eq. (6.2-11) and (6.2-12). For example, if $c = 4$, we obtain

$$\left(\frac{A_1}{A_2}\right)^{(1)} = \frac{1 - c - 2(\omega_1/\omega_{11})^2}{c} = 0.916$$

$$\left(\frac{A_1}{A_2}\right)^{(2)} = \frac{1 + c - 2(\omega_2/\omega_{11})^2}{c} = -2.04$$

The amplitudes for the two modes of vibration are shown in Fig. 6.2-5 for $c = 4$.

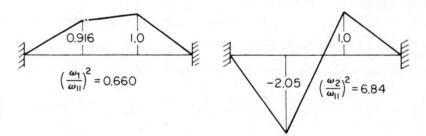

Fig. 6.2-5. The two normal modes of the system of Fig. 6.2-1 when $c = 4$.

EXAMPLE 6.2-3. In Fig. 6.2-6, the two simple pendulums are coupled together by means of a light spring k, which is unstrained when the two pendulums are in the vertical position. Determine the normal-mode vibrations.

Solution. Taking the counterclockwise angular displacements to be positive, and taking moments about the points of suspension, we obtain the

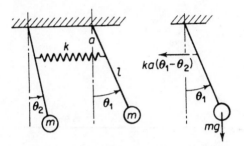

Fig. 6.2-6. Oscillations of a coupled pendulum.

following equations of motion for small oscillations:

$$ml^2\ddot{\theta}_1 = -mgl\theta_1 - ka^2(\theta_1 - \theta_2)$$

$$ml^2\ddot{\theta}_2 = -mgl\theta_2 + ka^2(\theta_1 - \theta_2)$$

Assuming the normal-mode solution as

$$\theta_1 = A_1 \cos \omega t$$

$$\theta_2 = A_2 \cos \omega t$$

and substituting into the differential equations, we obtain

$$(-\omega^2 ml^2 + mgl + ka^2)A_1 = ka^2 A_2$$

$$(-\omega^2 ml^2 + mgl + ka^2)A_2 = ka^2 A_1$$

The amplitude ratio then becomes

$$\frac{A_1}{A_2} = \frac{ka^2}{-\omega^2 ml^2 + mgl + ka^2} = \frac{-\omega^2 ml^2 + mgl + ka^2}{ka^2}$$

The frequency equation obtained from the above relationship is

$$\omega^4 - 2\left(\frac{g}{l} + \frac{ka^2}{ml^2}\right)\omega^2 + \left(\frac{g^2}{l^2} + 2\frac{ka^2 g}{ml^2 l}\right) = 0$$

or

$$\omega^2 = \frac{g}{l} + \frac{ka^2}{ml^2}(1 \pm 1)$$

The two natural frequencies of the system are then

$$\omega_1 = \sqrt{\frac{g}{l}}, \qquad \omega_2 = \sqrt{\frac{g}{l} + 2\frac{ka^2}{ml^2}}$$

The nature of the two modes is revealed on substituting ω_1 and ω_2 into the equation for the amplitude ratio.

Substituting ω_1,

$$\left(\frac{A_1}{A_2}\right)^{(1)} = +1.0$$

This result indicates that the two pendulums are moving in phase with equal amplitudes. The spring k does not, then, undergo extension or compression, and hence the two pendulums behave as if they were uncoupled and independent of each other. Since the frequency of a simple pendulum is $\sqrt{g/l}$, one would expect ω_1 to be the same.

Substituting ω_2, the amplitude ratio for this mode becomes

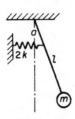

$$\left(\frac{A_1}{A_2}\right)^{(2)} = -1.0$$

and the two pendulums now vibrate in opposition with equal amplitudes. The spring k is now alternately elongated and compressed, and its stiffening effect on the pendulum is indicated by the higher frequency of vibration for this mode. Since the mid-point of the spring remains

Fig. 6.2-7.

stationary for this mode, the same conclusions could have been obtained by considering the system shown in Fig. 6.2-7.

6.3 Coordinate Coupling

In the spring-mass system of Fig. 6.2-1 the two systems $k_1 m_1$ and $k_2 m_2$ were coupled by a spring k_c, and the effect on the natural frequency of changing the strength of this coupling was studied in Example 6.2-2. Here we did not change the coordinates but varied the amount of coupling.

If we change the coordinates of the system, we would also expect a change in the coupling. For this discussion it is convenient to choose the system of Fig. 6.3-1 which shows a rigid bar supported by unequal springs k_1, k_2 with the center of mass not coinciding with its geometric center; i.e., $l_1 \neq l_2$.

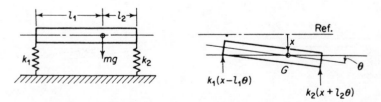

Fig. 6.3-1. Coupled translational and rotational vibration.

Using the equilibrium position as reference, we will let x be the linear displacement of its center of mass, and θ be the rotation of the bar. Letting J be the mass moment of inertia of the bar about its center of mass, the two equations of motion for small amplitudes can be readily written from the free-body diagram:

$$m\ddot{x} = -k_1(x - l_1\theta) - k_2(x + l_2\theta)$$
$$J\ddot{\theta} = k_1(x - l_1\theta)l_1 - k_2(x + l_2\theta)l_2$$

(6.3-1)

On rearranging, these equations may be presented in the form

$$\ddot{x} + \frac{1}{m}(k_1 + k_2)x + \frac{1}{m}(k_2l_2 - k_1l_1)\theta = 0$$

$$\ddot{\theta} + \frac{1}{J}(k_1l_1^2 + k_2l_2^2)\theta + \frac{1}{J}(k_2l_2 - k_1l_1)x = 0$$

(6.3-2)

or its matrix representation

$$\begin{bmatrix} 1 & 0 \\ 0 & 1 \end{bmatrix} \begin{Bmatrix} \ddot{x} \\ \ddot{\theta} \end{Bmatrix} + \begin{bmatrix} \dfrac{1}{m}(k_1 + k_2) & \dfrac{1}{m}(k_2l_2 - k_1l_1) \\ \dfrac{1}{J}(k_2l_2 - k_1l_1) & \dfrac{1}{J}(k_1l_1^2 + k_2l_2^2) \end{bmatrix} \begin{Bmatrix} x \\ \theta \end{Bmatrix} = 0 \quad (6.3\text{-}3)$$

It is evident that the coordinates x and θ are coupled by the term $(k_2l_2 - k_1l_1)$. If $k_1l_1 = k_2l_2$ the coupling term disappears and we obtain only the uncoupled x and θ vibrations; i.e., a force applied to the center of mass produces no rotation of the bar, whereas a torque applied to the bar produces no displacement of its center of mass. This also diagonalizes the stiffness matrix.

In the case $k_1l_1 \neq k_2l_2$ we would find that a force applied at G will produce both a linear displacement x and a rotation θ, whereas a torque applied to the bar will likewise cause both θ and x displacements. Consequently, the type of coupling found here is referred to as the static coupling, which is identified by the off-diagonal terms of the stiffness matrix.

We note now that there is some point C (see Fig. 6.3-2) along the bar where a force applied normal to the bar will result in pure translation; i.e., $k_1l_3 = k_2l_4$. If we define our x-coordinate through such a point and write the

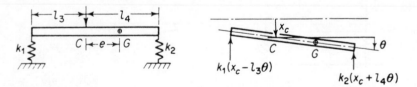

Fig. 6.3-2. Redefining of the coordinate x through point C.

equations of motion, we obtain

$$m(\ddot{x}_c + e\ddot{\theta}) = -k_1(x_c - l_3\theta) - k_2(x_c + l_4\theta)$$
$$J\ddot{\theta} + me\ddot{x}_c = k_1(x_c - l_3\theta)l_3 - k_2(x_c + l_4\theta)l_4 \qquad (6.3\text{-}4)$$

where $\ddot{x}_c + e\ddot{\theta}$ is the acceleration of G and the term $me\ddot{x}_c$ results from taking a moment about an accelerating point which does not coincide with the center of mass.*

Rearranging and noting that $k_1l_3 = k_2l_4$ these equations reduce to

$$m\ddot{x}_c + me\ddot{\theta} + (k_1 + k_2)x_c = 0$$
$$J_c\ddot{\theta} + me\ddot{x}_c - (k_1l_3^2 + k_2l_4^2)\theta = 0 \qquad (6.3\text{-}5)$$

or in matrix form

$$\begin{bmatrix} m & me \\ me & J_c \end{bmatrix}\begin{Bmatrix} \ddot{x}_c \\ \ddot{\theta} \end{Bmatrix} + \begin{bmatrix} (k_1 + k_2) & 0 \\ 0 & (k_1l_3^2 + k_2l_4^2) \end{bmatrix}\begin{Bmatrix} x_c \\ \theta \end{Bmatrix} = 0 \qquad (6.3\text{-}6)$$

It is evident that the coupling is due to the term e; however, the terms with e appear only in the off-diagonal terms of the mass matrix. Physically this implies that the coupling is *dynamic*; i.e., if the bar is displaced by a force at C and released, the inertia force which acts at G will result in a moment about C and produce a rotation superimposed on the translation. We have thus changed the form of coupling from static to dynamic coupling by redefining the coordinates of the system, and now the stiffness matrix is diagonalized while the mass matrix is not.

Fig. 6.3-3. Redefining of the coordinate x through point l.

In the previous two examples we chose the coordinates in such a manner that we obtained either static or dynamic coupling. It is, of course, possible to choose the coordinates differently than in the previous two cases so that both static and dynamic coupling will appear in the equations of motion. For example, if x, is chosen at the left end, as shown in Fig. 6.3-3, we would

---

* Moment about an accelerating point C not equal to the center of mass is

$$\Sigma M_c = J_c\ddot{\theta}\mathbf{1} + m\bar{r} \times a_c$$

$\mathbf{1}$ = unit vector along M_c

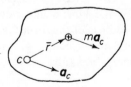

obtain the equations

$$m(\ddot{x}_1 + l_1\ddot{\theta}) = -k_1 x_1 - k_2(x_1 + l\theta)$$
$$J_1\ddot{\theta} + ml_1\ddot{x}_1 = -k_2(x_1 + l\theta)l \tag{6.3-7}$$

which, on rearranging, becomes

$$\begin{bmatrix} m_1 & ml_1 \\ ml_1 & J_1 \end{bmatrix}\begin{Bmatrix} \ddot{x}_1 \\ \ddot{\theta} \end{Bmatrix} + \begin{bmatrix} (k_1 + k_2) & k_2 l \\ k_2 l & k_2 l^2 \end{bmatrix}\begin{Bmatrix} x_1 \\ \theta \end{Bmatrix} = 0 \tag{6.3-8}$$

We have thus introduced both dynamic and static coupling as evidenced by the off-diagonal terms in both the mass and stiffness matrices.

We conclude now with some general statements:

(1) The coordinates x and θ are independent; i.e., x can be defined independently of θ.

(2) In general, the equations of motion will have both static and dynamic coupling, their general form being

$$\begin{bmatrix} m_{11} & m_{12} \\ m_{21} & m_{22} \end{bmatrix}\begin{Bmatrix} \ddot{x}_1 \\ \ddot{x}_2 \end{Bmatrix} + \begin{bmatrix} k_{11} & k_{12} \\ k_{21} & k_{22} \end{bmatrix}\begin{Bmatrix} x_1 \\ x_2 \end{Bmatrix} = 0 \tag{6.3-9}$$

where the terms m_{ij} and k_{ij} can be identified from the specific problem.

(3) The natural frequencies and mode shapes can be found by substituting $x_1 = A_1 \sin \omega t$ and $x_2 = A_2 \sin \omega t$ which results in the frequency equation

$$\begin{vmatrix} (k_{11} - m_{11}\omega^2) & (k_{12} - m_{12}\omega^2) \\ (k_{21} - m_{21}\omega^2) & (k_{22} - m_{22}\omega^2) \end{vmatrix} = 0 \tag{6.3-10}$$

and amplitude ratio

$$\frac{A_1}{A_2} = \frac{-(k_{12} - m_{12}\omega^2)}{(k_{11} - m_{11}\omega^2)} \tag{6.3-11}$$

where the two values of ω^2 found from the frequency equation are to be substituted. Although the form of these equations differs with different coordinates, the numerical values of the natural frequencies must be identical and independent of the choice of the coordinate system adopted. The amplitude ratio will of course differ with the choice of the coordinates, but the normal-mode shape of the physical system will be identical.

One further statement will be made here without proof. If the equations for the kinetic and potential energies of the system are examined, cross products of the form $\dot{x}_1\dot{x}_2$ in the kinetic energy equation indicate the presence of dynamic coupling, whereas cross products of the form $x_1 x_2$ in the potential energy equation indicate the presence of static coupling.

EXAMPLE 6.3-1. Express the kinetic and potential energies of the double pendulum of Fig. 6.3-4 in terms of the coordinates shown and establish the types of coupling present.

Solution. Using the angular coordinates θ_1 and θ_2 and assuming small amplitudes, the equations for the kinetic and potential energies are

$$T = \tfrac{1}{2}m_1 l_1^2 \dot{\theta}_1^2 + \tfrac{1}{2}m_2(l_1\dot{\theta}_1 + l_2\dot{\theta}_2)^2$$
$$= \tfrac{1}{2}(m_1 l_1^2 + m_2 l_1^2)\dot{\theta}_1^2 + m_2 l_1 l_2 \dot{\theta}_1 \dot{\theta}_2 + \tfrac{1}{2}m_2 l_2^2 \dot{\theta}_2^2$$
$$U = -m_1 g l_1(1 - \cos\theta_1) - m_2 g[l_1(1 - \cos\theta_1)$$
$$+ l_2(1 - \cos\theta_2)]$$
$$= -\tfrac{1}{2}(m_1 + m_2)g l_1 \theta_1^2 - \tfrac{1}{2}m_2 g l_2 \theta_2^2$$

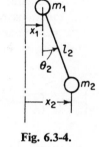

Fig. 6.3-4.

Examination of these equations indicates that the angular coordinates lead to dynamic coupling with zero static coupling.

If we set up the equations with x_1 and x_2 as coordinates we obtain

$$T_1 = \tfrac{1}{2}m_1 \dot{x}_1^2 + \tfrac{1}{2}m_2 \dot{x}_2^2$$

$$U = -\tfrac{1}{2}(m_1 + m_2)\frac{g}{l_1}x_1^2 - \tfrac{1}{2}m_2 \frac{g}{l_2}(x_2 - x_1)^2$$

It is evident then that the dynamic coupling has been eliminated and the system now has static coupling.

6.4 Principal Coordinates

From the discussion of the previous section, it is evident that the equations of motion of a system can be formulated in a number of different coordinate systems. Depending on the manner in which the coordinates are chosen, dynamic, static, or both forms of coupling could exist. The behavior of the system, however, is independent of the coordinate system adopted, and if we are looking for normal-mode oscillations, we need only to assume harmonic motion for the coordinate system adopted, which will result in the natural frequencies of the system irrespective of the coordinates chosen. The natural mode of oscillation will also be independent of the coordinate system chosen; however, its interpretation will obviously differ with the various coordinate systems adopted to describe its motion.

It is logical now to ask whether it is possible to describe the natural modes of oscillation in a coordinate system which will result in neither dynamic nor static coupling, so that the equations of motion will appear in a diagonalized form:

$$\begin{bmatrix} m_{11} & 0 \\ 0 & m_{22} \end{bmatrix} \begin{Bmatrix} \ddot{q}_1 \\ \ddot{q}_2 \end{Bmatrix} + \begin{bmatrix} k_{11} & 0 \\ 0 & k_{22} \end{bmatrix} \begin{Bmatrix} q_1 \\ q_2 \end{Bmatrix} = 0$$

If such is possible the two natural frequencies which previously required a simultaneous solution of the coupled equations can now be determined from the two uncoupled equations as

$$\omega_1 = \sqrt{k_{11}/m_{11}}, \qquad \omega_2 = \sqrt{k_{22}/m_{22}}$$

The answer to this question is that it is possible to find such a special kind of coordinate system called principal coordinates which will uncouple the equations of motion. However, the task of finding such coordinates is a mathematical one, except in the case of certain obviously simple problems, as in the following illustrative example. A more general discussion of the principal coordinates will be deferred at present in anticipation of the subject of generalized coordinates and Lagrange's equation which is taken up in Chapter 9.

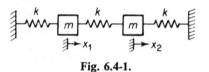

Fig. 6.4-1.

Consider again the symmetrical two-degrees-of-freedom system of Fig. 6.4-1 where the k'_s and m'_s are chosen to be equal. The equations of motion are now

$$m\ddot{x}_1 + kx_1 - k(x_2 - x_1) = 0$$
$$m\ddot{x}_2 + kx_2 + k(x_2 - x_1) = 0 \qquad (6.4\text{-}1)$$

If we add these equations, we obtain

$$m(\ddot{x}_1 + \ddot{x}_2) + k(x_1 + x_2) = 0 \qquad (6.4\text{-}2)$$

If we subtract the first from the second equation, we obtain

$$m(\ddot{x}_2 - \ddot{x}_1) + 3k(x_2 - x_1) = 0 \qquad (6.4\text{-}3)$$

Thus by letting $q_1 = x_1 + x_2$ and $q_2 = x_2 - x_1$, we have uncoupled the equations of motion which now appear as

$$\begin{bmatrix} m & 0 \\ 0 & m \end{bmatrix} \begin{Bmatrix} \ddot{q}_1 \\ \ddot{q}_2 \end{Bmatrix} + \begin{bmatrix} k & 0 \\ 0 & 3k \end{bmatrix} \begin{Bmatrix} q_1 \\ q_2 \end{Bmatrix} = 0 \qquad (6.4\text{-}4)$$

Thus the principal coordinates q_1 and q_2 of the system are found to be combinations of the previous coordinates x_1 and x_2 and the natural frequencies are found from inspection to be $\omega_1 = \sqrt{k/m}$ and $\omega_2 = \sqrt{3k/m}$ as was found previously in the normal-mode analysis of Example 6.2-1. It is further noted that x_1 and x_2 can be expressed in terms of the q_s and are equal to

$$x_1 = \tfrac{1}{2}(q_1 - q_2)$$
$$x_2 = \tfrac{1}{2}(q_1 + q_2)$$

$$(6.4\text{-}5)$$

In a more general case we can state that a linear transformation exists between the x and q coordinates such that

$$x_1 = c_{11}q_1 + c_{12}q_2 \qquad q_1 = h_{11}x_1 + h_{12}x_2$$
$$\text{or}$$
$$x_2 = c_{21}q_1 + c_{22}q_2 \qquad q_2 = h_{21}x_1 + h_{22}x_2$$

$$(6.4\text{-}6)$$

Furthermore, when principal coordinates are chosen the equations for the kinetic and potential energies will be free of cross products of the form $\dot{q}_1\dot{q}_2$ and q_1q_2 and contain only quadratic terms $\dot{q}_i^2$ and q_i^2.

6.5 Free Vibrations in Terms of Initial Conditions

For the two-degrees-of-freedom system, we found two natural frequencies ω_1 and ω_2 which satisfied the normal mode oscillations of the form

$$\begin{Bmatrix} x_1 \\ x_2 \end{Bmatrix} = \begin{Bmatrix} A_1 \\ A_2 \end{Bmatrix} \sin(\omega t + \psi)$$

$$(6.5\text{-}1)$$

The inclusion of the phase angle ψ only allows us the freedom of shifting the time origin and does not alter our findings regarding the normal modes of the system.

Using superscripts to identify the mode number, the two normal modes of the system can be represented as

$$\begin{Bmatrix} x_1 \\ x_2 \end{Bmatrix}^{(1)} = \begin{Bmatrix} A_1 \\ A_2 \end{Bmatrix}^{(1)} \sin(\omega_1 t + \psi_1)$$
$$\begin{Bmatrix} x_1 \\ x_2 \end{Bmatrix}^{(2)} = \begin{Bmatrix} A_1 \\ A_2 \end{Bmatrix}^{(2)} \sin(\omega_2 t + \psi_2)$$

$$(6.5\text{-}2)$$

where the amplitude ratio A_1/A_2 is specified for each mode.

Since each of these solutions for the normal modes satisfies the differential equations of motion, their sum must represent the general solution to the free vibration problem:

$$\begin{Bmatrix} x_1 \\ x_2 \end{Bmatrix} = \begin{Bmatrix} A_1 \\ A_2 \end{Bmatrix}^{(1)} \sin(\omega_1 t + \psi_1) + \begin{Bmatrix} A_1 \\ A_2 \end{Bmatrix}^{(2)} \sin(\omega_2 t + \psi_2)$$

$$(6.5\text{-}3)$$

The above equation has four arbitrary constants:

$$\begin{Bmatrix} A_1 \\ A_2 \end{Bmatrix}^{(1)}, \quad \begin{Bmatrix} A_1 \\ A_2 \end{Bmatrix}^{(2)}, \quad \psi_1 \quad \text{and} \quad \psi_2$$

which are sufficient to satisfy the four initial conditions

$$\begin{Bmatrix} x_1(0) \\ x_2(0) \end{Bmatrix} \quad \text{and} \quad \begin{Bmatrix} \dot{x}_1(0) \\ \dot{x}_2(0) \end{Bmatrix}$$

Thus the solution to the free-vibration problem, specified in terms of the initial conditions, is simply equal to the sum of the normal modes of the system adjusted in amount and phase.

EXAMPLE 6.5-1. If the system of Fig. 6.2-2 is started with initial conditions

$$\begin{Bmatrix} x_1(0) \\ x_2(0) \end{Bmatrix} \quad \text{and} \quad \begin{Bmatrix} \dot{x}_1(0) \\ \dot{x}_2(0) \end{Bmatrix}$$

determine its motion.

Solution. The normal modes of the system from Example 6.2-1 are:

$$\begin{Bmatrix} x_1 \\ x_2 \end{Bmatrix}^{(1)} = \begin{Bmatrix} 1 \\ 1 \end{Bmatrix} \sin (\sqrt{k/m}\, t + \psi_1)$$

$$\begin{Bmatrix} x_1 \\ x_2 \end{Bmatrix}^{(2)} = \begin{Bmatrix} -1 \\ 1 \end{Bmatrix} \sin (\sqrt{3k/m}\, t + \psi_2)$$

and the general solution in terms of the normal modes becomes

$$\begin{Bmatrix} x_1 \\ x_2 \end{Bmatrix} = \begin{Bmatrix} 1 \\ 1 \end{Bmatrix} \sin (\sqrt{k/m}\, t + \psi_1) + \begin{Bmatrix} -1 \\ 1 \end{Bmatrix} \sin (\sqrt{3k/m}\, t + \psi_2)$$

By differentiating, the velocities are obtained, and by letting $t = 0$ the unknowns of the equations can be expressed in terms of the initial conditions:

$$\begin{Bmatrix} x_1(0) \\ x_2(0) \end{Bmatrix} = \begin{Bmatrix} \sin \psi_1 & -\sin \psi_2 \\ \sin \psi_1 & +\sin \psi_2 \end{Bmatrix}$$

$$\begin{Bmatrix} \dot{x}_1(0) \\ \dot{x}_2(0) \end{Bmatrix} = \begin{Bmatrix} \sqrt{k/m} \cos \psi_1 & -\sqrt{3k/m} \cos \psi_2 \\ \sqrt{k/m} \cos \psi_1 & +\sqrt{3k/m} \cos \psi_2 \end{Bmatrix}$$

By adding and subtracting, these equations yield

$$\sin \psi_1 = \tfrac{1}{2}[x_1(0) + x_2(0)]$$

$$\sin \psi_2 = \tfrac{1}{2}[x_2(0) - x_1(0)]$$

$$\cos \psi_1 = \tfrac{1}{2}\sqrt{m/k}\, [\dot{x}_1(0) + \dot{x}_2(0)]$$

$$\cos \psi_2 = \tfrac{1}{2}\sqrt{m/3k}\, [\dot{x}_2(0) - \dot{x}_1(0)]$$

Expanding the trigonometric functions

$$\sin (\omega t + \psi) = \cos \psi \sin \omega t + \sin \psi \cos \omega t$$

the equations for x_1 and x_2, may be written as

$$x_1 = \tfrac{1}{2}\sqrt{m/k}\, [\dot{x}_1(0) + \dot{x}_2(0)] \sin \sqrt{k/m}\, t + \tfrac{1}{2}[x_1(0) + x_2(0)] \cos \sqrt{k/m}\, t$$
$$\quad - \tfrac{1}{2}\sqrt{k/3m}\, [\dot{x}_2(0) - \dot{x}_1(0)] \sin \sqrt{3k/m}\, t - \tfrac{1}{2}[x_2(0) - x_1(0)] \cos \sqrt{3k/m}\, t$$

$$x_2 = \tfrac{1}{2}\sqrt{m/k}\, [\dot{x}_1(0) + \dot{x}_2(0)] \sin \sqrt{k/m}\, t + \tfrac{1}{2}[x_1(0) + x_2(0)] \cos \sqrt{k/m}\, t$$
$$\quad + \tfrac{1}{2}\sqrt{m/3k}\, [\dot{x}_2(0) - \dot{x}_1(0)] \sin \sqrt{3k/m}\, t + \tfrac{1}{2}[x_2(0) - x_1(0)] \cos \sqrt{3k/m}\, t$$

Note that if the initial velocities are zero and that if $x_1(0) = x_2(0)$ we obtain the first normal mode, whereas if $x_1(0) = -x_2(0)$ we obtain the second normal mode.

6.6 Forced Oscillations—Harmonic Excitation

When a harmonically varying force acts on a system, the resulting vibration will have the same frequency as the excitation. As in the single-degree-of-freedom system, resonance will occur when the exciting frequency equals any of the natural frequencies of the system.

Consider again the system of Fig. 6.6-1 with a harmonic driving force $F_1 \sin \omega t$ acting on mass m_1. Its equation of motion, given in Sec. 6.2, is repeated here as

$$\begin{bmatrix} m_1 & 0 \\ 0 & m_2 \end{bmatrix} \begin{Bmatrix} \ddot{x}_1 \\ \ddot{x}_2 \end{Bmatrix} + \begin{bmatrix} (k_1 + k_c) & -k_c \\ -k_c & (k_2 + k_c) \end{bmatrix} \begin{Bmatrix} x_1 \\ x_2 \end{Bmatrix} = \begin{Bmatrix} F_1 \sin \omega t \\ 0 \end{Bmatrix} \qquad (6.6\text{-}1)$$

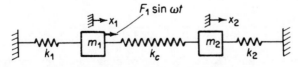

Fig. 6.6-1. Forced vibration at frequency ω.

With no damping, the motion of each mass will be either in-phase or out-of-phase with the driving force so that we can assume the solution to be of the form:

$$x_1 = X_1 \sin \omega t$$
$$x_2 = X_2 \sin \omega t \tag{6.6-2}$$

Substituting into the differential equations and factoring out the term $\sin \omega t$, we obtain the algebraic relationship between the amplitudes

$$\begin{bmatrix} (k_1 + k_c - m_1\omega^2) & -k_c \\ -k_c & (k_2 + k_c - m_2\omega^2) \end{bmatrix} \begin{Bmatrix} X_1 \\ X_2 \end{Bmatrix} = \begin{Bmatrix} F_1 \\ 0 \end{Bmatrix} \tag{6.6-3}$$

We recognize again the characteristic determinant which can be written in terms of the normal-mode frequencies ω_1 and ω_2 as

$$D = \begin{vmatrix} (k_1 + k_c - m_1\omega^2) & -k_c \\ -k_c & (k_2 + k_c - m_2\omega^2) \end{vmatrix} = m_1 m_2 (\omega_1^2 - \omega^2)(\omega_2^2 - \omega^2) \tag{6.6-4}$$

The equations for the amplitudes can be most conveniently written down in terms of Cramer's rule as follows:

$$X_1 = \frac{1}{D} \begin{vmatrix} F_1 & -k_c \\ 0 & (k_2 + k_c - m_2\omega^2) \end{vmatrix}$$

$$X_2 = \frac{1}{D} \begin{vmatrix} (k_1 + k_c - m_1\omega^2) & F_1 \\ -k_c & 0 \end{vmatrix} \tag{6.6-5}$$

and by substituting Eq. (6.6-4) for D we obtain

$$X_1 = \frac{F_1(k_2 + k_c - m_2\omega^2)}{m_1 m_2 (\omega_1^2 - \omega^2)(\omega_2^2 - \omega^2)}$$

$$X_2 = \frac{F_1 k_c}{m_1 m_2 (\omega_1^2 - \omega^2)(\omega_2^2 - \omega^2)} \tag{6.6-6}$$

We will first separate out the common factor $F_1/m_1 m_2$ and express the remaining parts in partial fractions as follows:

$$\frac{(k_2 + k_c - m_2\omega^2)}{(\omega_1^2 - \omega^2)(\omega_2^2 - \omega^2)} = \frac{C_1}{(\omega_1^2 - \omega^2)} + \frac{C_2}{(\omega_2^2 - \omega^2)}$$

$$\frac{k_c}{(\omega_1^2 - \omega^2)(\omega_2^2 - \omega^2)} = \frac{C_3}{(\omega_1^2 - \omega^2)} + \frac{C_4}{(\omega_2^2 - \omega^2)}$$

The constants C_1 and C_3 can be immediately found by multiplying the equations by $(\omega_1^2 - \omega^2)$ and letting $\omega^2 = \omega_1^2$. For example, multiplying through the first equation by $(\omega_1^2 - \omega^2)$ we obtain

$$\frac{(k_2 + k_c - m_2\omega^2)}{(\omega_2^2 - \omega^2)} = C_1 + \frac{C_2(\omega_1^2 - \omega^2)}{(\omega_2^2 - \omega^2)}$$

and on letting $\omega^2 = \omega_1^2$, the equation for C_1 becomes

$$C_1 = \frac{(k_2 + k_c - m_2\omega_1^2)}{(\omega_2^2 - \omega_1^2)}$$

Similarly, C_2 and C_4 are found by multiplying the equations by $(\omega_2^2 - \omega^2)$ and letting $\omega^2 = \omega_2^2$. Thus the equations for the remaining constants can be written down by inspection to be

$$C_2 = -\frac{(k_2 + k_c - m_2\omega_2^2)}{(\omega_2^2 - \omega_1^2)}$$

$$C_3 = \frac{k_c}{(\omega_2^2 - \omega_1^2)}$$

$$C_4 = \frac{-k_c}{(\omega_2^2 - \omega_1^2)}$$

We now recall that in the normal-mode vibrations the amplitude ratios for the two modes were equal to

$$\left(\frac{A_1}{A_2}\right)^{(1)} = \frac{(k_2 + k_c - m_2\omega_1^2)}{k_c}$$

$$\left(\frac{A_1}{A_2}\right)^{(2)} = \frac{(k_2 + k_c - m_2\omega_2^2)}{k_c} \tag{6.6-7}$$

and by normalizing to $A_2 = 1$ for each mode, the C_s can be written as

$$C_1 = \frac{k_c A_1^{(1)}}{(\omega_2^2 - \omega_1^2)} \qquad C_3 = \frac{k_c A_2^{(1)}}{(\omega_2^2 - \omega_1^2)}$$

$$C_2 = \frac{-k_c A_1^{(2)}}{(\omega_2^2 - \omega_1^2)} \qquad C_4 = \frac{-k_c A_2^{(2)}}{(\omega_2^2 - \omega_1^2)} \tag{6.6-8}$$

With these quantities the solution to the forced vibration problem can be written down immediately as

$$x_1 = \frac{F_1 k_c}{m_1 m_2 (\omega_2^2 - \omega_1^2)} \left\{ \frac{A_1^{(1)}}{\omega_1^2 - \omega^2} - \frac{A_1^{(2)}}{\omega_2^2 - \omega^2} \right\} \sin \omega t$$

$$x_2 = \frac{F_1 k_c}{m_1 m_2 (\omega_2^2 - \omega_1^2)} \left\{ \frac{A_2^{(1)}}{\omega_1^2 - \omega^2} - \frac{A_2^{(2)}}{\omega_2^2 - \omega^2} \right\} \sin \omega t \tag{6.6-9}$$

It appears then that we are able to express the equations for the forced vibration in terms of the normal-mode amplitudes and frequencies of the system and the resonance factors $(\omega_i^2 - \omega^2)$. For conciseness of interpretation we can present the above equations as

$$\begin{Bmatrix} x_1 \\ x_2 \end{Bmatrix} = \frac{F_1 k_c}{m_1 m_2 (\omega_2^2 - \omega_1^2)} \left[\begin{Bmatrix} A_1 \\ A_2 \end{Bmatrix}^{(1)} \frac{1}{(\omega_1^2 - \omega^2)} - \begin{Bmatrix} A_1 \\ A_2 \end{Bmatrix}^{(2)} \frac{1}{(\omega_2^2 - \omega^2)} \right] \sin \omega t$$

(6.6-10)

EXAMPLE 6.6-1. Determine the response of the symmetric system of Fig. 6.6-1 when excited by a harmonic force $F_1 \sin \omega t$ acting on mass 1.

Solution. Here $k_1 = k_2 = k_c = k$ and $m_1 = m_2 = m$, and the normal modes are characterized by

Mode 1   Mode 2

$\omega_1^2 = \dfrac{k}{m}, \quad \begin{Bmatrix} A_1 \\ A_2 \end{Bmatrix}^{(1)} = \begin{Bmatrix} 1 \\ 1 \end{Bmatrix} \qquad \omega_2^2 = \dfrac{3k}{m}, \quad \begin{Bmatrix} A_1 \\ A_2 \end{Bmatrix}^{(2)} = \begin{Bmatrix} -1 \\ 1 \end{Bmatrix}$

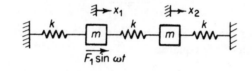

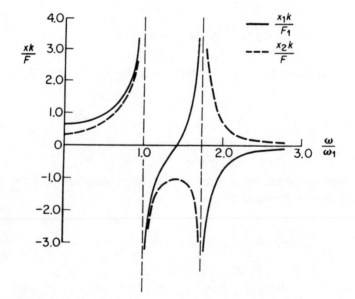

Fig. 6.6-2. Frequency response for system of Fig. 6.6-1. with $k_1 = k_2 = k_c = k$ and $m_1 = m_2 = m$.

The response from Eq. (6.6-10) is

$$x_1 = \frac{F_1}{2k}\left[\frac{1}{1-(\omega/\omega_1)^2} + \frac{1}{3-(\omega/3\omega_1)^2}\right]\sin \omega t$$

$$x_2 = \frac{F_1}{2k}\left[\frac{1}{1-(\omega/\omega_1)^2} - \frac{1}{3-(\omega/3\omega_1)^2}\right]\sin \omega t$$

The two equations are plotted in solid and dotted lines in Fig. 6.6-2.

6.7 Applications

There are many two-degrees-of-freedom systems which are of practical engineering interest. We will discuss several of these applications which illustrate the fundamentals of the two-degrees-of-freedom system.

(a) *Vehicle Suspension*. The automobile represents a complex system with many degrees of freedom. It is possible, however, to simplify the system as shown in Fig. 6.7-1 by considering only the major motions of the system:

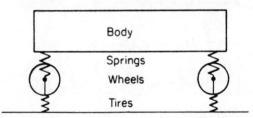

Fig. 6.7-1. Vibrational system representing an automobile.

(1) up-and-down linear motion of the body; (2) pitching angular motion of the body; and (3) up-and-down motion of the wheels.

The natural frequencies for the first two types of motion are of the same order of magnitude, and generally less than 1 c.p.s. The wheels, however, move up and down with greater rapidity, their natural frequencies being 6 to 10 times greater than that of the body motions. Because of this large separation of natural frequencies, the body motions (1) and (2) and the wheel motions (3) exist almost independently. For instance, the motions of the wheels are too rapid to affect the motions of the body appreciably, and the motions of the body are too slow to influence the motions of the wheels. It is evident, then, that we can study the body motions and the wheel motions separately, thereby greatly simplifying the problem.

To study the body motion, the system of Fig. 6.7-2 can be considered to be representative, where

m = mass of the body
J = mass moment of inertia of the body about the center of gravity
k_1, k_2 = stiffness of front and rear springs

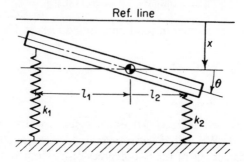

Fig. 6.7-2. Simplified vibratory system of automobiles.

Referring to Eq. (6.3-2) which applies to this problem, and letting

$$a = \frac{1}{m}(k_1 + k_2)$$

$$b = \frac{1}{m}(k_2 l_2 - k_1 l_1) \tag{a-1}$$

$$c = \frac{1}{J}(k_1 l_1^2 + k_2 l_2^2) = \frac{1}{mr^2}(k_1 l_1^2 + k_2 l_2^2)$$

the equations of motion are

$$\ddot{x} + ax + b\theta = 0 \tag{a-2}$$

$$\ddot{\theta} + c\theta + \frac{b}{r^2}x = 0 \tag{a-3}$$

It is evident from these equations that b is the coupling coefficient for the up-and-down and the pitching motion, and that the equations uncouple when $k_2 l_2 = k_1 l_1$. With zero coupling, a force applied to the center of gravity produces only translational motion x, while a torque applied to the body produces only rotational motion θ. It is also evident from the equations of motion that the uncoupled motions have the following frequencies

$$\omega_x = \sqrt{a} \tag{a-4}$$

$$\omega_\theta = \sqrt{c} \tag{a-5}$$

For the principal modes the x and θ motions are harmonic, and the solution can be expressed as follows:

$$x = X \cos \omega t$$
$$\theta = \Theta \cos \omega t \tag{a-6}$$

Substituting into the equations of motion, we obtain

$$(a - \omega^2)X + b\Theta = 0$$
$$\left(\frac{b}{r^2}\right)X + (c - \omega^2)\Theta = 0 \tag{a-7}$$

from which the amplitude ratio becomes

$$\frac{X}{\Theta} = -\frac{b}{a - \omega^2} = -\frac{c - \omega^2}{b/r^2} \tag{a-8}$$

Solving the above equation, we obtain the frequency equation for the principal modes:

$$\omega^4 - (a + c)\omega^2 + \left(ac - \frac{b^2}{r^2}\right) = 0$$

$$\omega_{1,2}^2 = \tfrac{1}{2}(a + c) \pm \sqrt{\tfrac{1}{4}(a + c)^2 - (ac - b^2/r^2)} \tag{a-9}$$

$$= \tfrac{1}{2}(a + c) \pm \sqrt{\tfrac{1}{4}(a - c)^2 + b^2/r^2}$$

The coupled frequencies given by the above equation always lie outside the uncoupled frequencies, that is, above the higher uncoupled frequency and below the lower uncoupled frequency. By expanding the radical of the above equation by the binomial theorem, the coupled frequencies for small coupling coefficients verify this statement, becoming

$$\omega_1^2 = a - \frac{b^2}{r^2(c - a)} \qquad \omega_2^2 = c + \frac{b^2}{r^2(c - a)} \tag{a-10}$$

EXAMPLE 3.7-1. Determine the principal modes of vibration of an automobile with the following data:

$$W = 3220 \text{ lb.} \qquad l_1 = 4.5 \text{ ft.} \qquad k_1 = 2400 \text{ lb./ft.}$$

$$r_2 = 16 \text{ ft.}^2 \qquad l_2 = 5.5 \text{ ft.} \qquad k_2 = 2600 \text{ lb./ft.}$$

$$l = 10 \text{ ft.} \qquad k_1 + k_2 = 5000 \text{ lb./ft.}$$

Solution. The constants a, b, and c are first determined as follows:

$$a = \frac{5000 \times 32.2}{3220} = 50.0$$

$$b = \left(\frac{14{,}300 - 10{,}800}{3220}\right) 32.2 = 35.0$$

$$c = \left(\frac{48{,}700 + 78{,}700}{3220 \times 16}\right) 32.2 = 79.7$$

$$\left(\frac{b}{r}\right)^2 = 76.7, \qquad c \pm a = \begin{cases} 129.7 \\ 29.7 \end{cases}$$

The coupled frequencies then become

$$\omega_{1,2}^2 = \tfrac{1}{2}(a + c) \pm \sqrt{\tfrac{1}{4}(a - c)^2 + (b/r)^2}$$

$$= \tfrac{1}{2} \times 129.7 \pm \sqrt{\tfrac{1}{4}(29.7)^2 + 76.7} = \begin{cases} 82.1 \\ 47.6 \end{cases}$$

$$\omega_{1,2} = \begin{cases} 9.06 \text{ rad./sec.} = 1.44 \text{ c.p.s.} \\ 6.90 \text{ rad./sec.} = 1.10 \text{ c.p.s.} \end{cases}$$

The amplitude ratios for the two frequencies are

$$\frac{X}{\Theta} = \frac{b}{\omega^2 - a} = \begin{cases} \dfrac{35}{47.6 - 50} = -14.6 \text{ ft./rad.} = -3.06 \text{ in./deg.} \\[2mm] \dfrac{35}{82.1 - 50} = +1.09 \text{ ft./rad.} = +0.228 \text{ in./deg.} \end{cases}$$

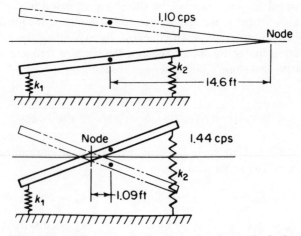

Fig. 6.7-3. Principal modes of vibration, showing node (points of zero motion) positions.

The position of the nodes then determines the type of motion, as illustrated by Fig. 6.7-3.

(b) *The Vibration Absorber.* A spring-mass system k_2, m_2 tuned to the frequency of the exciting force such that $\omega^2 = k_2/m_2$ will act as a vibration absorber. In Fig. 6.7-4, the original system k_1, m_1, is forced to vibrate under the harmonic excitation $F = F_0 \sin \omega t$. With the absorber system k_2, m_2 attached to the original system, the configuration becomes a two-degrees-of-freedom system whose equations of motion are

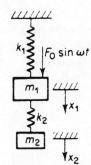

Fig. 6.7-4. Spring-mass system with absorber.

$$m_1 \ddot{x}_1 = k_2(x_2 - x_1) - k_1 x_1 + F_0 \sin \omega t \qquad (b\text{-}1)$$

$$m_2 \ddot{x}_2 = -k_2(x_2 - x_1) \qquad (b\text{-}2)$$

Making the following substitutions,

$$\omega_{11} = \sqrt{k_1/m_1} \qquad X_0 = F_0/k_1$$

$$\omega_{22} = \sqrt{k_2/m_2}$$

and assuming the solution to the forced vibration problem to be

$$x_1 = X_1 \sin \omega t$$

$$x_2 = X_2 \sin \omega t$$

the equations in terms of the amplitudes become

$$\left[1 + \frac{k_2}{k_1} - \left(\frac{\omega}{\omega_{11}}\right)^2\right] X_1 - \left(\frac{k_2}{k_1}\right) X_2 = X_0 \qquad (b\text{-}3)$$

$$-X_1 + \left[1 - \left(\frac{\omega}{\omega_{22}}\right)^2\right] X_2 = 0 \qquad (b\text{-}4)$$

The behavior of the system may be studied by examining the expressions for the amplitude of the two masses

$$\frac{X_1}{X_0} = \frac{[1 - (\omega/\omega_{22})^2]}{\left[1 + \dfrac{k_2}{k_1} - \left(\dfrac{\omega}{\omega_{11}}\right)^2\right]\left[1 - \left(\dfrac{\omega}{\omega_{22}}\right)^2\right] - \dfrac{k_2}{k_1}} \qquad (b\text{-}5)$$

$$\frac{X_2}{X_0} = \frac{1}{\left[1 + \dfrac{k_2}{k_1} - \left(\dfrac{\omega}{\omega_{11}}\right)^2\right]\left[1 - \left(\dfrac{\omega}{\omega_{22}}\right)^2\right] - \dfrac{k_2}{k_1}} \qquad (b\text{-}6)$$

It is evident from these equations that the amplitude X_1 of the mass 1 becomes zero when the exciting frequency ω coincides with the natural frequency ω_{22} of the absorber. For this frequency the amplitude X_2 of mass 2 is equal to

$$X_2 = -\frac{k_1}{k_2} X_0 = -\frac{F_0}{k_2} \qquad (b\text{-}7)$$

where the negative sign indicates that X_2 is out of phase with the impressed force. In fact, X_1 becomes zero at this frequency simply because the force $k_2 X_2$ exerted by the spring 2 on mass m_1 is equal and opposite to the impressed force F_0. This is the basis for the vibration absorber.

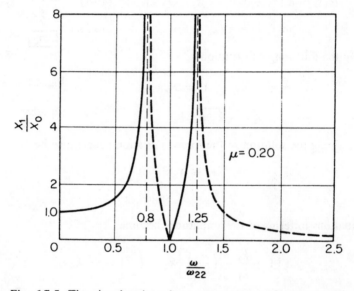

Fig. 6.7-5. The absorber introduces two resonant frequencies of the main mass.

The size of the absorber mass m_2 will depend on the magnitude of the disturbing force F_0, since the absorber must exert a force equal and opposite to the disturbing force which, in turn, will depend on the allowable deformation of the absorber spring, since

$$k_2 X_2 = \omega^2 m_2 X_2 = -F_0 \qquad (b\text{-}8)$$

Although the amplitude of the main mass m_1 is reduced to zero when $\omega = \omega_{22}$, there are two resonant frequencies at which the amplitude of m_1 becomes infinite, as shown in Fig. 6.7-5. These resonant frequencies are found on either side of ω_{22}, and can be obtained by setting the denominator of the amplitude equation to zero:

$$\left[1 + \frac{k_2}{k_1} - \left(\frac{\omega}{\omega_{11}}\right)^2\right]\left[1 - \left(\frac{\omega}{\omega_{22}}\right)^2\right] - \frac{k_2}{k_1} = 0 \qquad (b\text{-}9)$$

Replacing k_2/k_1 with $\dfrac{m_2}{m_1}\dfrac{k_2}{m_2}\dfrac{m_1}{k_1} = \mu(\omega_{22}/\omega_{11})^2$, where $\mu = m_2/m_1$, the above equation may be rearranged as

$$\left[\frac{\omega_{22}}{\omega_{11}}\right]^2\left(\frac{\omega}{\omega_{22}}\right)^4 - \left[1 + (1+\mu)\left(\frac{\omega_{22}}{\omega_{11}}\right)^2\right]\left(\frac{\omega}{\omega_{22}}\right)^2 + 1 = 0 \qquad (b\text{-}10)$$

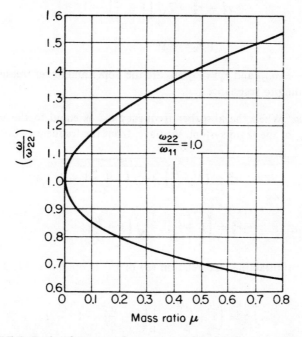

Fig. 6.7-6. Ratio of resonant frequency to absorber frequency plotted as a function of the mass ratio μ.

The two resonant frequencies can then be found from the above equation in (ω/ω_{22}), for any values of the parameters $(\omega_{22}/\omega_{11})$ and μ. Figure 6.7-6 shows one such curve for $(\omega_{22}/\omega_{11}) = 1$, the ratio of the resonant frequency to the absorber frequency being plotted as a function of the mass ratio μ.

The vibration absorber can be used only where the disturbing frequency is constant, since the absorber is effective only at the natural frequency of the absorber system. The vibration absorber is thus suited for synchronous machines and devices run on constant-frequency alternating currents. For gasoline engines and other variable-speed systems, the vibration absorber is not practical, since two resonant frequencies are introduced for each original resonant frequency.

EXAMPLE 6.7-2. For the vibration absorber with $\omega_{11} = \omega_{22}$, show that the natural frequencies of the system are related to the mass ratio by the expressions

$$r_1^2 + r_2^2 = 2\left(1 + \frac{\mu}{2}\right) \tag{a}$$

$$r_1^2 - r_2^2 = 2\sqrt{\left(1 + \frac{\mu}{2}\right)^2 - 1} \tag{b}$$

$$r_1 r_2 = 1.0 \tag{c}$$

where $r_1 = \omega_1/\omega_{22}$ and $r_2 = \omega_2/\omega_{22}$ are the upper and lower frequency ratios corresponding to resonance.

Solution. When the absorber frequency ω_{22} is equal to the natural frequency ω_{11} of the main system, we have, from the frequency equation

$$r_1^2 = \left(1 + \frac{\mu}{2}\right) + \sqrt{\left(1 + \frac{\mu}{2}\right)^2 - 1}$$

$$r_2^2 = \left(1 + \frac{\mu}{2}\right) - \sqrt{\left(1 + \frac{\mu}{2}\right)^2 - 1}$$

Adding or subtracting, the relation (a) or (b) is obtained. Multiplying, we obtain

$$r_1^2 r_2^2 = \left(1 + \frac{\mu}{2}\right)^2 - \left[\left(1 + \frac{\mu}{2}\right)^2 - 1\right] = 1.0$$

EXAMPLE 6.7-3. Because of unbalances in certain rotating machinery, a structure was found to vibrate excessively. A vibrometer record indicates that the vibration is composed of a large component at 1500 c.p.m. and a smaller one at 1280 c.p.m. Propose a procedure for the design of an absorber to be attached to the structure so that the natural frequencies will not be within 20 per cent of the disturbing frequencies.

Solution. For the natural frequencies to be at least 20 per cent from the disturbing frequencies, they must lie outside the region 1024 to 1800 c.p.m. If the absorber is tuned for 1500 c.p.m., the ratio of the natural frequencies to the absorber frequency becomes

$$r_1 = \frac{\omega_1}{\omega_{22}} = \frac{1800}{1500} = 1.2$$

$$r_2 = \frac{\omega_2}{\omega_{22}} = \frac{1024}{1500} = 0.683$$

and the mass ratio μ is established by the lower figure and the equation

$$r_2^2 = \left(1 + \frac{\mu}{2}\right) - \sqrt{\left(1 + \frac{\mu}{2}\right)^2 - 1}$$

Letting $x = \left(1 + \frac{\mu}{2}\right)$, the above equation becomes

$$r_2^4 - 2r_2^2 x + x^2 = x^2 - 1$$

$$x = \left(1 + \frac{\mu}{2}\right) = \frac{r_2^4 + 1}{2r_2^2} = \frac{0.683^4 + 1}{2 \times 0.683^2} = 1.31$$

$$\mu = 0.62 \text{ (see Fig. 6.7-6 for approximate check)}$$

Since the effective mass of the structure is not known it is necessary to use a test absorber of known weight W' tuned for 1500 c.p.m. and to note the lower natural frequency ratio r_2', from which μ' may be established. The proper mass can then be determined from the relation $W = W'(0.62/\mu')$.

(c) *The Centrifugal-Pendulum Vibration Absorber.* The vibration absorber discussed in the previous section is effective only when the disturbing frequency is equal to or near the fixed natural frequency of the absorber. Since such a system becomes resonant at two neighboring frequencies on each side of the absorber frequency, their range of effectiveness is narrowly limited,

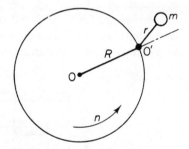

Fig. 6.7-7. Centrifugal pendulum.

making them useless for systems operating over a wide frequency range.

For torsional systems such as reciprocating engines, the disturbances are proportional to the rotational speed, which may vary over a wide range. To be effective, then, the absorber must have a natural frequency which will also vary with the speed. The pendulum-type absorber, schematically shown in Fig. 6.7-7 will satisfy this requirement of variable natural frequency. The pendulum is free to oscillate about a fixed point O' on a wheel or arm rotating with angular speed n rad./sec. The centrifugal field Rn^2 thus replaces the gravitational field g, and the natural frequency, which is equal to $\sqrt{g/l}$ for a gravity pendulum, now becomes equal to $n\sqrt{R/r}$ for the centrifugal pendulum.

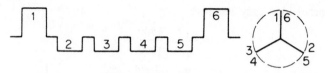

Fig. 6.7-8. Typical crankshaft of a six-cylinder engine.

For the natural frequency of the pendulum always to be equal to the disturbing frequency, the ratio R/r must be properly chosen. For a reciprocating engine, the frequency of the disturbing force which is proportional to the rotational speed will depend on the number of cylinders and whether the engine has a two- or four-stroke cycle. In a four-stroke cycle, each cylinder fires once every two revolutions, and therefore the number of impulses transmitted to the crankshaft per revolution is equal to $N/2$, where N is the number of cylinders. For the pendulum natural frequency to be equal to that of the disturbance, the ratio R/r must then be equal to $(N/2)^2$. For instance, the six-cylinder, four-stroke engine shown in Fig. 6.7-8 would fire three times per revolution, and R/r must equal 9. Since R is generally limited by the throw of the crank arm or the diameter of the flywheel, the geometric difficulties encountered in designing a conventional pendulum of sufficient mass and such small length r become evident. These difficulties, however, are overcome by a bifilar-type pendulum shown in Fig. 6.7-9. The U-shaped counter-weight fits loosely and rolls on two pins of diameter d_2 within two larger holes of equal diameters d_1. With respect to the crank, the counter-weight has a motion of translation, each point on it moving in a circular path

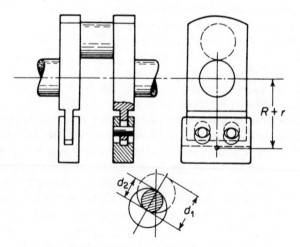

Fig. 6.7-9. Bifilar-type centrifugal pendulum.

of radius $r = d_1 - d_2$. Thus the pendulum arm r can be made very small in contrast to the distance $(R + r)$ measured from the center line of the crank to the center of gravity of the counterweight.

For the equation of motion for the centrifugal pendulum, we refer to Fig. 6.7-10. Placing axes through point O' parallel and normal to r, the acceleration of m relative to O' is

$$-r(\dot\theta + \dot\phi)^2 i + r(\ddot\theta + \ddot\phi)j$$

To this we add the acceleration of O', which is

$$(R\ddot\theta \sin\phi - R\dot\theta^2 \cos\phi)i$$
$$+ (R\ddot\theta \cos\phi + R\dot\theta^2 \sin\phi)j$$

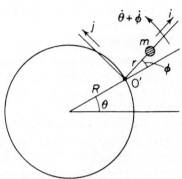

and obtain for the acceleration of m the equation

$$a_m = [R\ddot\theta \sin\phi - R\dot\theta^2 \cos\phi - r(\dot\theta + \dot\phi)^2]i \qquad \text{Fig. 6.7-10.}$$
$$+ [R\ddot\theta \cos\phi + R\dot\theta^2 \sin\phi + r(\ddot\theta + \ddot\phi)]j \qquad (c\text{-}1)$$

Since the moment about O' is zero, we have

$$r[R\ddot\theta \cos\phi + R\dot\theta^2 \sin\phi + r(\ddot\theta + \ddot\phi)] = 0 \qquad (c\text{-}2)$$

Assuming ϕ to be small, we let $\cos\phi = 1$ and $\sin\phi = \phi$, to obtain the equation

$$\ddot\phi + \frac{R}{r}\dot\theta^2\phi = -\frac{(R + r)}{r}\ddot\theta \qquad (c\text{-}3)$$

which indicates that the natural frequency of the centrifugal pendulum is $\dot\theta\sqrt{R/r}$ as stated before.

Although the system is one of two degrees of freedom with coordinates θ and ϕ, which requires an additional equation for the wheel, the exact equations are nonlinear and cannot be solved analytically without simplifying assumptions. If we assume θ to be composed of a steady rotation $\theta = nt$ plus an oscillatory component $\theta_0 \sin\omega t$, we have

$$\dot\theta = n + \omega\theta_0 \cos\omega t$$
$$\ddot\theta = -\omega^2\theta_0 \sin\omega t \qquad (c\text{-}4)$$

Thus, by assuming $\omega\theta_0$ to be small in comparison to n, the differential equation can be written as

$$\ddot\phi + \frac{n^2 R}{r}\phi = \omega^2\theta_0\left(\frac{R + r}{r}\right)\sin\omega t \qquad (c\text{-}5)$$

For the steady-state solution, $\phi = \phi_0 \sin(\omega t - \alpha)$, the relationship between the amplitudes becomes

$$\frac{\theta_0}{\phi_0} = \frac{n^2 \dfrac{R}{r} - \omega^2}{[(R + r)/r]\omega^2} \tag{c-6}$$

It is evident from this equation that when the disturbing frequency ω of the wheel coincides with the natural frequency $n\sqrt{R/r}$ of the pendulum, θ_0 becomes zero for a finite ϕ_0. Since the disturbing frequency ω is proportional to the rotational speed n, the centrifugal pendulum is effective for all speeds when properly designed.

Like the vibration absorber discussed in the previous section, the torque exerted by the centrifugal pendulum at $\omega = n\sqrt{R/r}$ is equal and opposite to the disturbing torque T_0. The required mass of the pendulum can therefore be found from the following equation

$$T_0 \cong m(R + r)n^2 R\phi \tag{c-7}$$

where $m(R + r)n^2$ is the force along r acting at O' and $R\phi$ is the moment arm for small ϕ.

(d) *Dynamics of Reciprocating Engines.* For the design of absorbers and dampers, it is necessary to know the nature of the excitation causing the vibration. In this section we discuss the dynamics of the reciprocating engines as a source of vibration excitation.

The moving parts of a reciprocating engine produce dynamic forces which result in undesirable vibrations. Rotating parts, such as the crankshaft, can be easily balanced; however, translating parts, such as the piston, and parts like the connecting rod with a more complex motion, cannot be so easily balanced.

Consider the connecting rod, the piston end of which is translating and the crank end rotating. For purposes of analysis it is desirable to replace the rod by an equivalent dynamical system composed of two concentrated masses, as shown in Fig. 6.7-11. For the two-mass system to be dynamically equivalent to the original connecting rod, it must satisfy the following requirements: (1) same total mass, (2) same center of gravity, and (3) same moment of inertia. These three conditions can be expressed by the equations

$$W = W_1 + W_2$$

$$W_2 h = Wc \tag{d-1}$$

$$Wk_G^2 = W_1 c^2 + W_2(h - c)^2$$

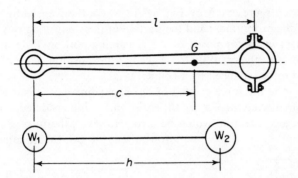

Fig. 6.7-11. Connecting rod and equivalent two-mass system.

where k_G is the radius of gyration of the connecting rod. Solving for W_1, W_2, and h,

$$W_1 = W\left(1 - \frac{c}{h}\right)$$

$$W_2 = W\frac{c}{h} \qquad (d\text{-}2)$$

$$h = c\left(1 + \frac{k_G^2}{c^2}\right) = \frac{k_1^2}{c}$$

In the above equations k_1 is the radius of gyration of the rod about the piston end and, therefore, h becomes equal to the distance to the center of percussion of the rod. Thus, in the dynamically equivalent rod, W_1 undergoes translation while W_2 is subjected to a more complex motion of translation and rotation. The force analysis on the rod, however, can be carried out by an analytical or a semigraphical procedure.

It is evident that, by proportioning the connecting rod with a slight extension beyond the crank end, it is possible to make $h = l$. The weight W_2 then coincides with the crank pin and its motion is that of pure rotation. The two concentrated masses are then expressed by the equations

$$W_1 = W\left(1 - \frac{c}{l}\right)$$

$$W_2 = W\frac{c}{l} \qquad (d\text{-}3)$$

We will analyze here only the simpler case where the connecting rod is replaced by concentrated masses expressed by the two equations above. The system is then represented by Fig. 6.7-12 where the translating mass m_t is the sum of the piston mass and the portion of the connecting rod W_1/g. The rotating mass m_r is composed of the remaining portion W_2/g and any unbalanced mass of the crankshaft assigned at this position, both of which will be assumed to be balanced by a counterweight.

Letting P be the force exerted by the expanding gas, it is evident that it would be an internal force when the entire engine is considered, and hence its resultant in any direction would be zero. However, the force P results in a torque about the crankshaft equal to

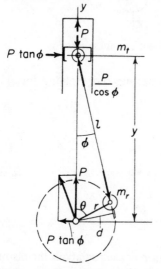

$$T_p = \frac{P}{\cos \phi} d = Py \tan \phi \qquad (d\text{-}4)$$

For the inertia forces of the moving parts we need only to consider the translating mass m_t since m_r is assumed to be balanced out. The force on m_t is in the vertical direction and equal to

$$F_y = m_t \ddot{s} \qquad (d\text{-}5)$$

where the acceleration $\ddot{s}$ of the piston is

$$\ddot{s} = r\omega^2 \left(\cos \omega t + \frac{r}{l} \cos 2\omega t + \cdots \right)$$

The equation for F_y now becomes

Fig. 6.7-12. Dynamic analysis of crank mechanism.

$$F_y = m_t r\omega^2 \cos \omega t + \left(m_t \frac{r}{l} \right) r\omega^2 \cos 2\omega t$$

$$(d\text{-}6)$$

We thus have, for the inertia force, a primary unbalance at a frequency equal to the rotational speed, and a secondary unbalance at a frequency equal to twice the rotational speed.

Since the gas pressure P which is in the same direction as F_y resulted in a torque about the crankshaft, the inertia force $-m_t \ddot{s}$ would also have a torque about the crankshaft equal to

$$T_i = -m_t \ddot{s} y \tan \phi$$

$$= -m_t r\omega^2 \left(\cos \omega t + \frac{r}{l} \cos 2\omega t \right) \left(l \cos \phi + r \cos \omega t \right) \tan \phi \qquad (d\text{-}7)$$

Replacing $\tan \phi$ and $\cos \phi$ with the following approximate relations

$$\tan \phi \cong \frac{r}{l} \sin \omega t$$

$$(d\text{-}8)$$

$$\cos \phi \cong 1.0$$

the equation for T_i becomes

$$T_i = -m_t r^2 \omega^2 \sin \omega t \left(\cos \omega t + \frac{r}{l} \cos 2\omega t \right) \left(1 + \frac{r}{l} \cos \omega t \right) \qquad (d\text{-}9)$$

Multiplying out and omitting higher powers of r/l,

$$T_i = -m_t r^2 \omega^2 \left(\sin \omega t \cos \omega t + \frac{r}{l} \sin \omega t \cos 2\omega t + \frac{r}{l} \sin \omega t \cos^2 \omega t \right) \qquad (d\text{-}10)$$

Using the trigonometric relations

$$\sin \omega t \cos 2\omega t = \tfrac{1}{2}(\sin 3\omega t - \sin \omega t)$$

$$\sin \omega t \cos \omega t = \tfrac{1}{2} \sin 2\omega t \qquad (d\text{-}11)$$

$$\cos^2 \omega t = \tfrac{1}{2}(1 + \cos 2\omega t)$$

the final equation for T_i reduces to

$$T_i = \tfrac{1}{2} m_t r^2 \omega^2 \left(\frac{r}{2l} \sin \omega t - \sin 2\omega t - \frac{3r}{2l} \sin 3\omega t \right) \qquad (d\text{-}12)$$

We can now summarize the unbalance of a single cylinder engine as follows:

(1) Inertia force F_y in the vertical direction, Eq. (d-6), with primary and secondary components.

(2) Inertia torque T_i about the crankshaft, Eq. (d-12), with primary and higher components.

(3) Power torque T_p about the crankshaft, Eq. (d-4), due to gas explosions in the cylinder. Since for the four-stroke cycle the engine fires once every two revolutions, the power torque results in half order components.

Multi-cylinder Engines. In the single-cylinder engine there will always be the unbalance due to the translating mass m_t. In the multi-cylinder engine the unbalance due to m_t can be canceled by the proper angular spacing of the cranks. The inertia unbalance of a counterbalanced multi-cylinder engine, from Eqs. (d-6) and (d-12), is

$$F_y = m_t r \omega^2 \sum_{n=1}^{j} \left[\cos (\omega t + \psi_n) + \frac{r}{l} \cos 2(\omega t + \psi_n) \right] \qquad (d\text{-}13)$$

$$T_i = \tfrac{1}{2} m_t r^2 \omega^2 \sum_{n=1}^{j} \left[\left(\frac{r}{2l} \right) \sin (\omega t + \psi_n) \right.$$

$$\left. - \sin 2(\omega t + \psi_n) - \frac{3r}{2l} \sin 3(\omega t + \psi_n) \right] \qquad (d\text{-}14)$$

where $\psi_1 = 0$, ψ_2, ψ_3, and so on, are angular positions of cranks 2, 3, and so on, with respect to crank 1.

Even if these forces and moments add up to zero, it is possible to have a pitching moment about a horizontal axis perpendicular to the center line of the engine. This moment can be found by summing the moments of the y forces about the first cylinder. If the distances from the nth cylinder to cylinder 1 is a_n, the pitching moment about the first cylinder becomes

$$M_x = m_t r \omega^2 \sum_{n=1}^{j} \left[a_n \cos (\omega t + \psi_n) + \frac{r}{l} a_n \cos 2(\omega t + \psi_n) \right] \quad (d\text{-}15)$$

where $a_1 = 0$.

(e) *Torsional Vibration Dampers.* Figure 6.7-13 represents a friction type of vibration damper, commonly known as the Lanchester damper, which has

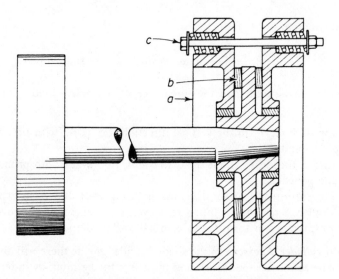

Fig. 6.7-13. Torsional vibration damper.

found practical use in torsional systems such as gas and diesel engines in limiting the amplitudes of vibration at critical speeds. The damper consists of two flywheels a free to rotate on the shaft and driven only by means of the friction rings b when the normal pressure is maintained by the spring-loaded bolts c.

When properly adjusted, the flywheels rotate with the shaft for small oscillations. However, when the torsional oscillations of the shaft tend to become large, the flywheels will not follow the shaft because of their large inertia, and energy is dissipated by friction due to the relative motion. The dissipation of energy thus limits the amplitude of oscillation, thereby preventing high torsional stresses in the shaft.

In spite of the simplicity of the torsional damper the mathematical analysis for its behavior is rather complicated. For instance, the flywheels may slip continuously, for part of the cycle, or not at all, depending on the pressure exerted by the spring bolts. If the pressure on the friction ring is either too great for slipping or zero, no energy is dissipated, and the damper becomes ineffective. Obviously, maximum energy dissipation takes place at some intermediate pressure, resulting in optimum damper effectiveness.

To obtain an insight to the problem, we consider briefly the case where the flywheels slip continuously. Assuming the shaft hub to be oscillating about its mean angular speed, as shown in Fig. 6.7-14, the flywheels will be acted upon by a constant frictional torque T while slipping. The acceleration of the flywheel, represented by the slope of the velocity curve, will hence be constant and equal to T/J, where J is the moment of inertia of the flywheels, and its velocity will be represented by a series of straight lines.

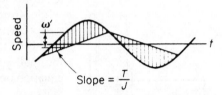

Fig. 6.7-14. Torsional damper under continuous slip.

The velocity of the flywheels will be increasing while the shaft speed is greater than that of the flywheels and decreasing when the shaft speed drops below that of the flywheels, as shown in the diagram.

The work done by the damper,

$$W = \int T \, d\theta = T \int \omega' \, dt \qquad (e\text{-}1)$$

where ω' is the relative velocity, is equal to the product of the torque T and the shaded area of Fig. 6.7-14. This shaded area, being small for large T and large for small T, the maximum energy is dissipated for some intermediate value of T.*

Obviously, the damper should be placed in a position where the amplitude of oscillation is the greatest. This position generally is found on the side of the shaft away from the main flywheel, since the node is usually near the largest mass.

(*f*) *The Untuned Viscous Vibration Damper.* In a rotating system such as an automobile engine, the disturbing frequencies for torsional oscillations are proportional to the rotational speed. However there is generally more than one such frequency, as was shown in Sec. *d*, and the centrifugal pendulum has the disadvantage that several pendulums tuned to the order number of the disturbance must be used. In contrast to the centrifugal pendulum, the

* J. P. Den Hartog and J. Ormondroyd, "Torsional-Vibration Dampers," *Trans. ASME* APM-52-13, pp. 133–152. (September–December, 1930).

untuned viscous torsional damper is effective over a wide operating range. It consists of a free rotational mass within a cylindrical cavity filled with viscous fluid, as shown in Fig. 6.7-15. Such a system is generally incorporated into the end pulley of a crankshaft which drives the fan belt, and is often referred to as the Houdaille damper.

We can examine the untuned viscous damper as a two-degrees-of-freedom system by considering the crankshaft, to which it is attached, as being fixed at one end with the damper at the other end. With the torsional stiffness of the shaft equal to K in. lb./rad. the damper can be considered to be excited by

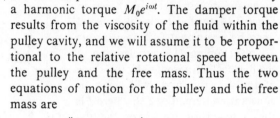

a harmonic torque $M_0 e^{i\omega t}$. The damper torque results from the viscosity of the fluid within the pulley cavity, and we will assume it to be proportional to the relative rotational speed between the pulley and the free mass. Thus the two equations of motion for the pulley and the free mass are

$$J\ddot{\theta} + K\theta + c(\dot{\theta} - \dot{\varphi}) = M_0 e^{i\omega t}$$

$$J_d \ddot{\varphi} - c(\dot{\theta} - \dot{\varphi}) = 0 \qquad (f\text{-}1)$$

Assuming the solution to be in the form

Fig. 6.7-15. Untuned viscous damper.

$$\theta = \theta_0 e^{i\omega t}$$

$$\varphi = \varphi_0 e^{i\omega t} \qquad (f\text{-}2)$$

where θ_0 and φ_0 are complex amplitudes, their substitution into the differential equations results in

$$\left[\left(\frac{K}{J} - \omega^2\right) + i\frac{c\omega}{J}\right]\theta_0 - \frac{ic\omega}{J}\varphi_0 = \frac{M_0}{J}$$

and

$$\left(-\omega^2 + i\frac{c\omega}{J_d}\right)\varphi_0 = \frac{ic\omega}{J_d}\theta_0 \qquad (f\text{-}3)$$

Eliminating φ_0 between the two equations, the expression for the amplitude θ_0 of the pulley becomes

$$\frac{\theta_0}{M_0} = \frac{(\omega^2 J_d - ic\omega)}{[\omega^2 J_d(K - J\omega^2)] + ic\omega[\omega^2 J_d - (K - J\omega^2)]} \qquad (f\text{-}4)$$

Letting $\omega_n^2 = K/J$ and $\mu = J_d/J$, the critical damping is

$$c_c = 2J\omega_n, \quad c = \frac{c}{c_c}2J\omega_n = 2\zeta J\omega_n$$

The amplitude equation then becomes

$$\left|\frac{K\theta_0}{M_0}\right| = \sqrt{\frac{\mu^2(\omega/\omega_n)^2 + 4\zeta^2}{\mu^2(\omega/\omega_n)^2(1 - \omega^2/\omega_n^2)^2 + 4\zeta^2[\mu(\omega/\omega_n)^2 - (1 - \omega^2/\omega_n^2)]^2}} \quad (f\text{-}5)$$

which indicates that $|K\theta_0/M_0|$ is a function of three parameters, ζ, μ, and (ω/ω_n).

If μ is held constant and $|K\theta_0/M_0|$ plotted as a function of (ω/ω_n), the curve for any ζ will appear somewhat similar to that of a single-degree-of-freedom system with a single peak. Of interest are the two extreme values of

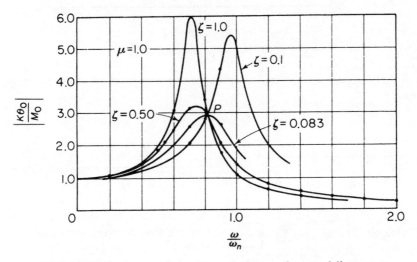

Fig. 6.7-16. Response of an untuned viscous damper (all curves pass through P).

$\zeta = 0$ and $\zeta = \infty$. When $\zeta = 0$, we have an undamped system with resonant frequency $\omega_n = \sqrt{K/J}$, and the amplitude will be infinite at this frequency. If $\zeta = \infty$, the damper mass and the wheel will move together as a single mass, and again we have an undamped system but with natural frequency of $\sqrt{k/(J + J_d)}$.

Thus, like the Lanchester damper of the previous section, there is an optimum damping ζ_0 for which the peak amplitude is a minimum as shown in Fig. 6.7-16. The result can be presented as a plot of the peak values as a function of ζ for any given μ, as shown in Fig. 6.7-17.

It can be shown that the optimum damping is equal to

$$\zeta_0 = \frac{1}{\sqrt{2(1 + \mu)(2 + \mu)}} \quad (f\text{-}6)$$

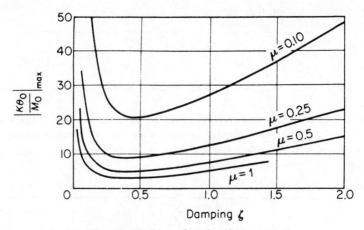

Fig. 6.7.-17. Plot of peak values versus ζ.

and that the peak amplitude for optimum damping is found at a frequency equal to

$$\frac{\omega}{\omega_n} = \sqrt{2/(2 + \mu)} \qquad (f\text{-}7)$$

These conclusions can be arrived at by observing that the curves of Fig. 6.7-16 all pass through a common point P, regardless of the numerical values of ζ. Thus, by equating the equation for $|K\theta_0/M|$ for $\zeta = 0$ and $\zeta = \infty$,

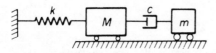

Fig. 6.7-18. Untuned viscous damper.

Eq. (f-7) is found. The curve for optimum damping then must pass through P with a zero slope, so that if we substitute $(\omega/\omega_n)^2 = 2/(2 + \mu)$ into Eq. (f-5) and equate it to the amplitude as found in the undamped curve for the same frequency, the expression for ζ_0 is found. It is evident that these conclusions apply also to the linear spring-mass sytem of Fig. 6.7-18, which is a special case of the damped vibration absorber with the damper spring equal to zero.

(g) *Gyroscopic Effect on Rotating Shafts.* A rotating wheel and shaft with angular momentum H can, under certain conditions, introduce a gyroscopic moment, thereby coupling the deflection and slope to produce a two-degrees-of-freedom problem. We will illustrate this effect in terms of a wheel rotating on an overhanging shaft, as shown in Fig. 6.7-19.

If the rotational speed of the shaft is ω, its components parallel and normal to the face of the wheel are $\omega \sin \theta$ and $\omega \cos \theta$. Thus the angular momentum in these directions are $J_d \omega \sin \theta$ and $J_p \omega \cos \theta$, where J_p and J_d are the moments of inertia of the wheel along the polar axis and its diameter. Resolving

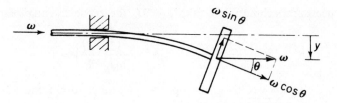

Fig. 6.7-19. Gyroscopic effect.

these vectors along the direction of ω and perpendicular to it, the component normal to ω is

$$H_n = J_p\omega \cos \theta \sin \theta - J_d\omega \sin \theta \cos \theta = (J_p - J_d)\omega \cos \theta \sin \theta \quad (g\text{-}1)$$

If the deflection plane of the shaft whirls with angular speed ω_1, a moment on the wheel equal to $H_n\omega_1$ is necessary, the component of H parallel to ω undergoing no change. Acting on the shaft is then an opposite moment:

$$M = -H_n\omega_1 = -(J_p - J_d)\omega\omega_1 \cos \theta \sin \theta \cong -(J_p - J_d)\omega\omega_1\theta \quad (g\text{-}2)$$

To take account of this moment on the deflection shape of the shaft, we can write the equations for the deflection and slope at the end of the shaft

$$y = F\frac{l^3}{3EI} + M\frac{l^2}{2EI}$$

$$\theta = F\frac{l^2}{2EI} + M\frac{l}{EI} \qquad (g\text{-}3)$$

where the coefficients of F and M are influence functions of deflection and slope due to unit force or unit moment acting on the shaft end, and F and M are the force and moment on the end of the shaft. The force F is simply $m\omega_1^2 y$, and M is the gyroscopic couple $-(J_p - J_d)\omega\omega_1\theta$ so that Eqs. (g-3) become

$$y = \left(m\omega_1^2 \frac{l^3}{3EI}\right)y - \left[(J_p - J_d)\omega\omega_1 \frac{l^2}{2EI}\right]\theta$$

$$\theta = \left(m\omega_1^2 \frac{l^2}{2EI}\right)y - \left[(J_p - J_d)\omega\omega_1 \frac{l}{EI}\right]\theta \qquad (g\text{-}4)$$

For a wheel or disk with an unbalance, we have found in Sec. 3.12 that the whirling speed ω_1 can be equal to ω. Thus the frequency equation may take the form

$$\left(1 - \frac{ma^2 l^3}{3EI}\right)\left[1 + (J_p - J_d)\omega^2 \frac{l}{EI}\right] + \left(m\omega^2 \frac{l^2}{2EI}\right)\left[(J_p - J_d)\omega^2 \frac{l^2}{2EI}\right] = 0$$

$$(g\text{-}5)$$

For a wheel approaching a thin disk, $J_p = 2J_d$, the frequency equation reduces to

$$\omega^4 + \frac{12EI}{mJ_d l^3}\left(\frac{ml^2}{3} - J_d\right)\omega^2 - \frac{12}{mJ_d}\left(\frac{EI}{l^2}\right)^2 = 0 \qquad (g\text{-}6)$$

Since, in the absence of the gyroscopic couple, the natural frequency of the system is $\omega_y = \sqrt{3EI/ml^3}$, we can rewrite the frequency equation as

$$\omega^4 + 4\omega_y^2\left(\frac{1}{\alpha} - 1\right)\omega^2 - \frac{4}{\alpha}\omega_y^4 = 0 \qquad (g\text{-}7)$$

where $\alpha = 3J_d/ml^2$ can be viewed as a coupling term. The relationship between $(\omega/\omega_y)^2$ and α is shown in Fig. 6.7-20. For very large values of α, the

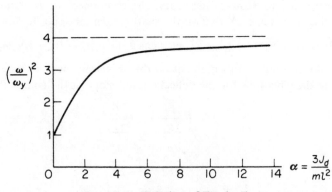

Fig. 6.7-20. Solution of Eq. (g-7).

ratio θ/y approaches zero and the natural frequency of the system tends to the value $\omega = \sqrt{12EI/ml^3}$.

PROBLEMS

1. Determine the equations of motion for the torsional system shown in Fig. 6-1. Express these equations in matrix form and compare with Eq. (6.2-4).

2. Determine the normal modes of the system of Prob. 1 when $K_1 = K_2$ and $J_1 = J_2$.

3. If $K_1 = 0$ in the torsional system of Prob. 1, the system becomes a degenerate two-degrees-of-freedom system

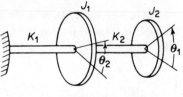

Fig. 6-1.

with only one natural frequency. Discuss the normal modes of this system. Show that the system can be treated as one of single degree of freedom by using the coordinate $\phi = \theta_1 - \theta_2$.

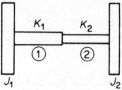

4. Determine the equation for the natural frequency of the torsional system shown in Fig. 6-2 and draw the normal-mode curve. Show that the nodal distance from J_2 is

Fig. 6-2.

$$l_2 \left[\frac{1 + K_2/K_1}{1 + J_2/J_1} \right].$$

5. Determine the natural frequency of the torsional system shown in Fig. 6-3 and draw the normal-mode curve.

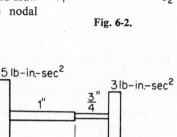

6. A gas engine has a flywheel weighing 600 lb. on each end, and their radius of gyration is 22 in. If the distance between the flywheels 30 is in. and the equivalent shaft between them has a diameter of 3

Fig. 6-3.

in., determine the natural frequency of torsional oscillation.

7. A shaft 2 in. in diameter and 24 in. long is attached at one end to a flywheel weighing 400 lb. with a radius of gyration of 10 in., and at the other end to a generator armature the moment of inertia of which is 0.6 that of the flywheel. Determine the natural frequency of torsional oscillation and locate the node.

8. In Fig. 6-4, two masses m_1 and m_2 are attached to a light string with tension T. Assuming that T remains unchanged when the masses are displaced normal to the string, write the equations of motion expressed in matrix form.

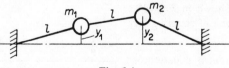

Fig. 6-4.

9. In Prob. 8, if the two masses are made equal, show that the normal-mode frequencies are $\omega_1 = \sqrt{T/ml}$ and $\omega_2 = \sqrt{3T/ml}$. Establish the configuration for these normal modes.

10. In Prob. 8, if $m_1 = 2m$, and $m_2 = m$, determine the normal-mode frequencies and mode shapes.

11. Determine the equations of motion for the system of Fig. 6-5 and establish the normal-mode frequencies and mode shapes.

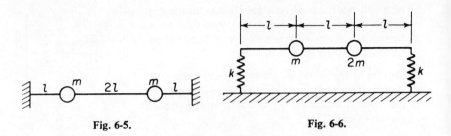

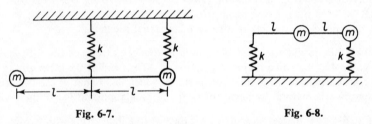

Fig. 6-5.   Fig. 6-6.

12. Set up the matrix equation of motion for the system shown in Fig. 6-6, using coordinates x_1 and x_2 at m and $2m$. Determine the equation for the normal-mode frequencies and describe the mode shapes.

13. In Prob. 12, if the coordinates x at m and θ are used, what form of coupling will they result in?

14. In Fig. 6-7, use coordinate x at the center of mass, and θ for the rotation of the bar, and set up the matrix equation of motion. Determine the normal-mode frequencies and locate the node of the bar for each mode.

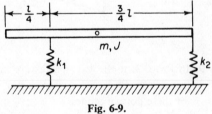

Fig. 6-7.   Fig. 6-8.

15. In Prob. 14, if the coordinates x_1 and x_2 at the ends of the bar are used, determine the form of the matrix equation and discuss the coupling.

16. Discuss the various coordinates which might be used to describe the motion of the system shown in Fig. 6-8, and state the type of coupling for each.

17. Determine the matrix equation of motion for the system shown in Fig. 6-9 when the coordinate x is placed at the center of mass of the uniform bar.

18. In Prob. 17, determine the co-ordinates which will result in dynamic coupling with no static coupling.

Fig. 6-9.

19. Determine the relationship between k_1 and k_2 which will eliminate the coupling in Prob. 17.

20. The following information is given for a certain automobile (see Fig. 6.7-2):

$$W = 3500 \text{ lb.} \qquad k_1 = 2000 \text{ lb./ft.}$$

$$l_1 = 4.4 \text{ ft.} \qquad k_2 = 2400 \text{ lb./ft.}$$

$$l_2 = 5.6 \text{ ft.} \qquad r = 4 \text{ ft.}$$

Determine the principal modes of vibration and locate the node for each mode.

21. The expansion joints of a concrete highway are 45 ft. apart. These joints cause a series of impulses at equal intervals to cars traveling at a constant speed. Determine the speeds at which pitching motion and up-and-down motion are most apt to arise for the automobile of Prob. 20.

22. Figure 6-10 shows an electric train made up of two cars of weight 50,000 lb. each. They are connected together by couplings of stiffness equal to 16,000 lb./in. Determine the natural frequency of the system.

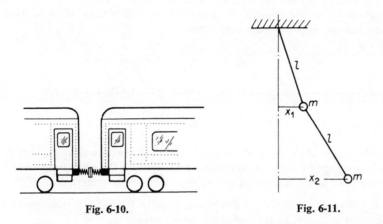

Fig. 6-10.   Fig. 6-11.

23. Assuming small amplitudes, set up the differential equation of motion for the double pendulum of Fig. 6-11, using the coordinates shown. Show that the natural frequencies of the system are given by the equation

$$\omega = \sqrt{\frac{g}{l}(2 \pm \sqrt{2})}$$

Determine the ratio of amplitudes x_1/x_2 and locate the nodes for the two modes of vibration.

24. Determine the natural frequencies of the system shown in Fig. 6-12 when

$$gm_1 = 3.86 \text{ lb.} \qquad k_1 = 20 \text{ lb./in.}$$

$$gm_2 = 1.93 \text{ lb.} \qquad k_2 = 10 \text{ lb./in.}$$

Describe the two modes of free vibration and locate the nodes.

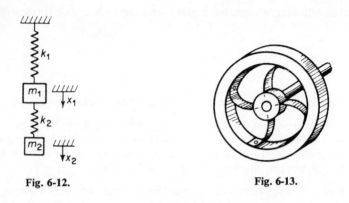

Fig. 6-12.   Fig. 6-13.

25. Figure 6-13 represents a torsional system composed of a shaft of stiffness K_1, a hub of radius r, and moment of inertia J_1, four leaf springs of stiffness k_2, and an outer wheel of radius R and moment of inertia J_2. Set up the differential equations for torsional oscillation, assuming one end of the shaft to be fixed. Show that the frequency equation reduces to

$$\omega^4 - \left(\omega_{11}^2 + \omega_{22}^2 + \frac{J_2}{J_1}\,\omega_{22}^2\right)\omega^2 + \omega_{11}^2\omega_{22}^2 = 0$$

where ω_{11} and ω_{22} are uncoupled frequencies given by the expressions

$$\omega_{11}^2 = \frac{K_1}{J_1}, \qquad \omega_{22}^2 = \frac{4k_2 R^2}{J_2}$$

26. For the system of Fig. 6-14, $W_1 = 200$ lb. and the absorber weight $W_2 = 50$ lb. If W_1 is excited by a 2-lb. in. unbalance rotating at 1800 r.p.m., determine the proper value of the absorber spring k_2. What will be the amplitude of W_2?

27. In Fig. 6-14, if a dashpot c is introduced between W_1 and W_2, determine the amplitude equations by the complex algebra method.

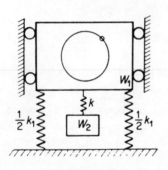

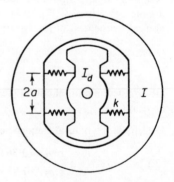

Fig. 6-14.   Fig. 6-15.

28. A flywheel of moment of inertia I has a torsional absorber of moment of inertia I_d free to rotate on the shaft and connected to the flywheel by four springs of stiffness k lb./in. as shown in Fig. 6-15. Set up the differential equations of motion for the system.

29. A bifilar-type centrifugal pendulum (Fig. 6.7-9) is proposed to eliminate a torsional disturbance of frequency equal to four times the rotational speed. If the distance R to the center of gravity of the pendulum mass is made equal to 4.0 in. and $d_1 = \frac{3}{4}$ in., what must be the diameter d_2 of the pins?

30. A jig used to size coal contains a screen that reciprocates with a frequency of 600 c.p.m. The jig weighs 500 lb. and has a fundamental frequency of 400 c.p.m. If an absorber weighing 125 lb. is to be installed to eliminate the vibration of the jig frame, determine the absorber spring stiffness. What will be the resulting two natural frequencies of the system?

31. In a certain refrigeration plant a section of pipe carrying the refrigerant vibrated violently at a compressor speed of 232 r.p.m. To eliminate this difficulty it was proposed to clamp a spring-mass system to the pipe to act as an absorber. For a trial test a 2.0-lb. absorber tuned to 232 c.p.m. resulted in two natural frequencies of 198 and 272 c.p.m. If the absorber system is to be designed so that the natural frequencies lie outside the region 160 to 320 c.p.m., what must be the weight and spring stiffness?

32. Figure 6-16 shows a type of damper frequently used on automobile crankshafts. J represents a solid disk free to spin on the shaft, and the space between the disk and case is filled with a silicone oil of coefficient of viscosity μ. The damping action results from any relative motion between the two. Derive an

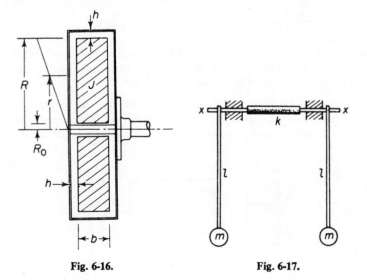

Fig. 6-16.   Fig. 6-17.

equation for the damping torque exerted by the disk on the case due to a relative velocity of ω.

33. Figure 6-17 shows two equal pendulums free to rotate about the x-x axis and coupled together by a rubber hose of torsional stiffness k lb. in./rad. Determine the natural frequencies for the principal modes of vibration, and describe how these motions may be started.

If $l = 19.3$ in., $mg = 3.86$ lb., and $k = 2.0$ lb. in./rad., determine the beat period for a motion started with $\theta_1 = 0$ and $\theta_2 = \theta_0$.

34. Set up the equations of motion for the system of Fig. 6-18.

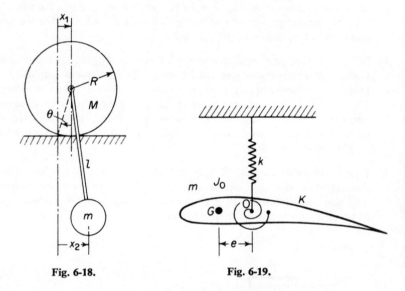

Fig. 6-18.   Fig. 6-19.

35. Figure 6-19 shows an airfoil section to be tested in a wind tunnel. If the center of gravity of the section is a distance e ahead of the point of support, determine the differential equations of motion of the system.

36. The rotor of Fig. 6-20 is mounted in bearings which are free to move in a single plane. The rotor is symmetrical about O with total mass M and moment of inertia J_0 about an axis perpendicular to the shaft. If a small unbalance mr

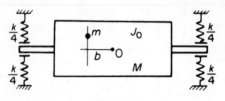

Fig. 6-20.

acts at an axial distance b from its center O, set up the equations of motion for a rotational speed ω.

37. For the Houdaille viscous damper with mass ratio $\mu = 0.25$, determine the optimum damping ζ_0 and the frequency at which the damper is most effective.

38. If the damping for the viscous damper of Prob. 37 is equal to $\zeta = 0.10$, determine the peak amplitude as compared to the optimum.

39. Establish the relationships given by Eqs. (f-7) and (f-6).

40. A two-story building is represented by the lumped mass system of Fig. 6-21 where $m_1 = \frac{1}{2}m_2$ and $k_1 = \frac{1}{2}k_2$. Show that its normal modes are

$$\left(\frac{X_1}{X_2}\right)^{(1)} = 2, \qquad \omega_1^2 = \frac{1}{2}\frac{k_1}{m_1}$$

$$\left(\frac{X_1}{X_2}\right)^{(2)} = -1, \qquad \omega_2^2 = 2\frac{k_1}{m_1}$$

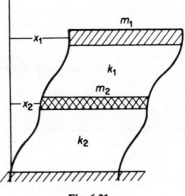

Fig. 6-21.

41. In Prob. 40, if a force is applied to m_1 to deflect it by unity, and the system is released from this position, determine the equation of motion of each mass by using the normal-mode-summation method.

42. In Prob. 41, determine the ratio of the maximum shear in the first and second stories.

43. Repeat Prob. 41 if the load is applied to m_2, displacing it by unity.

44. Show that the normal modes of the system shown in Fig. 6-22 are

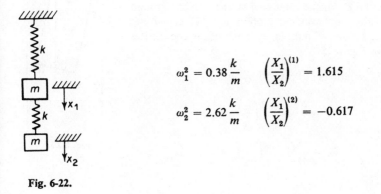

$$\omega_1^2 = 0.38\frac{k}{m} \qquad \left(\frac{X_1}{X_2}\right)^{(1)} = 1.615$$

$$\omega_2^2 = 2.62\frac{k}{m} \qquad \left(\frac{X_1}{X_2}\right)^{(2)} = -0.617$$

Fig. 6-22.

45. If the system of Prob. 44 is started with initial conditions,

$$x_1(0) = 0 \qquad x_2(0) = 1.0 \qquad \dot{x}_1(0) = \dot{x}_2(0) = 0$$

show that the equations of motion are

$$x_1(t) = 0.448 \cos \omega_1 t - 0.450 \cos \omega_2 t$$
$$x_2(t) = 0.723 \cos \omega_1 t + 0.279 \cos \omega_2 t$$

46. Determine the normal modes of the system shown in Fig. 6-23 which are to be used in Prob. 47.

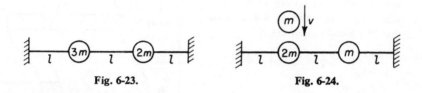

Fig. 6-23.   Fig. 6-24.

47. A lump of putty of mass m strikes $m_1 = 2m$ with velocity v and sticks to it. See Fig. 6-24. Determine its subsequent motion, using the normal modes of Prob. 46.

48. If the two-story building of Prob. 40 is excited by a harmonic force $F_1 = F_0 \sin \omega t$, applied to m_1, determine the equation for the steady-state vibration of each floor.

49. If the ground in Prob. 40 is vibrating with displacement $x_0(t) = X_0 \sin \omega t$, determine the equations for the steady-state vibration of each floor.

50. To simulate the effect of an earthquake on a rigid building, the base is assumed to be connected to the ground through two springs, K_h for the translational stiffness and K_r for the rotational stiffness. If the ground is now given a harmonic motion $y_g = Y_G \sin \omega t$, set up the equations of motion in terms of the coordinates shown in Fig. 6-25.

51. Solve the equations of Prob. 50 by letting

$$\omega_h = \frac{K_h}{M} \qquad \left(\frac{\rho_c}{l_0}\right)^2 = \frac{1}{3}$$

$$\omega_r = \frac{K_r}{M\rho_c^2} \qquad \left(\frac{\omega_r}{\omega_h}\right)^2 = 4$$

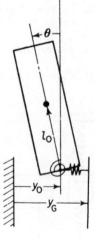

Fig. 6-25.

The first natural frequency and mode shape are

$$\omega_1/\omega_h = 0.734 \quad \text{and} \quad \frac{Y_G}{l_0\theta} = -3.16$$

which indicate a motion which is predominantly translation. Establish the second natural frequency and its mode.

52. The response and mode configuration for Prob. 50 and 51 are shown in Fig. 6-26. Verify the mode shapes for several values of the frequency ratio.

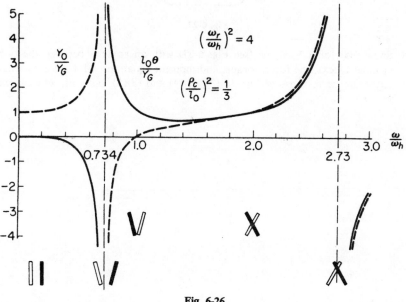

Fig. 6-26.

53. Determine the center of percussion of the connecting rod shown in Fig. 1-6, Ch. 1.

54. Show that the 4-cylinder engine with crank angles of 0°, 180°, 180°, and 0° has balanced primary forces, torques, and moments, and unbalanced secondary forces, torques and moments.

55. If the crank angles of the 4-cylinder engine are 0°, 90°, 270°, and 180°, show that the primary moments are unbalanced and the remaining primary and secondary quantities are balanced.

56. Investigate the state of inertia unbalance of the 6-cylinder engine with crank angles of 0°, 120°, 240°, 240°, 120°, and 0°.

57. Investigate the state of inertia unbalance of an 8-cylinder in-line engine with crank angles of 0°, 180°, 90°, 270°, 270°, 90°, 180°, 0°.

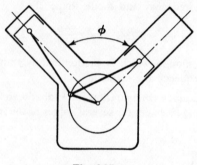

Fig. 6-27.

58. Show that for a V-engine (see Fig. 6-27) with an angle ϕ between the two cylinder blocks, the force, torque, and moment equations will have additional terms similar to those of Eqs. (d-13), (d-14) and (d-15), but with ωt replaced by $\omega t + \phi$.

7

Numerical Methods for Multi-degree-of-Freedom Systems

7.1 Introduction

Although all of the vibration theory for the multi-degrees-of-freedom system can be demonstrated by the two-degrees-of-freedom system, the problem of obtaining numerical results for systems of higher degrees of freedom requires special consideration. The recent development of the high-speed computer has revised the approach to many of these problems, and the emphasis here will be that of formulation for efficient numerical computation. Although these computations can be programmed directly from the recurrent dynamical equations, there are obvious advantages for presenting the problem in the logical matrix form. The matrix formulation, besides being convenient for computer programming, gives a clear conceptual presentation of the problem.

7.2 Influence Coefficients

We have shown so far that the equations of motion for any dynamical system can be formulated in terms of several different coordinate systems. For a lumped mass system it is convenient to choose coordinates with their origin fixed at the equilibrium position of each mass, as shown in

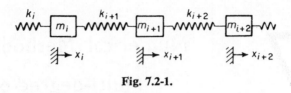

Fig. 7.2-1.

Fig. 7.2-1. The force summation method then leads to a dynamically un-coupled but a statically coupled set of equations of the form

$$\begin{bmatrix} m_1 & 0 & 0 & \cdots & \cdots & \\ 0 & m_2 & 0 & \cdots & \cdots & \\ 0 & & m_3 & & & \\ \cdot & & & \cdot & & \\ \cdot & & & & \cdot & \\ \cdot & & & & & \cdot \\ 0 & & & & & m_n \end{bmatrix} \begin{Bmatrix} \ddot{x}_1 \\ \ddot{x}_2 \\ \ddot{x}_3 \\ \cdot \\ \cdot \\ \cdot \\ \ddot{x}_n \end{Bmatrix} + \begin{bmatrix} k_{11} & k_{12} & \cdots & k_{1n} \\ k_{21} & k_{22} & & \cdot \\ & & \cdot & \cdot \\ & & & \cdot \\ & & & \cdot \\ & & & k_{nn} \end{bmatrix} \begin{Bmatrix} x_1 \\ x_2 \\ \cdot \\ \cdot \\ \cdot \\ \cdot \\ x_n \end{Bmatrix} = \begin{Bmatrix} 0 \\ 0 \\ \cdot \\ \cdot \\ \cdot \\ \cdot \\ 0 \end{Bmatrix}$$

(7.2-1)

In this equation, the k_{ij} are stiffness influence coefficients and although they can be simply identified from the equations of motion, $-k_{ij}$ can be inter-preted as the force at i due to a unit displacement at j where all the mass stations are held fixed except the jth station.

Dividing $\ddot{x}_i$ by m_i, the equation reduces to

$$-\begin{Bmatrix} \ddot{x}_1 \\ \ddot{x}_2 \\ \cdot \\ \cdot \\ \cdot \end{Bmatrix} = \begin{bmatrix} k_{11}/m_1 & k_{12}/m_1 & \cdots \\ k_{21}/m_2 & k_{22}/m_2 & \cdots \\ \cdot & & \\ \cdot & & \\ \cdot & & \end{bmatrix} \begin{Bmatrix} x_1 \\ x_2 \\ \cdot \\ \cdot \\ \cdot \end{Bmatrix}.$$

(7.2-2)

If the system is executing harmonic motion at frequency ω, we can replace $\ddot{x}$ with $-\omega^2 x$ and rewrite the equation as

$$\begin{Bmatrix} x_1 \\ x_2 \\ \cdot \\ \cdot \\ \cdot \end{Bmatrix} = \frac{1}{\omega^2} \begin{bmatrix} k_{11}/m_1 & k_{12}/m_1 & \cdots \\ k_{21}/m_2 & k_{22}/m_2 & \cdots \\ \cdot & & \\ \cdot & & \\ \cdot & & \end{bmatrix} \begin{Bmatrix} x_1 \\ x_2 \\ \cdot \\ \cdot \\ \cdot \end{Bmatrix}$$

(7.2-3)

We will have occasion to refer to this equation again in Sec. 7.5 when the subject of matrix iteration is discussed.

It is also possible to formulate the equations of motion in terms of the flexibility influence coefficients a_{ij} which is defined as the displacement at i due to a unit force applied at j. For example, in the three-mass system of Fig. 7.2-2, with forces F_1, F_2, and F_3 acting at stations 1, 2, 3, the displacements in terms of the flexibility influence coefficients are

$$x_1 = a_{11}F_1 + a_{12}F_2 + a_{13}F_3$$
$$x_2 = a_{21}F_1 + a_{22}F_2 + a_{23}F_3 \qquad (7.2\text{-}4)$$
$$x_3 = a_{31}F_1 + a_{32}F_2 + a_{33}F_3$$

For the free vibration problem we can replace the forces F_1, F_2, F_3 with the inertia forces $-m_1\ddot{x}_1$, $-m_2\ddot{x}_2$, $-m_3\ddot{x}_3$, and if the system is vibrating

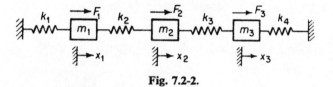

Fig. 7.2-2.

harmonically with frequency ω we can replace the term $-m_i\ddot{x}_i$ with $m_i\omega^2 x_i$, in which case the above equations become

$$x_1 = a_{11}(m_1\omega^2 x_1) + a_{12}(m_2\omega^2 x_2) + a_{13}(m_3\omega^2 x_3)$$
$$x_2 = a_{21}(m_1\omega^2 x_1) + a_{22}(m_2\omega^2 x_2) + a_{23}(m_3\omega^2 x_3) \qquad (7.2\text{-}5)$$
$$x_3 = a_{31}(m_1\omega^2 x_1) + a_{32}(m_2\omega^2 x_2) + a_{33}(m_3\omega^2 x_3)$$

On factoring out the ω^2 from the right side, the equations can be expressed in matrix form as

$$\begin{Bmatrix} x_1 \\ x_2 \\ x_3 \end{Bmatrix} = \omega^2 \begin{bmatrix} a_{11}m_1 & a_{12}m_2 & a_{13}m_3 \\ a_{21}m_1 & a_{22}m_2 & a_{23}m_3 \\ a_{31}m_1 & a_{32}m_2 & a_{33}m_3 \end{bmatrix} \begin{Bmatrix} x_1 \\ x_2 \\ x_3 \end{Bmatrix} \qquad (7.2\text{-}6)$$

which is similar in form to Eq. 7.2-3 except that now ω^2 appears in place of $1/\omega^2$.

It should be mentioned at this point that the method of influence coefficients can also be used for the forced vibration problem by letting $F_1 = f_1(t) - m_1\ddot{x}_1$, $F_2 = f_2(t) - m_2\ddot{x}_2$, etc.

For a system with n degrees of freedom, the number of influence coefficients will be n^2; however, only $n(n + 1)/2$ will have different values, since $a_{ij} = a_{ji}$. Moreover, a_{ii} being the deflection at i due to a unit load at i, influence coefficients with similar subscripts are equal to the reciprocal of the stiffness k_{ii} of the given point. Dividing through Eq. 7.2-5 by ω^2, the set of equations may be rearranged as follows:

$$\left(a_{11}m_1 - \frac{1}{\omega^2}\right)x_1 + (a_{12}m_2)x_2 + (a_{13}m_3)x_3 + \cdots = 0$$

$$(a_{21}m_1)x_1 + \left(a_{22}m_2 - \frac{1}{\omega^2}\right)x_2 + (a_{23}m_3)x_3 + \cdots = 0 \qquad (7.2\text{-}7)$$

$$(a_{31}m_1)x_1 + (a_{32}m_2)x_2 + \left(a_{33}m_3 - \frac{1}{\omega^2}\right)x_3 + \cdots = 0$$

These equations are satisfied if the determinant of the equations vanishes:

$$\begin{vmatrix} \left(a_{11}m_1 - \dfrac{1}{\omega^2}\right) & (a_{12}m_2) & (a_{13}m_3) & \cdots \\[2ex] (a_{21}m_1) & \left(a_{22}m_2 - \dfrac{1}{\omega^2}\right) & (a_{23}m_3) & \cdots \\[2ex] (a_{31}m_1) & (a_{32}m_2) & \left(a_{33}m_3 - \dfrac{1}{\omega^2}\right) & \cdots \\[2ex] \vdots \end{vmatrix} = 0 \qquad (7.2\text{-}8)$$

The above determinant called the characteristic equation will result in an equation of nth degree in $1/\omega^2$, leading to n natural frequencies of the system. The solution of the frequency equation, however, is laborious for n greater than 2, and an iteration procedure, discussed in Sec. 7.5, is recommended.

EXAMPLE 7.2-1. Determine the influence coefficients for the uniform cantilever beam of Fig. 7.2-3 and express the equation of motion in matrix form.

Solution. The influence coefficients may be determined by placing a unit load at (1), (2), and (3), as shown, and by making use of the area moment method* for the determination of the deflections. With the tangent line drawn at the wall, the deflection at the various points is equal to the moment of the M/EI curve about the point in question. As an example, the value of $a_{21} = a_{12}$ is found as follows:

$$a_{21} = \frac{1}{EI}\left[(l \times 2l) \times l + \frac{1}{2}(2l \times 2l) \times \frac{2}{3}(2l)\right] = \frac{14}{3}\frac{l^3}{EI}$$

* Timoshenko and MacCullough, *Elements of Strength of Materials*, 2nd Ed., D. Van Nostrand Co., Inc., 1940, pp. 158–160.

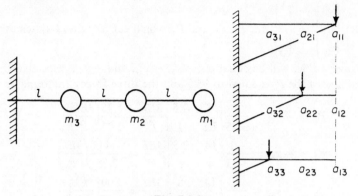

Fig. 7.2-3.

The other values determined as above are

$$a_{11} = \frac{27l^3}{3EI}, \qquad a_{21} = a_{12} = \frac{14}{27} a_{11}$$

$$a_{22} = \frac{8}{27} a_{11}, \qquad a_{23} = a_{32} = \frac{2.5}{27} a_{11}$$

$$a_{33} = \frac{1}{27} a_{11}, \qquad a_{13} = a_{31} = \frac{4}{27} a_{11}$$

The matrix equation can now be written as

$$\begin{bmatrix} y_1 \\ y_2 \\ y_3 \end{bmatrix} = \frac{\omega^2 a_{11}}{27} \begin{bmatrix} 27 & 14 & 4 \\ 14 & 8 & 2.5 \\ 4 & 2.5 & 1 \end{bmatrix} \begin{bmatrix} m_1 & 0 & 0 \\ 0 & m_2 & 0 \\ 0 & 0 & m_3 \end{bmatrix} \begin{bmatrix} y_1 \\ y_2 \\ y_3 \end{bmatrix}$$

EXAMPLE 7.2-2. Formulate the matrix equation for the system of Fig. 7.2-4 in terms of the stiffness influence coefficients.

Solution. Place loads P_1, P_2, P_3 at (1), (2), and (3) as shown in Fig. 7.2-4. By using the influence coefficients of the previous problem the deflections at (1), (2), (3), can be written as

$$\left(\frac{l^3}{3EI}\right)(27P_1 + 14P_2 + 4P_3) = y_1$$

$$\left(\frac{l^3}{3EI}\right)(14P_1 + 8P_2 + 2.5P_3) = y_2$$

$$\left(\frac{l^3}{3EI}\right)(4P_1 + 2.5P_2 + 1.0P_3) = y_3$$

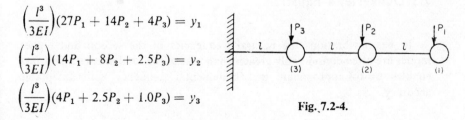

Fig. 7.2-4.

If we let $\left\{ \begin{matrix} y_1 \\ y_2 \\ y_3 \end{matrix} \right\} = \left\{ \begin{matrix} 1 \\ 0 \\ 0 \end{matrix} \right\}$, the P_1, P_2, and P_3, which can be solved systematically
by Cramer's rule, will be equal to $-k_{11}$, $-k_{21}$, and $-k_{31}$, respectively. The
determinant of the P_s is equal to 3.25 and the three forces are found as

$$P_1 = -k_{11} = \frac{1}{3.25} \begin{vmatrix} 8 & 2.5 \\ 2.5 & 1 \end{vmatrix} \frac{3EI}{l^3} = \frac{1.75}{3.25}\left(\frac{3EI}{l^3}\right)$$

$$P_2 = -k_{21} = \frac{-1}{3.25} \begin{vmatrix} 14 & 2.5 \\ 4 & 1 \end{vmatrix} \frac{3EI}{l^3} = \frac{-4.00}{3.25}\left(\frac{3EI}{l^3}\right)$$

$$P_3 = -k_{31} = \frac{1}{3.25} \begin{vmatrix} 14 & 8 \\ 4 & 2.5 \end{vmatrix} \frac{3EI}{l^3} = \frac{3.00}{3.25}\left(\frac{3EI}{l^3}\right)$$

By letting the deflections be $\left\{ \begin{matrix} 0 \\ 1 \\ 0 \end{matrix} \right\}$ and $\left\{ \begin{matrix} 0 \\ 0 \\ 1 \end{matrix} \right\}$ the remaining stiffness coefficients
are found to be

$$-k_{12} = \frac{-4}{3.25}\left(\frac{3EI}{l^3}\right) \qquad -k_{13} = \frac{3}{3.25}\left(\frac{3EI}{l^3}\right)$$

$$-k_{22} = \frac{11.0}{3.25}\left(\frac{3EI}{l^3}\right) \qquad -k_{23} = \frac{-13.5}{3.25}\left(\frac{3EI}{l^3}\right)$$

$$-k_{32} = \frac{-13.5}{3.25}\left(\frac{3EI}{l^3}\right) \qquad -k_{33} = \frac{20.0}{3.25}\left(\frac{3EI}{l^3}\right)$$

and we find that the relationship $k_{ij} = k_{ji}$ is verified. The matrix equation of
motion is then

$$\begin{bmatrix} m_1 & 0 & 0 \\ 0 & m_2 & 0 \\ 0 & 0 & m_3 \end{bmatrix} \begin{Bmatrix} \ddot{y}_1 \\ \ddot{y}_2 \\ \ddot{y}_3 \end{Bmatrix} - \left(\frac{3EI}{3.25 l^3}\right) \begin{bmatrix} 1.75 & -4 & 3 \\ -4 & 11 & -13.5 \\ 3 & -13.5 & 20 \end{bmatrix} \begin{Bmatrix} y_1 \\ y_2 \\ y_3 \end{Bmatrix} = 0$$

Essentially we have inverted the flexibility matrix into the stiffness matrix
without the formal use of matrix theory.

7.3 Dunkerley's Equation

In beam vibrations the natural frequencies of the second and higher
modes are often considerably greater than that of the fundamental. This fact
enables us to approximate the fundamental frequency with acceptable
accuracy.

We recall first a certain known fact of algebra which states that if the coefficient of the highest term of the nth-degree equation is reduced to unity, the coefficient of the second highest term will be equal to the sum of the roots of the equation. For instance, consider the characteristic equation (7.2-8) when $n = 3$. The frequency equation then becomes

$$\left(\frac{1}{\omega^2}\right)^3 - [a_{11}m_1 + a_{22}m_2 + a_{33}m_3]\left(\frac{1}{\omega^2}\right)^2$$

$$- [a_{12}m_2 a_{21}m_1 + \cdots]\frac{1}{\omega^2} - [\qquad] = 0 \quad (7.3\text{-}1)$$

If the roots of the equation are $1/\omega_1$, $1/\omega_2$, and $1/\omega_3$, the above equation can be factored into the following form:

$$\left(\frac{1}{\omega^2} - \frac{1}{\omega_1^2}\right)\left(\frac{1}{\omega^2} - \frac{1}{\omega_2^2}\right)\left(\frac{1}{\omega^2} - \frac{1}{\omega_3^2}\right) = 0$$

$$= \left(\frac{1}{\omega^2}\right)^3 - \left[\frac{1}{\omega_1^2} + \frac{1}{\omega_2^2} + \frac{1}{\omega_3^2}\right]\left(\frac{1}{\omega^2}\right)^2 \quad (7.3\text{-}2)$$

$$- [\qquad]\frac{1}{\omega^2} - [\qquad] = 0$$

Comparison of Eqs. (7.3-1) and (7.3-2) indicates that

$$\frac{1}{\omega_1^2} + \frac{1}{\omega_2^2} + \frac{1}{\omega_3^2} = a_{11}m_1 + a_{22}m_2 + a_{33}m_3$$

$$= \frac{m_1}{k_1} + \frac{m_2}{k_2} + \frac{m_3}{k_3} \quad (7.3\text{-}3)$$

$$= \frac{1}{\omega_{11}^2} + \frac{1}{\omega_{22}^2} + \frac{1}{\omega_{33}^2}$$

where the terms ω_{11}, ω_{22}, and ω_{33} are the natural frequencies of the system, with each mass acting separately in the absence of the other masses. Since ω_2 and ω_3 are natural frequencies corresponding to the higher modes and are larger than the fundamental, all terms on the left side of the equation except the first may be omitted for the approximate determination of the fundamental frequency.

The method can be extended to any number of degrees of freedom, and the approximate equation for the fundamental frequency can be written as

$$\frac{1}{\omega_1^2} \cong \frac{1}{\omega_{11}^2} + \frac{1}{\omega_{22}^2} + \frac{1}{\omega_{33}^2} + \frac{1}{\omega_{44}^2} + \cdots \quad (7.3\text{-}4)$$

This is known as *Dunkerley's equation* and has many useful applications, some of which are given in the following examples.

EXAMPLE 7.3-1. Dunkerley's equation is useful for estimating the fundamental frequency of a structure undergoing vibration testing. Natural frequencies of structures are often determined by attaching to the structure an eccentric mass exciter, and noting the frequencies corresponding to the maximum amplitude. The frequencies so measured represent those of the structure plus exciter and may deviate considerably from the natural frequencies of the structure itself when the mass of the exciter is a substantial percentage of the total mass. In such cases the fundamental frequency of the structure by itself may be determined by the following equation:

$$\frac{1}{\omega_1^2} = \frac{1}{\omega_{11}^2} + \frac{1}{\omega_{22}^2} \tag{a}$$

where ω_1 = fundamental frequency of structure plus exciter,

ω_{11} = fundamental frequency of the structure by itself,

ω_{22} = natural frequency of exciter mounted on the structure in the absence of other masses.

It is sometimes convenient to express this equation in another form, for instance,

$$\frac{1}{\omega_1^2} = \frac{1}{\omega_{11}^2} + a_{22}m_2 \tag{b}$$

where m_2 is the mass of the concentrated weight or exciter and a_{22} the influence coefficient of the structure at the point of attachment of the exciter.

EXAMPLE 7.3-2. An airplane rudder tab showed a resonant frequency of 30 c.p.s. when vibrated by an eccentric mass shaker weighing 1.5 lb. By attaching an additional weight of 1.5 lb. to the shaker, the resonant frequency was lowered to 24 c.p.s. Determine the true natural frequency of the tab.

Solution. The measured resonant frequencies are those due to the total mass of the tab and shaker. Letting f_{11} be the true natural frequency of the tab, and substituting into Eq. (b) of Example 7.3-1, we obtain

$$\frac{1}{(2\pi \times 30)^2} = \frac{1}{(2\pi f_{11})^2} + \frac{1.5}{386} a_{22}$$

$$\frac{1}{(2\pi \times 24)^2} = \frac{1}{(2\pi f_{11})^2} + \frac{3.0}{386} a_{22}$$

Eliminating a_{22}, the true natural frequency is

$$f_{11} = 45.3 \text{ c.p.s.}$$

The rigidity or stiffness of the tab at the point of attachment of the shaker may be determined from $1/a_{22}$ which from the same equations is found to be

$$k_2 = \frac{1}{a_{22}} = \frac{1}{0.00407} = 246 \text{ lb./in.}$$

EXAMPLE 7.3-3. Determine the fundamental frequency of a uniformly loaded cantilever beam with a concentrated mass M at the end, equal to the mass of the uniform beam (see Fig. 7.3-1).

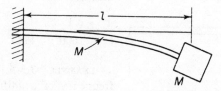

Fig. 7.3-1.

Solution. The frequency equation for the uniformly loaded beam by itself is, from Example 1.5-3, Chapter 1,

$$\omega_{11}^2 = 3.515^2\left(\frac{EI}{Ml^3}\right)$$

For the concentrated mass by itself attached to a weightless cantilever beam, we have

$$\omega_{22}^2 = 3.00\left(\frac{EI}{Ml^3}\right)$$

Substituting into Dunkerley's formula rearranged in the following form, the natural frequency of the system is determined as

$$\omega_1^2 = \frac{\omega_{11}^2\omega_{22}^2}{\omega_{11}^2 + \omega_{22}^2} = \frac{3.515^2 \times 3.0}{3.515^2 + 3.0}\left(\frac{EI}{Ml^3}\right) = 2.41\left(\frac{EI}{Ml^3}\right)$$

This result may be compared to the frequency equation obtained by Rayleigh's method (see Example 1.5-3), which is

$$\omega_1^2 = \frac{3EI}{(1 + \frac{33}{140})Ml^3} = 2.43\left(\frac{EI}{Ml^3}\right)$$

EXAMPLE 7.3-4. The natural frequency of a given airplane wing in torsion is 1600 c.p.m. What will be the new torsional frequency if a 1000-lb. bomb is hung at a position one-sixth of the semi-span from the center line of the airplane such that its moment of inertia about the torsional axis is 1800 lb. in. sec.^{2}? The torsional stiffness of the wing at this point is 60×10^6 lb. in./rad.

Solution. The frequency of the bomb attached to the weightless wing is

$$f_{22} = \frac{1}{2\pi} \sqrt{\frac{60 \times 10^6}{1800}} = 29.1 \text{ c.p.s.} = 1745 \text{ c.p.m.}$$

The new torsional frequency with the bomb, from Eq. (*a*) of Example 7.3-1 then becomes

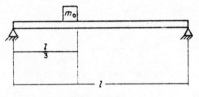

Fig. 7.3-2.

$$\frac{1}{f_1^2} = \frac{1}{1600^2} + \frac{1}{1745^2}, \quad f_1 = 1180 \text{ c.p.m.}$$

EXAMPLE 7.3-5. The fundamental frequency of a uniform beam of mass M, simply supported as in Fig. 7.3-2 is equal to $\pi^2 \sqrt{EI/Ml^3}$. If a lumped mass m_0 is attached to the beam at $x = l/3$, determine the new fundamental frequency.

Solution. Starting with Eq. (*b*) of Example 7.3-1, we let ω_{11} be the fundamental frequency of the uniform beam and ω_1 the new fundamental frequency with m_0 attached to the beam. Multiplying through Eq. (*b*) by ω_1^2, we have

or

$$1 = \left(\frac{\omega_1}{\omega_{11}}\right)^2 + a_{22} m_0 \omega_{11}^2 \left(\frac{\omega_1}{\omega_{11}}\right)^2$$

$$\left(\frac{\omega_1}{\omega_{11}}\right)^2 = \frac{1}{1 + a_{22} m_0 \omega_{11}^2}$$

The quantity a_{22} is the influence coefficient at $x = l/3$ due to a unit load applied at the same point. It can be found from the beam formula in Example 1.5-4 to be

$$a_{22} = \frac{8}{6 \times 81} \frac{l^3}{EI}$$

Substituting $\omega_{11}^2 = \pi^4 EI/Ml^3$ together with a_{22}, we obtain the convenient formula

$$\left(\frac{\omega_1}{\omega_{11}}\right)^2 = \frac{1}{1 + \dfrac{8\pi^4}{6 \times 81} \dfrac{m_0}{M}} = \frac{1}{1 + 1.6 \dfrac{m_0}{M}}$$

7.4 Orthogonality of Principal Modes

When only static coupling appears in the dynamical equation, the equation for the ith mass has the form

$$-m_i \ddot{x}_i = k_{i1}x_1 + k_{i2}x_2 + \cdots k_{in}x_n \qquad (7.4\text{-}1)$$

Thus, for the principal mode (s), the above equation becomes

$$m_i \omega_s^2 A_i^{(s)} = k_{i1}A_1^{(s)} + k_{i2}A_2^{(s)} + \cdots k_{in}A_n^{(s)} = \sum_{j=1}^{n} k_{ij}A_j^{(s)} \qquad (7.4\text{-}2)$$

whereas for the principal mode (p) we obtain

$$m_i \omega_p^2 A_i^{(p)} = k_{i1}A_1^{(p)} + k_{i2}A_2^{(p)} + \cdots k_{in}A_n^{(p)} = \sum_{j=1}^{n} k_{ij}A_j^{(p)} \qquad (7.4\text{-}3)$$

On multiplying Eq. (7.4-2) by $A_i^{(p)}$ and Eq. (7.4-3) by $A_i^{(s)}$ we have

$$m_i \omega_s^2 A_i^{(s)} A_i^{(p)} = k_{i1}A_1^{(s)}A_i^{(p)} + k_{i2}A_2^{(s)}A_i^{(p)} + \cdots k_{in}A_n^{(s)}A_i^{(p)} = \sum_{j=1}^{n} k_{ij}A_j^{(s)}A_i^{(p)} \qquad (7.4\text{-}4)$$

$$m_i \omega_p^2 A_i^{(s)} A_i^{(p)} = k_{i1}A_1^{(p)}A_i^{(s)} + k_{i2}A_2^{(p)}A_i^{(s)} + \cdots k_{in}A_n^{(p)}A_i^{(s)} = \sum_{j=1}^{n} k_{ij}A_j^{(p)}A_i^{(s)} \qquad (7.4\text{-}5)$$

The above equations apply to every mass point i which takes on the values $i = 1$ to n, so that there are n such equations corresponding to each of the above Eqs. (7.4-4) and (7.4-5) which may be written as

$$\omega_s^2 \sum_{i=1}^{n} m_i A_i^{(s)} A_i^{(p)} = \sum_{i=1}^{n} \sum_{j=1}^{n} k_{ij}A_j^{(s)}A_i^{(p)} \qquad (7.4\text{-}6)$$

$$\omega_p^2 \sum_{i=1}^{n} m_i A_i^{(s)} A_i^{(p)} = \sum_{i=1}^{n} \sum_{j=1}^{n} k_{ij}A_j^{(p)}A_i^{(s)} \qquad (7.4\text{-}7)$$

The right sides of these equations, however, can be shown to be identical to each other since the stiffness coefficients $k_{ij} = k_{ji}$. (Try writing out these terms for $n = 2$ and $n = 3$.) Thus by subtracting we obtain the equation

$$(\omega_s^2 - \omega_p^2) \sum_{i=1}^{n} m_i A_i^{(s)} A_i^{(p)} = 0 \qquad (7.4\text{-}8)$$

and since we will assume that the natural frequencies differ for the two modes, we arrive at the result

$$\sum_{i=1}^{n} m_i A_i^{(s)} A_i^{(p)} = 0 \qquad (p \neq s) \qquad (7.4\text{-}9)$$

which we call the *orthogonality relation* between the principal modes of vibration. Although it is possible in some cases to have multiple roots where $\omega_s = \omega_p$, their mode shapes will differ and the above relation will be found to be applicable.

When $p = s$ the sum, as represented by Eq. 7.4-9, will be a constant whose value depends on the normal-mode amplitudes. Generally, the principal modes are normalized to unity at some station, and since the A_i are then nondimensional, the sum as expressed by Eq. 7.4-9 will have the dimensions of mass. Thus when $p = s$ the orthogonality equation can be presented as

$$\sum_{i=1}^{n} m_i (A_i^2)^{(s)} = M_s \qquad (7.4\text{-}10)$$

where M_s is called the generalized mass for mode s.

When dynamic coupling appears along with static coupling the orthogonality equation, which can be derived in a similar manner, is more complicated. The equation then takes the form

$$\sum_{i=1}^{n} \sum_{j=1}^{n} m_{ij} A_i^{(s)} A_j^{(p)} = 0 \qquad \text{for} \quad p \neq s$$
$$= M_s \quad \text{for} \quad p = s \qquad (7.4\text{-}11)$$

7.5 Method of Matrix Iteration

The equations of motion, formulated either on the basis of the stiffness equation or the flexibility equation, are similar in form and appear as

$$\begin{Bmatrix} x_1 \\ x_2 \\ \cdot \\ \cdot \\ \cdot \\ x_n \end{Bmatrix} = \lambda \begin{bmatrix} \alpha_{11} & \alpha_{12} & \cdots & \alpha_{1n} \\ \alpha_{21} & \alpha_{22} & \cdots & \\ \cdot & & & \\ \cdot & & & \\ \cdot & & & \\ \alpha_{n1} & \alpha_{n2} & \cdots & \alpha_{nn} \end{bmatrix} \begin{Bmatrix} x_1 \\ x_2 \\ \cdot \\ \cdot \\ \cdot \\ x_n \end{Bmatrix} \qquad (7.5\text{-}1)$$

where λ is equal to $1/\omega^2$ for the stiffness formulation, and ω^2 for the flexibility formulation.

The iteration procedure is started by assuming a set of deflections for the right column of Eq. (7.5-1) and performing the indicated operation, which results in a column of numbers. This is then normalized by making one of the amplitudes equal to unity and dividing each term of the column by the particular amplitude which was normalized. The procedure is then repeated with the normalized column until the amplitudes stabilize to a definite pattern.

As will be shown in Sec. 7.6, the iteration process converges to the lowest value of λ so that for the equation formulated on the flexibility influence coefficients, the fundamental or the lowest mode of vibration is found. Likewise, for the equation formulated on the basis of the stiffness influence coefficients, the convergence is to the highest mode which corresponds to the lowest value of $\lambda = 1/\omega^2$.

EXAMPLE 7.5-1. The uniform beam of Fig. 7.5-1, free to vibrate in the plane shown, has two concentrated weights $W_1 = 500$ lb. and $W_2 = 100$ lb. Determine the fundamental frequency of the system.

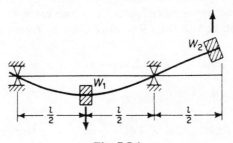

Fig. 7.5-1.

Solution. The influence coefficients for this problem, determined from deflection equations of beams by placing a unit load at positions 1 and 2, are

$$a_{11} = \frac{l^3}{48EI} = \tfrac{1}{6}a_{22}, \qquad a_{12} = a_{21} = \frac{l^3}{32EI} = \tfrac{1}{4}a_{22}, \qquad a_{22} = \frac{l^3}{8EI}$$

Substituting into Eq. (7.2-6),

$$\begin{bmatrix} x_1 \\ x_2 \end{bmatrix} = \frac{\omega^2 l^3}{8EIg} \begin{bmatrix} \frac{500}{6} & \frac{100}{4} \\ \frac{500}{4} & 100 \end{bmatrix} \begin{bmatrix} x_1 \\ x_2 \end{bmatrix}$$

Starting with $x_1 = x_2 = 1.0$ for the right column, we obtain

$$\begin{bmatrix} x_1 \\ x_2 \end{bmatrix} = \frac{\omega^2 l^3}{8EIg} \begin{bmatrix} 108.3 \\ 225.0 \end{bmatrix} = \frac{108.3\omega^2 l^3}{8EIg} \begin{bmatrix} 1.00 \\ 2.08 \end{bmatrix}$$

If the procedure is repeated with $x_1 = 1.0$ and $x_2 = 2.08$, the second result is

$$\begin{bmatrix} x_1 \\ x_2 \end{bmatrix} = \frac{\omega^2 l^3}{8EIg} \begin{bmatrix} 135.3 \\ 333.0 \end{bmatrix} = \frac{135.3\omega^2 l^3}{8EIg} \begin{bmatrix} 1.00 \\ 2.46 \end{bmatrix}$$

By repeating the procedure a few more times the deflections will converge to

$$\begin{bmatrix} 1.00 \\ 2.60 \end{bmatrix} = \frac{148.3\omega^2 l^3}{8EIg} \begin{bmatrix} 1.00 \\ 2.60 \end{bmatrix}$$

The fundamental frequency from the above equation using $g = 386$ in/.sec.[2] is

$$\omega = \sqrt{8EIg/148.3l^3} = 4.56\sqrt{EI/l^3}$$

and the amplitude ratio is found to be

$$\frac{x_1}{x_2} = \frac{1.0}{2.60}$$

If only the fundamental frequency is of interest, sufficient accuracy can be obtained from the results of the first and second iterations. From the first iteration the inertia forces are $500\omega^2/g$ and $208\omega^2/g$. These forces produce deflections obtained in the second iteration, which are $x_1 = 135.3\omega^2 l^3/8EIg = 16.92\omega^2 l^3/EIg$ and $x_2 = 2.46x_1$. The work done by these forces is then

$$U = \tfrac{1}{2}(500 + 208 \times 2.46)\frac{\omega^2}{g} x_1 = \tfrac{1}{2} \times 1012 \times \frac{\omega^2 x_1}{g}$$

and the corresponding kinetic energy is

$$T = \tfrac{1}{2}(500 + 100 \times 2.46^2)\frac{\omega^2}{g} x_1^2 = \tfrac{1}{2} \times 1105 \times \frac{\omega^2 x_1^2}{g}$$

Equating the two, the fundamental frequency is found as

$$\omega = \sqrt{\frac{1012 \times 386}{1105 \times 16.92}\frac{EI}{l^3}} = 4.57\sqrt{\frac{EI}{l^3}}$$

7.6 Calculation of Higher Modes

When the equations of motion are formulated in terms of the flexibility influence coefficients, the iteration procedure converges to the lowest mode present in the assumed deflection. It is evident that if the lowest mode was absent in the assumed defection, the iteration technique must converge to the next lowest or the second mode.

Letting the assumed curve for the second mode be expressed by the sum of the normal modes,

$$\begin{Bmatrix} x_1 \\ x_2 \\ \cdot \\ \cdot \\ \cdot \end{Bmatrix} = C_1 \begin{Bmatrix} A_1 \\ A_2 \\ \cdot \\ \cdot \\ \cdot \end{Bmatrix}^{(1)} + C_2 \begin{Bmatrix} A_1 \\ A_2 \\ \cdot \\ \cdot \\ \cdot \end{Bmatrix}^{(2)} + C_3 \begin{Bmatrix} A_1 \\ A_2 \\ \cdot \\ \cdot \\ \cdot \end{Bmatrix}^{(3)} + \cdots \quad (7.6\text{-}1)$$

we will impose the condition that $C_1 = 0$ to remove the first mode from the assumed deflection. For this we introduce the orthogonality relationship which in matrix notation for $n = 3$ is*

$$\sum_{i=1}^{n} m_i A_i^{(r)} A_i^{(s)} = [A_1 \quad A_2 \quad A_3]^{(r)} \begin{bmatrix} m_1 & 0 & 0 \\ 0 & m_2 & 0 \\ 0 & 0 & m_3 \end{bmatrix} \begin{Bmatrix} A_1 \\ A_2 \\ A_3 \end{Bmatrix}^{(s)} = \begin{cases} 0 & \text{for} \quad r \neq s \\ 1 & \text{for} \quad r = s \end{cases}$$

(7.6-2)

By premultiplying each term of Eq. (7.6-1) by the row matrix corresponding to mode 1 and the diagonal mass matrix, i.e.,

$$[A_1 \quad A_2 \quad A_3]^{(1)} \begin{bmatrix} m_1 & 0 & 0 \\ 0 & m_2 & 0 \\ 0 & 0 & m_2 \end{bmatrix} \qquad (7.6-3)$$

the right side of the equation, after imposing Eq. (7.6-2), becomes equal to C_1:

$$C_1 = [A_1 \quad A_2 \quad A_3]^{(1)} \begin{bmatrix} m_1 & 0 & 0 \\ 0 & m_2 & 0 \\ 0 & 0 & m_3 \end{bmatrix} \begin{Bmatrix} x_1 \\ x_2 \\ x_3 \end{Bmatrix}$$

$$= m_1 A_1^{(1)} x_1 + m_2 A_2^{(1)} x_2 + m_3 A_3^{(1)} x_3 \qquad (7.6\text{-}4)$$

Equating C_1 to zero and solving for the assumed mode,

$$x_1 = -\frac{m_2}{m_1}\left(\frac{A_2}{A_1}\right)^{(1)} x_2 - \frac{m_3}{m_1}\left(\frac{A_3}{A_1}\right)^{(1)} x_3$$

$$x_2 = x_2$$

$$x_3 = x_3$$

where the equations for x_2 and x_3 are introduced as identities. This set of equations is then expressed in the matrix form

$$\begin{Bmatrix} x_1 \\ x_2 \\ x_3 \end{Bmatrix} = \begin{bmatrix} 0 & -\dfrac{m_2}{m_1}\left(\dfrac{A_2}{A_1}\right)^{(1)} & -\dfrac{m_3}{m_1}\left(\dfrac{A_3}{A_1}\right)^{(1)} \\ 0 & 1 & 0 \\ 0 & 0 & 1 \end{bmatrix} \begin{Bmatrix} x_1 \\ x_2 \\ x_3 \end{Bmatrix} \qquad (7.6\text{-}5)$$

* The multiplication is started from the right end, resulting in a column to be multiplied by the row matrix which can be considered as a square matrix with the remaining rows of zero elements.

Since this equation is the result of $C_1 = 0$, the first mode has been swept out of the assumed deflection. Consequently, the above matrix is referred to as the *sweeping matrix* and is designated as

$$\{x\} = [S]\{x\} \qquad (7.6\text{-}6)$$

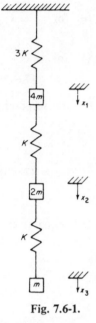

Fig. 7.6-1.

When Eq. (7.6-6) is applied to the original matrix equation, the result is

$$\{x\} = \omega^2[am][S]\{x\} \qquad (7.6\text{-}7)$$

The iteration procedure applied to Eq. (7.6-7) will then converge to the second mode.

For the third and higher modes, the sweeping procedure is repeated, making $C_1 = C_2 = 0$, etc. This procedure reduces the order of the matrix equation by one each time; however, the convergence for higher modes becomes more critical if impurities are introduced through the sweeping matrices. It is well to check the highest mode by the inversion of the original matrix equation which should be equal to the equation formulated in terms of the stiffness influence coefficients.

EXAMPLE 7.6-1. Write the matrix equation, based on flexibility coefficients, for the system of Fig. 7.6-1 and determine all the natural modes.

Solution. The influence coefficients are found by applying a unit load, one at a time, to points (1), (2), and (3):

$$a_{11} = a_{21} = a_{31} = a_{12} = a_{13} = \frac{1}{3k}$$

$$a_{22} = a_{32} = a_{23} = \left(\frac{1}{3k} + \frac{1}{k}\right) = \frac{4}{3k}$$

$$a_{33} = \frac{1}{3k} + \frac{1}{k} + \frac{1}{k} = \frac{7}{3k}$$

The equations of motion in matrix form are then

$$\begin{Bmatrix} x_1 \\ x_2 \\ x_3 \end{Bmatrix} = \frac{\omega^2 m}{3k} \begin{bmatrix} 4 & 2 & 1 \\ 4 & 8 & 4 \\ 4 & 8 & 7 \end{bmatrix} \begin{Bmatrix} x_1 \\ x_2 \\ x_3 \end{Bmatrix} \qquad (a)$$

Starting with arbitrary values of x_1, x_2, x_3, the above equation converges to the first mode as:

$$\begin{Bmatrix} x_1 \\ x_2 \\ x_3 \end{Bmatrix} = \frac{\omega^2 m}{3k} 14.32 \begin{Bmatrix} 0.25 \\ 0.79 \\ 1.00 \end{Bmatrix} \tag{b}$$

The fundamental frequency is then found to be

$$\omega_1 = \sqrt{3k/14.32m} = 0.457\sqrt{k/m} \tag{c}$$

To determine the second mode, we sweep out the first mode by letting $C_1 = 0$. The resulting sweeping matrix is

$$\begin{bmatrix} 0 & -\frac{1}{2}\left(\frac{0.79}{0.25}\right) & -\frac{1}{4}\left(\frac{1.00}{0.25}\right) \\ 0 & 1 & 0 \\ 0 & 0 & 1 \end{bmatrix} = \begin{bmatrix} 0 & -1.58 & -1 \\ 0 & 1 & 0 \\ 0 & 0 & 1 \end{bmatrix} \tag{d}$$

and the new equation for the second mode iteration is

$$\begin{Bmatrix} x_1 \\ x_2 \\ x_3 \end{Bmatrix} = \frac{\omega^2 m}{3k} \begin{bmatrix} 4 & 2 & 1 \\ 4 & 8 & 4 \\ 4 & 8 & 7 \end{bmatrix} \begin{bmatrix} 0 & -1.58 & -1 \\ 0 & 1 & 0 \\ 0 & 0 & 1 \end{bmatrix} \begin{Bmatrix} x_1 \\ x_2 \\ x_3 \end{Bmatrix}$$

$$= \frac{\omega^2 m}{3k} \begin{bmatrix} 0 & -4.32 & -3.0 \\ 0 & 1.67 & 0 \\ 0 & 1.67 & 3.0 \end{bmatrix} \begin{Bmatrix} x_1 \\ x_2 \\ x_3 \end{Bmatrix} \tag{e}$$

Starting the iteration process with arbitrary amplitudes, Eq. (e) converges to the second mode

$$\begin{Bmatrix} x_1 \\ x_2 \\ x_3 \end{Bmatrix} = \frac{\omega^2 m}{3k} 3 \begin{Bmatrix} -1.0 \\ 0 \\ 1.0 \end{Bmatrix} \tag{f}$$

The natural frequency of the second mode is therefore found to be

$$\omega_2 = \sqrt{k/m} \tag{g}$$

For the determination of the third mode we impose the conditions, $C_1 = C_2 = 0$:

$$C_1 = \sum_{i=1}^{3} m_i A_i^{(1)} x_i = 4(0.25)x_1 + 2(0.79)x_2 + 1(1.0)x_3 = 0$$

$$C_2 = \sum_{i=1}^{3} m_i A_i^{(2)} x_i = 4(-1.0)x_1 + 2(0)x_2 + 1(1.0)x_3 = 0$$

From these two equations we obtain

$$x_1 = 0.25x_3, \qquad x_2 = -0.79x_3$$

so that a sweeping matrix devoid of the first two modes is

$$\begin{bmatrix} 0 & 0 & 0.25 \\ 0 & 0 & -0.79 \\ 0 & 0 & 1.00 \end{bmatrix} \qquad (h)$$

Applying this to the original equation, we obtain

$$\begin{Bmatrix} x_1 \\ x_2 \\ x_3 \end{Bmatrix} = \frac{\omega^2 m}{3k} \begin{bmatrix} 4 & 2 & 1 \\ 4 & 8 & 4 \\ 4 & 8 & 7 \end{bmatrix} \begin{bmatrix} 0 & 0 & 0.25 \\ 0 & 0 & -0.79 \\ 0 & 0 & 1.00 \end{bmatrix} \begin{Bmatrix} x_1 \\ x_2 \\ x_3 \end{Bmatrix}$$

$$= \frac{\omega^2 m}{3k} \begin{bmatrix} 0 & 0 & 0.42 \\ 0 & 0 & -1.32 \\ 0 & 0 & 1.68 \end{bmatrix} \begin{Bmatrix} x_1 \\ x_2 \\ x_3 \end{Bmatrix} \qquad (i)$$

The above equation results immediately in the third mode, which is

$$\begin{Bmatrix} x_1 \\ x_2 \\ x_3 \end{Bmatrix} = \frac{\omega^2 m}{3k} 1.68 \begin{Bmatrix} 0.25 \\ -0.79 \\ 1.00 \end{Bmatrix} \qquad (j)$$

The natural frequency of the third mode is then found as

$$\omega_3 = \sqrt{3k/1.68m} = 1.34\sqrt{k/m} \qquad (k)$$

7.7 The Holzer-Type Problem

When only one coordinate is associated with each lumped mass of the multi-degrees-of-freedom system, we have a Holzer-type problem which can be solved by proceeding numerically from one end of the system to the other. By presenting such problems in matrix form it is possible to give a clear description of the computational procedure. Both the linear spring-mass system and the torsional lumped mass system will be treated.

(a) *The Spring-Mass System.* Figure 7.7-1 shows a part of a linear spring-mass system. Isolating one of the masses, we designate the quantities to its left and right by superscripts L and R.

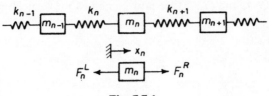

Fig. 7.7-1.

The first equation is an identity for the displacement

$$x_n^L = x_n^R \tag{7.7-1}$$

The second equation is the force equation for the mass m_n:

$$m_n \ddot{x}_n = F_n^R - F_n^L$$

which for harmonic motion becomes

$$F_n^R = -\omega^2 m_n x_n^L + F_n^L \tag{7.7-2}$$

Since either the superscript L or R could be placed on x_n, we use L for the above equation. Equations (7.7-1) and (7.7-2) are now combined into a point matrix as follows:

$$\begin{bmatrix} x \\ F \end{bmatrix}_n^R = \begin{bmatrix} 1 & 0 \\ -\omega^2 m & 1 \end{bmatrix} \begin{bmatrix} x \\ F \end{bmatrix}_n^L \tag{7.7-3}$$

where the column $\begin{bmatrix} x \\ F \end{bmatrix}$ is called a state vector.

Next we isolate the spring k_n, as shown in Fig. 7.7-2. Since the spring is massless, the two end forces are equal:

$$F_{n-1}^R = F_n^L \tag{7.7-4}$$

The force and the relative displacement are related by the spring modulus

$$x_n^L - x_{n-1}^R = \frac{F_{n-1}^R}{k_n} \tag{7.7-5}$$

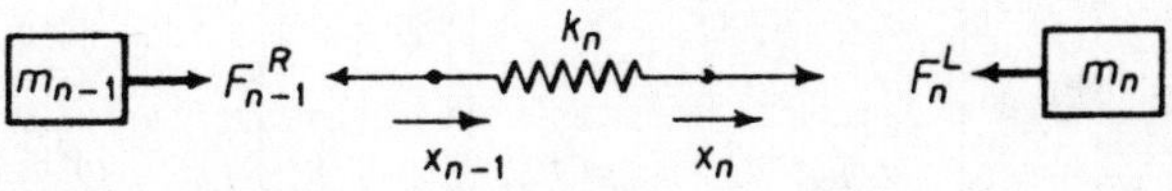

Fig. 7.7-2.

Equations (7.7-4) and (7.7-5) are now combined into a field matrix:

$$
\begin{bmatrix} x \\ F \end{bmatrix}_n^L = \begin{bmatrix} 1 & \dfrac{1}{k} \\ 0 & 1 \end{bmatrix} \begin{bmatrix} x \\ F \end{bmatrix}_{n-1}^R
\tag{7.7-6}
$$

To relate the quantities at station n in terms of quantities at station $n - 1$, we substitute Eq. (7.7-6) into Eq. (7.7-3) to obtain the transfer matrix:

$$
\begin{bmatrix} x \\ F \end{bmatrix}_n^R = \begin{bmatrix} 1 & 0 \\ -\omega^2 m & 1 \end{bmatrix} \begin{bmatrix} 1 & \dfrac{1}{k} \\ 0 & 1 \end{bmatrix} \begin{bmatrix} x \\ F \end{bmatrix}_{n-1}^R
$$

$$
\rightarrow = \begin{bmatrix} 1 & \dfrac{1}{k} \\ -\omega^2 m & \left(1 - \dfrac{\omega^2 m}{k}\right) \end{bmatrix}_n \begin{bmatrix} x \\ F \end{bmatrix}_{n-1}^R
\tag{7.7-7}
$$

Thus with known values of the state vector $\begin{bmatrix} x \\ F \end{bmatrix}_1$ at station 1, and a chosen value of ω^2, it is possible to compute progressively the state vector $\begin{bmatrix} x \\ F \end{bmatrix}_n$ at station n. Depending on the boundary conditions, either x_n or F_n can be plotted as a function of ω^2, and the natural frequencies of the system are established when the boundary conditions are satisfied.

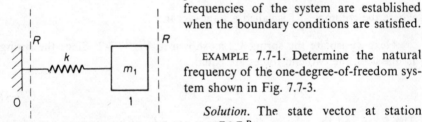

Fig. 7.7-3.

EXAMPLE 7.7-1. Determine the natural frequency of the one-degree-of-freedom system shown in Fig. 7.7-3.

Solution. The state vector at station 0 is $\begin{bmatrix} 0 \\ F_0 \end{bmatrix}_0^R$. From Eq. (7.7-7) the state vector at station 1 becomes

$$
\begin{bmatrix} x \\ F \end{bmatrix}_1^R = \begin{bmatrix} 1 & \dfrac{1}{k} \\ -\omega^2 m & \left(1 - \dfrac{\omega^2 m}{k}\right) \end{bmatrix} \begin{bmatrix} 0 \\ F_0 \end{bmatrix} = \begin{bmatrix} \dfrac{F_0}{k} \\ \left(1 - \dfrac{\omega^2 m}{k}\right) F_0 \end{bmatrix}
$$

Since $F_1^R = 0$, we have from the above equation

$$1 - \frac{\omega^2 m}{k} = 0$$

or

$$\omega = \sqrt{k/m}$$

(b) *Torsional System.* The development for the transfer matrix for the torsional system is identical to that of the linear spring-mass system. We isolate the nth disk and write for the torque equation (see Fig. 7.7-4)

$$-\omega^2 J_n \theta_n = T_n^R - T_n^L$$

The identity in the displacement is again

$$\theta_n^L = \theta_n^R$$

and the two equations for the point matrix

$$\begin{bmatrix} \theta \\ T \end{bmatrix}_n^R = \begin{bmatrix} 1 & 0 \\ -\omega^2 J & 1 \end{bmatrix} \begin{bmatrix} \theta \\ T \end{bmatrix}_n^L \quad (7.7\text{-}8)$$

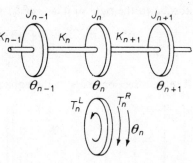

Fig. 7.7-4.

We next isolate the shaft of torsional stiffness K_n, and write for its two equations (see Fig. 7.7-5)

$$\theta_n^L - \theta_{n-1}^R = \frac{T_{n-1}^R}{K_n}$$

$$T_n^L = T_{n-1}^R$$

The field matrix is then formed as

$$\begin{bmatrix} \theta \\ T \end{bmatrix}_n^L = \begin{bmatrix} 1 & \dfrac{1}{K} \\ 0 & 1 \end{bmatrix}_n \begin{bmatrix} \theta \\ T \end{bmatrix}_{n-1}^R \quad (7.7\text{-}9)$$

Substituting Eq. (7.7-9) into (7.7-8), the transfer matrix relating the state vectors at stations n and $n - 1$ is

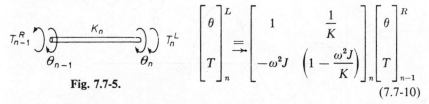

Fig. 7.7-5.

$$\begin{bmatrix} \theta \\ T \end{bmatrix}_n^L = \begin{bmatrix} 1 & \dfrac{1}{K} \\ -\omega^2 J & \left(1 - \dfrac{\omega^2 J}{K}\right) \end{bmatrix}_n \begin{bmatrix} \theta \\ T \end{bmatrix}_{n-1}^R \quad (7.7\text{-}10)$$

We thus note that each of Eqs. (7.7-8), (7.7-9), and (7.7-10) is identical to those of the linear spring-mass system.

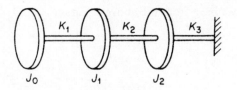

Fig. 7.7-6.

The Holzer method was actually developed for a torsional system such as Fig. 7.7-6. Starting at a free end 0 with zero torque and unit amplitude, the torque to the right of 0 is $-\omega^2 J_0 1$ so that we can start with the state vector

$$\begin{bmatrix} \theta \\ T \end{bmatrix}_0^R = \begin{bmatrix} 1 \\ -\omega^2 J_0 \end{bmatrix}$$

The next section leads to the equation

$$\begin{bmatrix} \theta \\ T \end{bmatrix}_1^R = \begin{bmatrix} 1 & \dfrac{1}{K_1} \\ -\omega^2 J_1 & \left(1 - \dfrac{\omega^2 J_1}{K_1}\right) \end{bmatrix} \begin{bmatrix} 1 \\ -\omega^2 J_0 \end{bmatrix} = \begin{bmatrix} \left(1 - \dfrac{\omega^2 J_0}{K_1}\right) \\ -\omega^2 J_1 - \omega^2 J_0 \left(1 - \dfrac{\omega^2 J_1}{K_1}\right) \end{bmatrix}$$

The amplitude θ_1 is thus the amplitude $\theta_0 = 1$ minus the twist in shaft 1, which is $\omega^2 J_0 / K_1$ or $\theta_1 = (1 - \omega^2 J_0 / K_1)$. The torque T_1^R is the torque to the right of disk 1, which is the torque carried by shaft 2. It is equal to the inertia torque of both disks 0 and 1, or

$$-\omega^2 J_0 \theta_0 - \omega^2 J_1 \theta_1 = -\omega^2 J_0 - \omega^2 J_1 \left(1 - \frac{\omega^2 J_0}{K_1}\right)$$

$$= -\omega^2 J_1 - \omega^2 J_0 \left(1 - \frac{\omega^2 J_1}{K_1}\right)$$

as indicated in the matrix equation above.

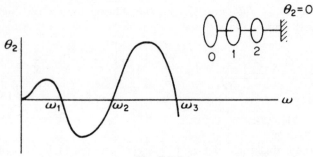

Fig. 7.7-7.

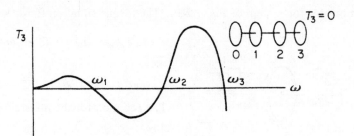

Fig. 7.7-8.

The computation is carried out numerically with chosen values of ω^2 and the results are plotted as in Figs. 7.7-7 or 7.7-8, depending on the boundary conditions at station n.

In the systems shown so far, the stations were numbered in increasing order from left to right. Thus in Eqs. (7.7-7) and (7.7-10) an arrow under the equal sign is used to denote this direction. Some prefer to number the stations from right to left, since then the configuration of the matrices conform with the configuration of the system. Thus for the system shown in Fig. 7.7-9 the transfer matrix is

$$\begin{bmatrix} \theta \\ T \end{bmatrix}_{n+1}^{R} \underset{\leftarrow}{=} \begin{bmatrix} \left(1 - \dfrac{\omega^2 J}{K}\right) & -\dfrac{1}{K} \\ \omega^2 J & 1 \end{bmatrix}_{n} \begin{bmatrix} \theta \\ T \end{bmatrix}_{n}^{R} \tag{7.7-11}$$

which is somewhat simpler to associate with the physical system. Again the arrow under the equal sign denotes the direction of increasing n. Note, however, that the terms of the matrix are now rearranged; the student should verify this equation, starting from the free-body development.

(c) *Systems with Damping.* When damping is considered, we find that the transfer matrix is not altered, but that the mass and stiffness elements are replaced by complex quantities. This can be easily shown by writing the

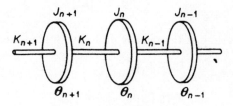

Fig. 7.7-9.

equations for the system of Fig. 7.7-10. The torque equation for disk n is

$$-\omega^2 J_n \theta_n = T_n^R - T_n^L - i\omega C_n \theta_n$$

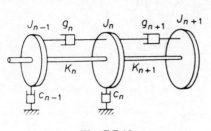

or

$$(i\omega C_n - \omega^2 J_n)\theta_n = T_n^R - T_n^L$$

The equation for the nth shaft is

$$T_n^L = K_n(\theta_n - \theta_{n-1})$$
$$+ i\omega g_n(\theta_n - \theta_{n-1})$$

Fig. 7.7-10.

$$= (K_n + i\omega g_n)(\theta_n - \theta_{n-1})$$

Thus the point matrix and the field matrix for the damped system become

$$\begin{bmatrix} \theta \\ T \end{bmatrix}_n^R = \begin{bmatrix} 1 & 0 \\ (i\omega C - \omega^2 J) & 1 \end{bmatrix}_n \begin{bmatrix} \theta \\ T \end{bmatrix}_n^L \qquad (7.7\text{-}12)$$

$$\begin{bmatrix} \theta \\ T \end{bmatrix}_n^L = \begin{bmatrix} 1 & \dfrac{1}{K + i\omega g} \\ 0 & 1 \end{bmatrix}_n \begin{bmatrix} \theta \\ T \end{bmatrix}_{n-1}^R \qquad (7.7\text{-}13)$$

which are identical to the undamped case except for the mass and stiffness elements which are now complex.

EXAMPLE 7.7-2. Determine the natural frequencies and the mode shapes of the system shown in Fig. 7.7-11.

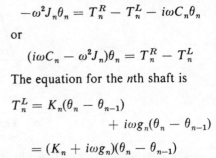

$K_2 = 10^6 \qquad K_3 = 2 \times 10^6 \qquad$ lb-in./rad.

$J_1 = 50 \qquad J_2 = 100 \qquad J_3 = 200 \qquad$ ib-in.-sec^2

Fig. 7.7-11.

Solution. Starting at the right of disk 1, with the state vector

$$\begin{bmatrix} \theta \\ T \end{bmatrix}_1^R = \begin{bmatrix} 1.00 \\ -50\omega^2 \end{bmatrix}$$

the state vectors at station 2 and station 3 are

$$\begin{bmatrix} \theta \\ T \end{bmatrix}_2^R = \begin{bmatrix} 1 & 10^{-6} \\ -100\omega^2 & \left(1 - \dfrac{100\omega^2}{10^6}\right) \end{bmatrix}_2 \begin{bmatrix} 1.0 \\ -50\omega^2 \end{bmatrix}_1$$

$$\begin{bmatrix} \theta \\ T \end{bmatrix}_3^R = \begin{bmatrix} 1 & \frac{1}{2} \times 10^{-6} \\ -200\omega^2 & \left(1 - \dfrac{200\omega^2}{2 \times 10^6}\right) \end{bmatrix}_3 \begin{bmatrix} \theta \\ T \end{bmatrix}_2^R$$

Thus by assuming different values of ω, the quantities θ and T can be found, first for station 2 and then for station 3. Since the end 3 is free, the frequencies which result in $T_3^R = 0$ are the natural frequencies of the system. The following table gives the state vectors at each station for three values of ω, and Fig.

ω rad./sec.	$\begin{bmatrix} \theta \\ T \end{bmatrix}_1^R$	$\begin{bmatrix} \theta \\ T \end{bmatrix}_2^R$	$\begin{bmatrix} \theta \\ T \end{bmatrix}_3^R$
126	1.00 0.794 × 10⁶	0.206 1.121 × 10⁶	−0.355 −0.009 × 10⁶
150	1.00 1.126 × 10⁶	−0.126 0.842 × 10⁶	−0.547 −1.618 × 10⁶
210	1.00 2.205 × 10⁶	−1.205 −3.104 × 10⁶	0.347 −0.044 × 10⁶

7.7-12 is a plot of T_3^R indicating that the natural frequencies of the system are $\omega_1 = 126$ rad./sec. and $\omega_2 = 210$ rad./sec. The torques T_3 for $\omega = 126$ and 210 are not zero, but are close enough to zero to approximate the condition

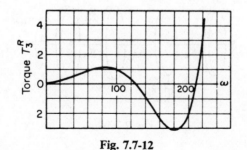

Fig. 7.7-12

$T_3 = 0$. The mode shapes for the two natural frequencies are also shown in Fig. 7.7-13.

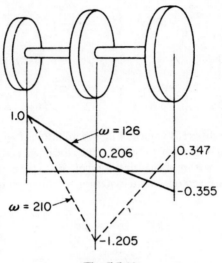

Fig. 7.7-13.

EXAMPLE 7.7-3. The torsional system of Fig. 7.7-14 is excited by a harmonic torque at a point to the right of disk 4. Determine the torque-frequency curve and establish the first natural frequency of the system.

$$J_1 = J_2 = 500 \text{ lb. in. sec.}^2$$
$$J_3 = J_4 = 1000 \text{ lb. in. sec.}^2$$
$$K_2 = K_3 = K_4 = 10^6 \text{ lb. in./rad.}$$
$$c_2 = 10^4 \text{ lb. in. sec./rad.}$$
$$g_4 = 2 \times 10^4 \text{ lb. in. sec./rad.}$$

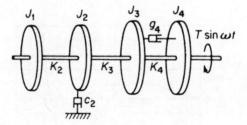

Fig. 7.7-14.

Solution. The numerical computations for $\omega^2 = 1000$ are shown in the first accompanying table. The complex mass and stiffness terms are first tabulated for each station n. Substituting into the point and field matrices, i.e., Eqs. (7.7-12) and (7.7-13), the complex amplitude and torque for each

n	$(\omega^2 J_n - i\omega c_n)10^{-6}$	$(K_n + i\omega g_n)10^{-6}$
1	$0.50 + 0.0i$	
2	$0.50 - 0.316i$	$1.0 + 0.0i$
3	$1.0 + 0.0i$	$1.0 + 0.0i$
4	$1.0 + 0.0i$	$1.0 + 0.635i$

station are found, as in the second table here.

n	θ_n	T_n^R (for $\omega^2 = 1000$)
1	$1.0 + 0.0i$	$(0.50 + 0.0i) \times 10^6$
2	$0.50 + 0.0i$	$(0.750 + 0.158i) \times 10^6$
3	$-0.250 + 0.158i$	$(0.50 + 0.0i) \times 10^6$
4	$-0.607 + 0.384i$	$(-0.107 + 0.384i) \times 10^6$

The above computations are repeated for a sufficient number of frequencies to plot the torque-frequency curve of Fig. 7.7-15. The plot shows the real and imaginary parts of T_4^R as well as their resultant, which in this problem is the exciting torque. For example, the resultant torque at $\omega^2 = 1000$ is

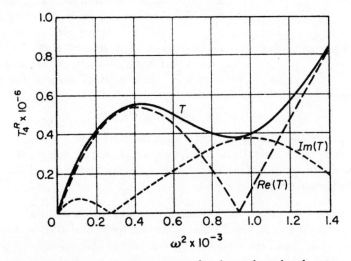

Fig. 7.7-15. Torque-frequency curve for damped torsional system of Fig. 7.7-14.

$10^6\sqrt{0.107^2 + 0.384^2} = 0.394 \times 10^6$ in. lb. The first natural frequency of the system from this diagram is found to be approximately $\omega = \sqrt{930} = 30.5$ rad./sec., where the natural frequency is defined as that frequency which requires no torque to sustain the motion of the undamped system.

EXAMPLE 7.7-4. In Fig. 7.7-14 if $T = 2000$ in. lb. and $\omega = 31.6$ rad./sec., determine the amplitude of the second disk.

Solution. The table above indicates that a torque of 394,000 in. lb. will produce an amplitude of $\theta_2 = 0.50$ radian. Since amplitude is proportional to torque, the amplitude of the second disk for the specified torque is $0.50 \times \frac{2}{394} = 0.00254$ radian.

7.8 Geared and Branched Systems

Geared System. Consider the geared torsional system of Fig. 7.8-1 where the speed ratio of shaft 2 to shaft 1 is n. If the oscillatory amplitude of shaft 1 is θ_1, the oscillatory amplitude of shaft 2 will be $\theta_2 = n\theta_1$. The oscillatory kinetic energy of disk 2 is then

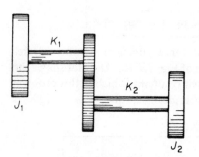

$$T = \tfrac{1}{2}J_2\dot{\theta}_2^2 = \tfrac{1}{2}n^2 J_2 \dot{\theta}_1^2 \qquad (7.8\text{-}1)$$

and the equivalent inertia of disk 2 referred to shaft 1 is $n^2 J_2$.

To determine the equivalent stiffness of shaft 2 referred to shaft 1, assume disk 2 to be fixed and apply a twisting torque to shaft 1, rotating gear 1 through an angle θ_1. The gear on shaft 2 will then rotate through an angle $n\theta_1$ and, since this is the twist of shaft 2, the poten-

Fig. 7.8-1. A geared torsional system.

tial energy of shaft 2 becomes equal to

$$U = \tfrac{1}{2}K_2(n\theta_1)^2 = \tfrac{1}{2}(n^2 K_2)\theta_1^2 \qquad (7.8\text{-}2)$$

The second shaft referred to shaft 1 must therefore have a stiffness of $n^2 K_2$.

The rule for geared systems is thus quite simple and is: *multiply all stiffness and inertias of the geared shaft by n^2*, where n is the speed ratio of the geared shaft to the reference shaft.

EXAMPLE 7.8-1. Determine by the matrix method, the natural frequency of the system shown in Fig. 7.8-2(a), where the speed of shaft 2 is n times the speed of shaft 3, and the masses of the gears are considered to be negligible.

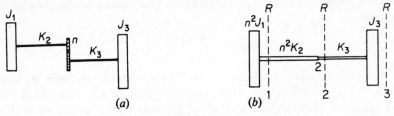

Fig. 7.8-2.

Solution. Choosing shaft 3 as reference, we multiply J_1 and K_2 by n^2 to form the equivalent system of Fig. 7.8-2(b). We can start with the state vector at $1R$, which is

$$\begin{bmatrix} \theta \\ T \end{bmatrix}_1^R = \begin{bmatrix} 1 \\ -\omega^2 n^2 J_1 \end{bmatrix}$$

Going from $1R$ to $2R$, we let $J_2 = 0$ for the field matrix and write

$$\begin{bmatrix} \theta \\ T \end{bmatrix}_2^R = \begin{bmatrix} 1 & \dfrac{1}{n^2 K_2} \\ 0 & 1 \end{bmatrix} \begin{bmatrix} 1 \\ -\omega^2 n^2 J_1 \end{bmatrix}$$

Going from $2R$ to $3R$, we have

$$\begin{bmatrix} \theta \\ T \end{bmatrix}_3^R = \begin{bmatrix} 1 & \dfrac{1}{K_3} \\ -\omega^2 J_3 & \left(1 - \dfrac{\omega^2 J_3}{K_3}\right) \end{bmatrix} \begin{bmatrix} 1 & \dfrac{1}{n^2 K_2} \\ 0 & 1 \end{bmatrix} \begin{bmatrix} 1 \\ -\omega^2 n J_1 \end{bmatrix}$$

$$= \begin{bmatrix} 1 & \left(\dfrac{1}{n^2 K_2} + \dfrac{1}{K_3}\right) \\ -\omega^2 J_3 & \left(1 - \dfrac{\omega^2 J_3}{K_3} - \dfrac{\omega^2 J_3}{n^2 K_2}\right) \end{bmatrix} \begin{bmatrix} 1 \\ -\omega^2 n^2 J_1 \end{bmatrix}$$

Since $T_3^R = 0$,

$$-\omega^2 J_3 - \omega^2 n^2 J_1 \left(1 - \dfrac{\omega^2 J_3}{K_3} - \dfrac{\omega^2 J_3}{n^2 K_2}\right) = 0$$

Solving for ω^2, the natural frequency is found to be

$$\omega^2 = \frac{(n^2 J_1 + J_3) K_2 K_3}{(K_3 + n^2 K_2) J_1 J_3}$$

The result can be checked from the equation for the natural frequency of a two-disk system connected by a single shaft, by replacing K with the equivalent stiffness

$$K_{eff} = \frac{n^2 K_2 K_3}{n^2 K_2 + K_3}$$

Branched Systems. Branched systems are frequently encountered, some common examples being the dual propeller system of a marine installation and the drive shaft and differential of the automobile, as shown in Fig. 7.8-3.

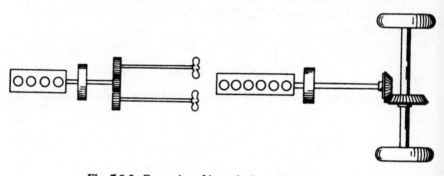

Fig. 7.8-3. Examples of branched torsional systems.

Such systems can be reduced to the form shown in Fig. 7.8-4, with a one-to-one gear, by multiplying all the inertias and stiffnesses of the branches by the squares of their speed ratios. How such systems are handled by the matrix method is illustrated in Example 7.8-2.

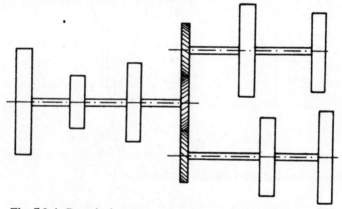

Fig. 7.8-4. Branched system reduced to common speeds by 1 to 1 gears.

EXAMPLE 7.8-2. Outline the matrix procedure for solving the torsional branched system of Fig. 7.8-5(a).

Solution. Again we convert to a system having one-to-one gears by multiplying the stiffness and inertia of branch B by n^2, as shown in Fig. 7.8-5(b).

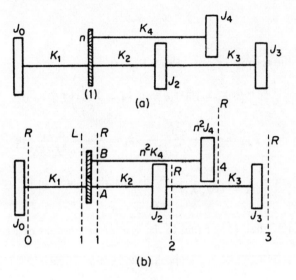

Fig. 7.8-5.

We can immediately write three equations:

$$\begin{bmatrix} \theta \\ T \end{bmatrix}_1^L = \begin{bmatrix} 1 & \dfrac{1}{K_1} \\ 0 & 1 \end{bmatrix}\begin{bmatrix} 1 \\ -\omega^2 J_0 \end{bmatrix} \tag{a}$$

$$\begin{bmatrix} \theta \\ T \end{bmatrix}_2^R = \begin{bmatrix} 1 & \dfrac{1}{K_2} \\ -\omega^2 J_2 & \left(1 - \dfrac{\omega^2 J_2}{K_2}\right) \end{bmatrix}\begin{bmatrix} \theta_A \\ T_A \end{bmatrix}_1^R \tag{b}$$

$$\begin{bmatrix} \theta \\ T \end{bmatrix}_4^R = \begin{bmatrix} 1 & \dfrac{1}{n^2 K_4} \\ -\omega^2 n^2 J_4 & \left(1 - \dfrac{\omega^2 J_4}{K_4}\right) \end{bmatrix}\begin{bmatrix} \theta_B \\ T_B \end{bmatrix}_1^R \tag{c}$$

and we need only to relate the state vector

$$\begin{bmatrix} \theta_A \\ T_A \end{bmatrix}_1^R \quad \text{in terms of} \quad \begin{bmatrix} \theta \\ T \end{bmatrix}_1^L$$

in order to proceed along shaft A. We can do this by first noting that

$$\theta_{A1}^R = \theta_1^L$$
$$\theta_{B1}^R = -\theta_1^L \tag{d}$$

and that the torque $T_4^R = 0$. From Eq. (c),

$$T_4^R = -\omega^2 n^2 J_4 \theta_{B1}^R + \left(1 - \frac{\omega^2 J_4}{K_4}\right) T_{B1}^R = 0$$

$$T_{B1}^R = \frac{\omega^2 n^2 J_4}{(1 - \omega^2 J_4/K_4)} \theta_{B1}^R = \frac{-\omega^2 n^2 J_4}{(1 - \omega^2 J_4/K_4)} \theta_1^L$$

and letting $\alpha = \dfrac{\omega^2 n^2 J_4}{(1 - \omega^2 J_4/K_4)}$, $T_{B_1}^R$ is linearly related to θ_1^L by the equation

$$T_{B1}^R = -\alpha \theta_1^L \tag{e}$$

We also note that T_1^L is equal to the sum of the torques carried by shafts A and B, or

$$T_1^L = T_{A1}^R + T_{B1}^R \tag{f}$$

Eliminating $T_{B_1}^R$ from Eqs. (e) and (f), we obtain

$$T_{A1}^R = T_1^L + \alpha \theta_1^L \tag{g}$$

Thus from Eqs. (d) and (g) the state vector in question becomes

$$\begin{bmatrix} \theta_A \\ T_A \end{bmatrix}_1^R = \begin{bmatrix} 1 & 0 \\ \alpha & 1 \end{bmatrix} \begin{bmatrix} \theta \\ T \end{bmatrix}_1^L \tag{h}$$

Substituting this state vector back into Eq. (b), we obtain the transfer vector from $1L$ to $2R$

$$\begin{bmatrix} \theta \\ T \end{bmatrix}_2^R = \begin{bmatrix} 1 & \dfrac{1}{K_2} \\ -\omega^2 J_2 & \left(1 - \dfrac{\omega^2 J_2}{K_2}\right) \end{bmatrix} \begin{bmatrix} 1 & 0 \\ \alpha & 1 \end{bmatrix} \begin{bmatrix} \theta \\ T \end{bmatrix}_1^L \tag{i}$$

Thus, with $\begin{bmatrix} \theta \\ T \end{bmatrix}_1^L$ known from Eq. (a), we can proceed to station $2R$ and points farther along on shaft A in the usual manner.

7.9 Beams

When a beam is replaced by lumped masses connected by massless beam sections, a method developed by N. O. Myklestad and M. A. Prohl can be used to compute progressively the deflection, slope, moment, and shear from one station to the next, in a manner similar to the Holzer method. Again it is advantageous, from the point of view of conciseness and efficiency of computation, to express these equations in matrix form.

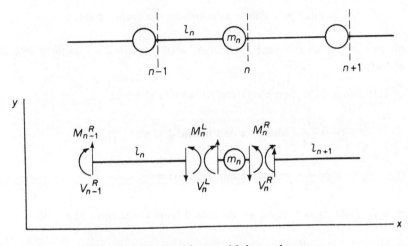

Fig. 7.9-1. Idealized beam with lumped masses.

(a) *Uncoupled Flexural Vibrations.* Figure 7.9-1 shows a typical section of an idealized beam with lumped masses. Examining the nth section, the forces and moments acting on the mass and beam section are indicated by the free-body diagram. From them the equations for the shear and moment are found to be

$$M_n^L = M_n^R \qquad\qquad V_{n-1}^R = V_n^L$$
$$V_n^L = V_n^R - \omega^2 m_n y_n \qquad M_{n-1}^R = M_n^L - V_n^L l_n \qquad (7.9\text{-}1)$$

For the elastic deformation of the nth beam section, we refer to Fig. 7.9-2. The deflection and slope at the ends are then related by the equations

$$y_n = y_{n-1} + l_n \theta_{n-1} + M_n^L \frac{l_n^2}{2(EI)_n} - V_n^L \frac{l_n^3}{3(EI)_n} \qquad (7.9\text{-}2)$$

$$\theta_n = \theta_{n-1} + \frac{M_n^L l_n}{(EI)_n} - \frac{V_n^L l^2}{2(EI)_n}$$

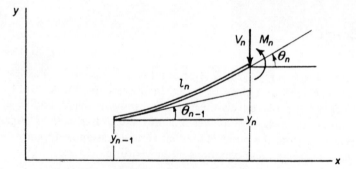

Fig. 7.9-2. Elastic deformation of a beam section.

where the various influence coefficients used are based on uniform section, and are:

(1)   Slope of station n referred to tangent at $n-1$

$$= \frac{l}{EI} \quad \text{due to a unit moment at } n$$

$$= \frac{l^2}{2EI} \quad \text{due to a unit shear at } n$$

(2)   Deflection of station n measured from the tangent at $n-1$

$$= \frac{l^2}{2EI} \quad \text{due to a unit moment at } n$$

$$= \frac{l^3}{3EI} \quad \text{due to a unit shear at } n$$

Expressing M_n and V_n in Eq. (7.9-2) in terms of M_{n-1} and V_{n-1} from Eq. (7.9-1), these equations can be rewritten as

$$y_n = y_{n-1} + l_n \theta_{n-1} + \frac{M_{n-1}^R l_n^2}{2(EI)_n} + \frac{V_{n-1}^R l_n^3}{6(EI)_n}$$

$$\theta_n = 0 \quad + \theta_{n-1} + \frac{M_{n-1}^R l_n}{(EI)_n} + \frac{V_{n-1}^R l_n^2}{2(EI)_n} \qquad (7.9\text{-}3)$$

$$M_n^L = 0 \quad + 0 \quad + M_{n-1}^R + V_{n-1}^R l_n$$

$$V_n^L = 0 \quad + 0 \quad + 0 \quad + V_{n-1}^R$$

and expressed in the matrix form

$$
\begin{bmatrix} y \\ \theta \\ M \\ V \end{bmatrix}_n^L = \begin{bmatrix} 1 & l & \dfrac{l^2}{2EI} & \dfrac{l^3}{6EI} \\ 0 & 1 & \dfrac{l}{EI} & \dfrac{l^2}{2EI} \\ 0 & 0 & 1 & l \\ 0 & 0 & 0 & 1 \end{bmatrix}_n \begin{bmatrix} y \\ \theta \\ M \\ V \end{bmatrix}_{n-1}^R
\tag{7.9-4}
$$

For the point mass m_n, we have

$$
y_n^R = y_n^L \qquad M_n^R = M_n^L
$$
$$
\theta_n^R = \theta_n^L \qquad V_n^R = V_n^L + \omega^2 m_n y_n^L
$$

which leads to the point matrix

$$
\begin{bmatrix} y \\ \theta \\ M \\ V \end{bmatrix}_n^R = \begin{bmatrix} 1 & 0 & 0 & 0 \\ 0 & 1 & 0 & 0 \\ 0 & 0 & 1 & 0 \\ \omega^2 m & 0 & 0 & 1 \end{bmatrix}_n \begin{bmatrix} y \\ \theta \\ M \\ V \end{bmatrix}_n^L
\tag{7.9-5}
$$

Substituting Eq. (7.9-4) for the column on the right side of Eq. (7.9-5), the final equation relating the state vectors at n and $n-1$ is found to be

$$
\begin{bmatrix} y \\ \theta \\ M \\ V \end{bmatrix}_n^R = \begin{bmatrix} 1 & 0 & 0 & 0 \\ 0 & 1 & 0 & 0 \\ 0 & 0 & 1 & 0 \\ m\omega^2 & 0 & 0 & 1 \end{bmatrix}_n \begin{bmatrix} 1 & l & \dfrac{l^2}{2EI} & \dfrac{l^3}{6EI} \\ 0 & 1 & \dfrac{l}{EI} & \dfrac{l^2}{2EI} \\ 0 & 0 & 1 & l \\ 0 & 0 & 0 & 1 \end{bmatrix}_n \begin{bmatrix} y \\ \theta \\ M \\ V \end{bmatrix}_{n-1}^R
$$

$$
= \begin{bmatrix} 1 & l & \dfrac{l^2}{2EI} & \dfrac{l^3}{6EI} \\ 0 & 1 & \dfrac{l}{EI} & \dfrac{l^2}{2EI} \\ 0 & 0 & 1 & l \\ \omega^2 m & \omega^2 ml & \dfrac{\omega^2 ml^2}{2EI} & 1 + \omega^2 m \dfrac{l^3}{6EI} \end{bmatrix}_n \begin{bmatrix} y \\ \theta \\ M \\ V \end{bmatrix}_{n-1}^R
\tag{7.9-6}
$$

For any frequency ω, Eq. (7.9-6) enables us to start at the left boundary 0 and proceed to the right boundary N, these quantities being linearly related by the equation:

$$
\begin{bmatrix} y \\ \theta \\ M \\ V \end{bmatrix}_N = \begin{bmatrix} u_{11} & u_{12} & u_{13} & u_{14} \\ u_{21} & u_{22} & u_{23} & u_{24} \\ u_{31} & u_{32} & u_{33} & u_{34} \\ u_{41} & u_{42} & u_{43} & u_{44} \end{bmatrix} \begin{bmatrix} y \\ \theta \\ M \\ V \end{bmatrix}_0
\tag{7.9-7}
$$

Generally, two of the boundary conditions at each end are known, so that the frequencies which satisfy these conditions are the natural frequencies of the beam.

EXAMPLE 5.9-1. A cantilever beam, fixed at the left end, is represented by several lumped masses. Determine the boundary equations leading to the natural frequencies.

Solution. At station 0, $y_0 = \theta_0 = 0$, and from Eq. (7.9-7) we obtain

$$M_N = u_{33}M_0 + u_{34}V_0$$
$$V_N = u_{43}M_0 + u_{44}V_0$$

where M_0 and V_0 are unknown and M_N and V_N must be zero. The boundary conditions are then satisfied if the determinant of the equation is zero, or

$$
\begin{vmatrix} u_{33} & u_{34} \\ u_{43} & u_{44} \end{vmatrix} = 0
$$

so that this quantity may be plotted as a function of ω to establish the natural frequencies of the beam.

(b) *Rotating Beams.* In this section we will examine rotating beams, such as propellers and turbine blades, for vibration perpendicular to the plane of rotation. Due to centrifugal force, we will need to consider terms in addition to the beam analysis of the previous section.

The centrifugal force, which is normal to the axis of rotation, is shown in Fig. 7.9-3 and is equal to $m_n\Omega^2 x_n$ for mass m_n. The additional quantity which must be introduced is then

$$F_n^L = F_n^R + m_n\Omega^2 x_n \tag{7.9-8}$$

where

$$F_n^L = \sum_{i=n}^{N} m_i\Omega^2 x_i \tag{7.9-9}$$

Because of this term, the moment equation is modified to

$$M_{n-1}^R = M_n^L - V_n^L l_n - F_n^L(y_n^L - y_{n-1}^R)$$

The deflection and slope are also influenced by F_n^L, and we account for it by considering only the component of F_n^L normal to the beam as a shear load:

$$y_n^L = y_{n-1}^R + l_n\theta_{n-1}^R - \theta_n^L F_n^L \frac{l^3}{3(EI)_n} + M_n^L \frac{l_n^2}{2(EI)_n} - V_n^L \frac{l_n^3}{3(EI)_n}$$

$$\theta_n^L = \theta_{n-1}^R + \frac{M_n^L l_n}{(EI)_n} - \frac{V_n^L l_n^2}{2(EI)_n} - \frac{\theta_n^L F_n^L l^2}{2(EI)_n}$$

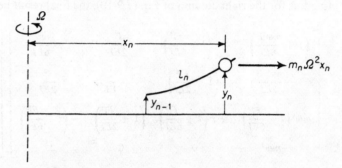

Fig. 7.9-3. Centrifugal force on a rotating beam.

These equations can now be rearranged for calculation to proceed from right to left as follows:

$$y_{n-1}^R = y_n^L - \theta_n^L\left[l_n + \frac{F_n^L l_n^3}{6(EI)_n}\right] + M_n^L \frac{l_n^2}{2(EI)_n} - V_n^L \frac{l^3}{6(EI)_n}$$

$$\theta_{n-1}^R = \theta_n^L\left[1 + \frac{F_n^L l^2}{2(EI)_n}\right] - M_n^L \frac{l_n}{(EI)_n} + V_n^L \frac{l^2}{2(EI)_n}$$

$$M_{n-1}^R = M_n^L\left[1 + \frac{F_n^L l_n^2}{2(EI)_n}\right] - V_n^L\left[l_n + \frac{F_n^L l_n^3}{6(EI)_n}\right] - \theta_n^L F_n^L\left[l_n + \frac{F_n^L l_n^3}{6(EI)_n}\right]$$

$$V_{n-1}^R = V_n^L$$

Arranged in matrix form, these equations appear as

$$
\begin{bmatrix} y \\ \theta \\ M \\ V \end{bmatrix}_{n-1}^{R}
=
\overleftarrow{
\begin{bmatrix}
1 & -\left(l + \dfrac{Fl^3}{6EI}\right) & \dfrac{l^2}{2EI} & -\dfrac{l^3}{6EI} \\[2ex]
0 & \left(1 + \dfrac{Fl^2}{2EI}\right) & -\dfrac{l}{EI} & \dfrac{l^2}{2EI} \\[2ex]
0 & -F\left(l + \dfrac{Fl^3}{6EI}\right) & \left(1 + \dfrac{Fl^2}{2EI}\right) & -\left(l + \dfrac{Fl^3}{6EI}\right) \\[2ex]
0 & 0 & 0 & 1
\end{bmatrix}_{n}
}
\begin{bmatrix} y \\ \theta \\ M \\ V \end{bmatrix}_{n}^{L}
$$

$$(7.9\text{-}10)$$

To complete the problem, we need the point matrix for mass m_n:

$$
\begin{bmatrix} y \\ \theta \\ M \\ V \end{bmatrix}_n^L =
\begin{bmatrix} 1 & 0 & 0 & 0 \\ 0 & 1 & 0 & 0 \\ 0 & 0 & 1 & 0 \\ -m\omega^2 & 0 & 0 & 1 \end{bmatrix}
\begin{bmatrix} y \\ \theta \\ M \\ V \end{bmatrix}_n^R
\tag{7.9-11}
$$

Substituting this for the right column of Eq. (7.9-10), the final result becomes

$$
\begin{bmatrix} y \\ \theta \\ M \\ V \end{bmatrix}_{n-1}^R =
\begin{bmatrix}
\left(1 + \dfrac{m\omega^2 l^3}{6EI}\right) & -\left(l + \dfrac{Fl^3}{6EI}\right) & \dfrac{l^2}{2EI} & -\dfrac{l^3}{6EI} \\[2mm]
-\dfrac{m\omega^2 l^2}{2EI} & \left(1 + \dfrac{Fl^2}{2EI}\right) & -\dfrac{l}{EI} & \dfrac{l^2}{2EI} \\[2mm]
m\omega^2\left(l + \dfrac{Fl^3}{6EI}\right) & -F\left(l + \dfrac{Fl^3}{6EI}\right) & \left(1 + \dfrac{Fl^2}{2EI}\right) & -\left(l + \dfrac{Fl^2}{6EI}\right) \\[2mm]
-m\omega^2 & 0 & 0 & 1
\end{bmatrix}_n
\begin{bmatrix} y \\ \theta \\ M \\ V \end{bmatrix}_n^R
$$

$$\tag{7.9-12}$$

(c) *Coupled Flexure-Torsion Vibration.* Natural modes of vibration of airplane wings and other beam structures are often coupled flexure-torsion modes which for higher modes differ considerably from those of uncoupled modes. To treat such problems it is necessary to introduce one additional influence coefficient, h_n, defined as the angle of twist of station n relative to station $n - 1$, due to a unit torque at n. Referring to the beam section of

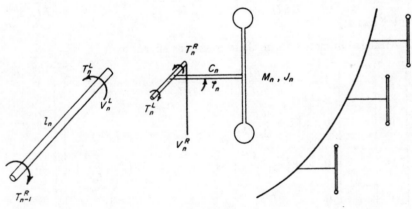

Fig. 7.9-4.

Fig. 7.9-4, the equations pertaining to the torque T are

$$T_n^R - T_n^L = J_n\ddot{\varphi}_n + m_n c_n \ddot{y}_n$$
$$= -J_n\omega^2\varphi_n - m_n c_n \omega^2 y_n$$
$$\varphi_n^L - \varphi_{n-1}^R = T_n^L h_n$$
$$\varphi_n^R = \varphi_n^L$$

where $J_n = J_{ncg} + m_n c_n^2$ is the moment of inertia of the nth section about the elastic axis of the beam. The shear across the mass is

$$V_n^L - V_n^R = -m_n\omega^2(y_n + c_n\varphi_n)$$

and the point matrix across m can be written as

$$
\begin{bmatrix} y \\ \theta \\ M \\ V \\ \varphi \\ T \end{bmatrix}_n^R =
\left[\begin{array}{cccc:cc}
1 & 0 & 0 & 0 & 0 & 0 \\
0 & 1 & 0 & 0 & 0 & 0 \\
0 & 0 & 1 & 0 & 0 & 0 \\
m\omega^2 & 0 & 0 & 1 & m\omega^2 c & 0 \\
\hdashline
0 & 0 & 0 & 0 & 1 & 0 \\
-mc\omega^2 & 0 & 0 & 0 & -J\omega^2 & 1
\end{array}\right]
\begin{bmatrix} y \\ \theta \\ M \\ V \\ \varphi \\ T \end{bmatrix}_n^L
\qquad (7.9\text{-}13)
$$

The field matrix between station $(n-1)R$ and $(n)L$ is the same as Eq. (7.9-4) with two additional equations:

$$
\begin{bmatrix} y \\ \theta \\ M \\ V \\ \varphi \\ T \end{bmatrix}_n^L =
\left[\begin{array}{cccc:cc}
1 & l & \dfrac{l^2}{2EI} & \dfrac{l^3}{6EI} & 0 & 0 \\
0 & 1 & \dfrac{l}{EI} & \dfrac{l^2}{2EI} & 0 & 0 \\
0 & 0 & 1 & l & 0 & 0 \\
0 & 0 & 0 & 1 & 0 & 0 \\
\hdashline
0 & 0 & 0 & 0 & 1 & h \\
0 & 0 & 0 & 0 & 0 & 1
\end{array}\right]_n
\begin{bmatrix} y \\ \theta \\ M \\ V \\ \varphi \\ T \end{bmatrix}_{n-1}^R
\qquad (7.9\text{-}14)
$$

Thus by substituting Eq. (7.9-14) for the right column of Eq. (7.9-13), the state vectors at station $(n)R$ are related to the state vectors at station $(n-1)R$.

EXAMPLE 7.9-2. Figure 7.9-5 shows the mass breakdown for a fighter-plane wing and fuselage for flexure-torsion vibration. Find the boundary equations for the determination of the symmetric flexure-torsion modes.

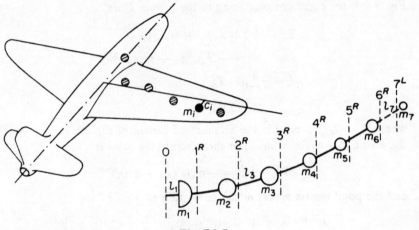

Fig. 7.9-5.

Solution. To use the matrix equation (7.9-14), we let the center line of the airplane be station 0 and let half of the mass and mass moment of inertia of the fuselage about the elastic axis be m_1 and J_1, with $l_1 = 0$. At the wing tip we put in station 7 with $m_7 = l_7 = 0$ to make use of the column matrix $[\;\;]_7^L$ which now becomes equal to $[\;\;]_6^R$.

For the symmetric modes, the bending slope θ_0^R, the shear V_0^R, and the twisting torque T_0^R are zero at the center line, whereas at the wing tip the moment, shear, and the torque are zero. The boundary equations from Eq. (7.9-14) then appear as

$$
\begin{bmatrix} - \\ - \\ M \\ V \\ - \\ T \end{bmatrix}^L_7
=
\begin{bmatrix} & & & & \\ & & & & \\ & & u_{ij} & & \\ & & & & \\ & & & & \end{bmatrix}
\begin{bmatrix} y \\ 0 \\ M \\ 0 \\ \varphi \\ 0 \end{bmatrix}^R_0
$$

which may be rewritten as

$$
\begin{bmatrix} M \\ V \\ T \end{bmatrix}^L_7
=
\begin{bmatrix} u_{31} & u_{33} & \mu_{35} \\ u_{41} & u_{43} & u_{45} \\ u_{61} & u_{63} & u_{65} \end{bmatrix}
\begin{bmatrix} y \\ M \\ \varphi \end{bmatrix}^R_0
$$

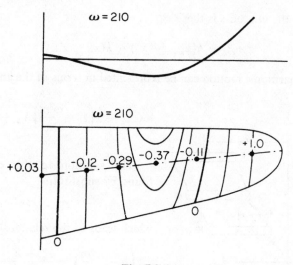

Fig. 7.9-6.

The quantities y_0, M_0, and φ_0 at the center line are unknown; however, the column matrix at the left for the moment, shear, and torque at the wing tip are zero. Thus the determinant of the u_{ij} which is a function of ω must be zero to satisfy the boundary conditions.

By plotting the quantity

$$D(\omega) = \begin{vmatrix} u_{31} & u_{33} & u_{35} \\ u_{41} & u_{43} & u_{45} \\ u_{61} & u_{63} & u_{65} \end{vmatrix}$$

against ω, the natural frequencies for the symmetric modes are established. The mode shapes are then established for the natural frequencies found, by computing y_n, θ_n, and φ_n. Figure 7.9-6 shows a typical contour plot for the second symmetric mode of a particular fighter plane.

7.10 Difference Equation

Frequently in a dynamical system identical sections are repeated several times. The equations of motion can then be treated with advantage by the difference equation.

As an example of repeating sections, consider the N-story building shown in Fig. 7.10-1, where the mass of each floor is m and the lateral or shear stiffness of each section between floors is k lb./in. The equation of motion for the nth mass is then

$$m\ddot{x}_n = k(x_{n+1} - x_n) - k(x_n - x_{n-1}) \qquad (7.10\text{-}1)$$

which for harmonic motion can be represented in terms of the amplitudes as

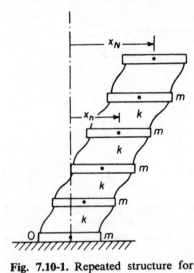

$$X_{n+1} - 2\left(1 - \frac{\omega^2 m}{2k}\right)X_n + X_{n-1} = 0$$
$$(7.10\text{-}2)$$

The solution to this equation is found by substituting

$$X_n = e^{i\beta n} \qquad (7.10\text{-}3)$$

which leads to the relationship

$$\left(1 - \frac{\omega^2 m}{2k}\right) = \frac{e^{i\beta} + e^{-i\beta}}{2} = \cos\beta$$

$$\frac{\omega^2 m}{k} = 2(1 - \cos\beta) = 4\sin^2\frac{\beta}{2}$$
$$(7.10\text{-}4)$$

The general solution for X_n is

$$X_n = A\cos\beta n + B\sin\beta n \qquad (7.10\text{-}5)$$

Fig. 7.10-1. Repeated structure for difference equation analysis.

where A and B are evaluated from the boundary conditions. At the base, $n = 0$, the amplitude $X_0 = 0$, so that $A = 0$. At the top story, $n = N$, the equation of motion is

$$m\ddot{x}_N = -k(x_N - x_{N-1})$$

which, in terms of the amplitude, becomes

$$X_{N-1} = \left(1 - \frac{\omega^2 m}{k}\right)X_N \qquad (7.10\text{-}6)$$

Substituting from the general solution, we obtain the following relationship for the evaluation of β:

$$\sin\beta(N - 1) = [1 - 2(1 - \cos\beta)]\sin\beta N$$

This result can be reduced to the product form

$$2\cos\beta(N + \tfrac{1}{2})\sin\frac{\beta}{2} = 0 \qquad (7.10\text{-}7)$$

which is satisfied by

$$\sin \frac{\beta}{2} = 0, \quad \frac{\beta}{2} = 0, \pi, 2\pi, 3\pi \cdots$$

$$\cos \beta(N + \tfrac{1}{2}) = 0, \quad \frac{\beta}{2} = \frac{\pi}{2(2N + 1)}, \frac{3\pi}{2(2N + 1)}, \frac{5\pi}{2(2N + 1)}, \cdots$$

(7.10-8)

The natural frequencies are then available from Eq. 7.10-4 as

$$\omega = 2\sqrt{k/m} \sin \frac{\beta}{2} \qquad (7.10\text{-}9)$$

which lead to

$$\omega_1 = 2\sqrt{\frac{k}{m}} \sin \frac{\pi}{2(2N + 1)}$$

$$\omega_2 = 2\sqrt{\frac{k}{m}} \sin \frac{3\pi}{2(2N + 1)}$$

$$\vdots$$

$$\omega_N = 2\sqrt{\frac{k}{m}} \sin \frac{(2N - 1)\pi}{2(2N + 1)}$$

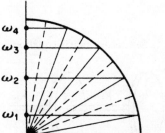

Fig. 7.10-2. Natural frequencies of a repeated structure with $N = 4$.

Figure 7.10-2 shows a graphical representation of these natural frequencies when $N = 4$.

The method of difference equation presented here is applicable to many other dynamical systems where repeating sections are present. The natural frequencies are always given by Eq. (7.10-9); however, the quantity β must be established for each problem from its boundary conditions.

REFERENCES

1. Dunkerley, S., "On the Whirling and Vibration of Shafts." *Phil. Trans. Roy. Soc.* A. vol. 185 (1895) pp. 269–360.

2. Frazer, R. A., W. J. Duncan, and A. R. Collar, *Elementary Matrices.* (London: Cambridge Univ. Press, 1963.)

3. Pestel, E. C., and F. A. Leckie, *Matrix Methods in Elastomechanics.* (New York: McGraw-Hill Book Company, 1963.)

4. Pipes, L. A., *Applied Mathematics for Engineers and Physicists*, 2nd Ed. (New York: McGraw-Hill Book Company, 1958.)

5. Prohl, M. A., "A General Method for Calculating Critical Speeds of Flexible Rotors," *Trans ASME* A-142 (Sept., 1945).

6. Myklestad, N. O., "A New Method of Calculating Natural Modes of Uncoupled Bending Vibration of Airplane Wings and Other Types of Beams," *Jour. Aero. Sci.*, pp. 153–162 (Apr., 1944).

7. Timoshenko, S., and G. H. MacCullough, *Elements of Strength of Materials*, 2nd Ed. (Princeton: D. Van Nostrand Co., Inc., 1940.)

8. Thomson, W. T., "Matrix Solution for the Vibration of Nonuniform Beams," *J. Appl'd. Mech.*, pp. 337–339 (Sept., 1950).

PROBLEMS

1. Set up the matrix equation for the system shown in Fig. 7-1, in the form

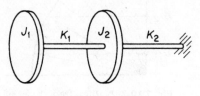

$$\{\theta\} = \omega^2[a][J]\{\theta\}$$

Fig. 7-1.

2. Determine the influence co-efficients for the spring-mass system of Fig. 7-2.

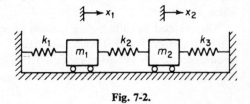

Fig. 7-2.

3. Write the kinetic and potential energy expressions for the system of Prob. 2, Fig. 7-2, when

$$k_1 = k, \qquad m_1 = m,$$
$$k_2 = 3k, \qquad m_2 = 2m$$
$$k_3 = 2k,$$

and determine the equation for ω^2 by equating the two energies. Letting $x_2/x_1 = n$, plot ω^2 versus n. Pick off the maximum and minimum values of ω^2 and the corresponding values of n, and show that they represent the two natural modes of the system.

4. Determine the influence coefficients for the two-mass cantilever beam of Fig. 7-3, and write its equation of motion in matrix form.

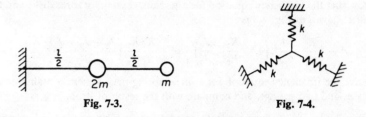

Fig. 7-3.   Fig. 7-4.

5. Three equal springs of stiffness k lb./in. are joined at one end, the other ends being arranged symmetrically at 120° from each other, as shown in Fig. 7-4. Show that the influence coefficients of the junction in a direction making an angle θ with any spring is independent of θ and equal to $1/1.5k$.

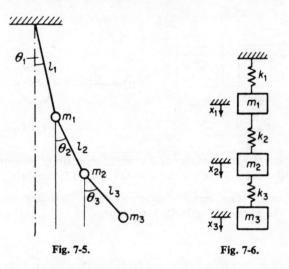

Fig. 7-5.   Fig. 7-6.

6. Determine the influence coefficients for the triple pendulum and the spring-mass system of Figs. 7-5 and 7-6.

7. Determine the influence coefficients for the deflection and slope of the shaft shown in Fig. 7-7, which is fixed at one end and terminating with a disk, with moment of inertia J about its diameter, at the other end.

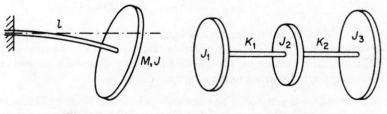

Fig. 7-7.   Fig. 7-8.

8. Show that the frequency equation for a torsional system of three disks and two shafts, shown in Fig. 7-8 is

$$\omega^4 - \left[\frac{K_1}{J_1} + \frac{K_2}{J_2}\left(1 + \frac{K_1}{K_2} + \frac{J_2}{J_3}\right)\right]\omega^2 + \frac{K_1 K_2}{J_1 J_2}\left(\frac{J_1 + J_2 + J_3}{J_3}\right) = 0$$

9. Derive the frequency equation for a linear spring-mass system containing three masses and two springs, and compare with the result of Prob. 8.

10. Figure 7-9 represents an equivalent torsional system of a propeller, radial engine, and supercharger. Determine the two natural frequencies by using matrix formulation.

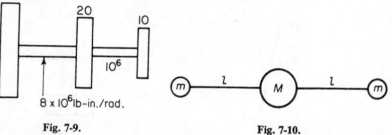

100 lb-in.-sec^2

20

10

10^6

8 x 10^6 lb-in./rad.

Fig. 7-9.

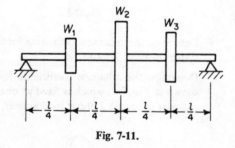

Fig. 7-10.

11. Determine the natural modes of the simplified model of an airplane shown in Fig. 7-10 where $M/m = n$ and the beam of length l is uniform.

12. Determine the fundamental frequency of the three-mass cantilever beam of Fig. 7.2.3, using Dunkerley's equation.

13. Using Dunkerley's equation, determine the fundamental frequency of the beam shown in Fig. 7-11 for

$$W_1 = W, \quad W_2 = 4W, \quad W_3 = 2W$$

W_1　W_2　W_3

$\frac{l}{4}$ | $\frac{l}{4}$ | $\frac{l}{4}$ | $\frac{l}{4}$

Fig. 7-11.

14. A load of 1000 lb. at the wing tip of a fighter plane produced a corresponding deflection of 0.78 in. If the fundamental bending frequency of the same wing is 622 c.p.m., approximate the new bending frequency when a 320-lb. fuel tank (including fuel) is attached to the wing tip.

15. A given beam was vibrated by an eccentric weight shaker weighing 12 lb. at the mid-span, and resonance was found at 435 c.p.s. With an additional weight of

10 lb., the resonant frequency was lowered to 398 c.p.s. Determine the natural frequency of the beam.

16. Determine the two natural modes of the system shown in Fig. 7-9 and show that they are orthogonal.

17. Determine the normal modes of the cantilever beam of Fig. 7-3 and verify their orthogonality.

18. For the system shown in Fig. 7-6, let

$$k_1 = 3k, \quad m_1 = 4m$$

$$k_2 = k, \quad m_2 = 2m$$

$$k_3 = k, \quad m_3 = m$$

Set up the matrix equation and determine the three principal modes by iteration. Check the orthogonality of the modes found.

19. Using Dunkerley's equation, calculate the fundamental frequency for Prob. 18 and compare with the results of matrix iteration.

20. Determine the three principal modes of the beam shown in Fig. 7-11 when $W_1 = W_2 = W_3$. Check the fundamental frequency by Dunkerley's equation.

21. Show that Dunkerley's equation always results in a fundamental frequency that is lower than the true value.

22. Using Holzer's method in matrix form, determine the first two natural frequencies and normal modes of the torsional system shown in Fig. 7-12 with the following values of J and K:

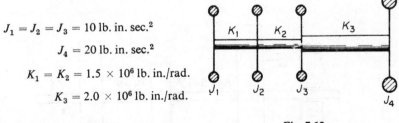

$$J_1 = J_2 = J_3 = 10 \text{ lb. in. sec.}^2$$

$$J_4 = 20 \text{ lb. in. sec.}^2$$

$$K_1 = K_2 = 1.5 \times 10^6 \text{ lb. in./rad.}$$

$$K_3 = 2.0 \times 10^6 \text{ lb. in./rad.}$$

Fig. 7-12.

23. A fighter-plane wing is reduced to a series of disks and shafts for Holzer's analysis as shown in Fig. 7-13. Determine the first two natural frequencies for symmetric and antisymmetric torsional oscillations of the wings, and plot the torsional mode corresponding to each.

n	J lb. in. sec.^{2}	K lb. in./rad.
1	50	15×10^6
2	138	30
3	145	22
4	181	36
5	260	120
6	$\frac{1}{2} \times 140,000$	

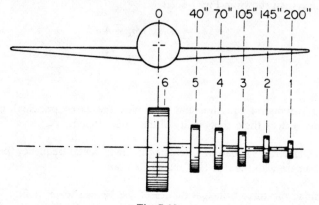

Fig. 7-13.

24. If a harmonic torque of 10,000 in. lb. at $\omega = 150$ rad./sec. is applied to disk 3 of the system shown in Fig. 7.7-11, determine the amplitude and phase of each disk.

25. Figure 7-14 represents a torsional system with a torsional damper. Determine the torque-frequency curve for the system.

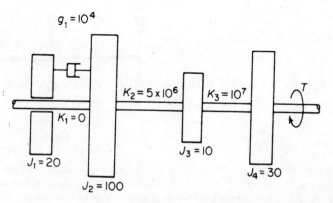

Fig. 7-14.

26. For the system of Example 7.7-3, Fig. 7.7-14, determine the amplitude and phase of each disk at $\omega^2 = 600$ when an impressed torque of 0.040×10^6 in. lb. is impressed on disk 4.

27. Figure 7-15 represents a linear system with damping between mass 1 and 2. Carry out a complex tabular analysis for numerical values assigned by the instructor, and determine the amplitude and phase of each mass at a specified frequency.

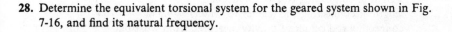

Fig. 7-15.

28. Determine the equivalent torsional system for the geared system shown in Fig. 7-16, and find its natural frequency.

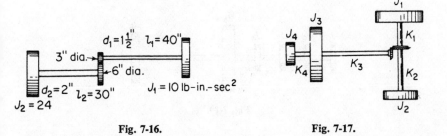

Fig. 7-16.   Fig. 7-17.

29. If the small and large gears of the system shown in Fig. 7-16 have the following inertias, $J' = 2, J'' = 6$, determine the equivalent single-shaft system and establish the natural frequencies.

30. Determine the two lowest natural frequencies of the torsional system shown in

Fig. 7-17 for the following values of J, K, and n:

$J_1 = 15$ lb. in. sec.^{2} $K_1 = 2 \times 10^6$ lb. in./rad.

$J_2 = 10$ lb. in. sec.^{2} $K_2 = 1.6 \times 10^6$ lb. in./rad.

$J_3 = 18$ lb. in. sec.^{2} $K_3 = 1 \times 10^6$ lb. in./rad.

$J_4 = 6$ lb. in. sec.^{2} $K_4 = 4 \times 10^6$ lb. in./rad.

Speed ratio of drive shaft to axle = 4 to 1.

What are the amplitude ratios of J_2 to J_1 at the natural frequencies?

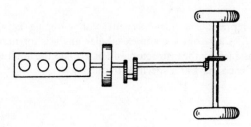

Fig. 7-18.

31. Determine the equivalent disk inertia $J_1 = J_2 = J_3 = J_4$ of the 4-cylinder engine shown in Fig. 7-18 with the following information:
    piston weight = 2.0 lb. each
    stroke = 4.2 in.
    weight of connecting rod = 3.4 lb. each
    length of connecting rod (center to center) = 9.5 in.
    connecting rod center of gravity measured from wrist pin = 6.5 in.
    Assume crankshaft to be dynamically balanced.

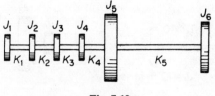

Fig. 7-19.

32. Reduce the system of Fig. 7-18 to the equivalent torsional system of Fig. 7-19. The necessary information is given as follows:
    J of each wheel = 9.2 lb. in. sec.^{2}
    J of flywheel = 12.3 lb. in. sec.^{2}
    transmission speed ratio (drive shaft to engine speed) = 1.0 to 3.0
    differential speed ratio (axle to drive shaft) = 1.0 to 3.5

axle dimensions $= 1\frac{1}{4}$ in. diameter, 25 in. long (each)
drive-shaft dimensions $= 1\frac{1}{2}$ in. diameter, 74 in. long
stiffness of crankshaft between cylinders measured experimentally $= 6.1 \times 10^6$ lb. in./rad.
stiffness of crankshaft between cylinder 4 and flywheel $= 4.5 \times 10^6$ lb. in./rad.

33. Determine the equations of motion for the torsional system of Fig. 7-20 and arrange them into the matrix iteration form. Solve for the principal modes of oscillation.

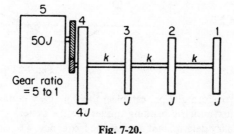

Fig. 7-20.

34. Apply the matrix method of Sec. 7.9(a) to a cantilever beam of length l and mass m at the end, and show that the natural frequency equation is directly obtained.

35. Apply the matrix method to a cantilever beam with two equal masses spaced equally a distance l. Show that the boundary conditions of zero slope and deflection lead to the equation

$$\theta_1 = \frac{\frac{1}{2}m\omega^2 lK(5 + \frac{1}{6}m\omega^2 l^2 K)}{1 + \frac{1}{2}l^2 Km\omega^2} = \frac{1 + \frac{3}{2}m\omega^2 l^2 K + (\frac{1}{6}m\omega^2 l^2 K)^2}{2l + \frac{1}{6}m\omega^2 l^3 K}$$

where $K = l/EI$.
Obtain the frequency equation from the above relationship and determine the two natural frequencies.

36. Solve Prob. 34 by the method of Sec. 7.9(b) when the beam is rotated about an axis through the fixed end with an angular speed Ω.

37. Determine the natural frequencies of the cantilever beam of Example 7.2-1 by the method of Sec. 7.9(a).

38. From the boundary equation, Eq. (7.9-7), establish the boundary determinant $D(\omega)$ for a simply supported beam.

39. Determine the boundary determinant $D(\omega)$ for a clamped-clamped beam.

40. Determine the boundary determinant $D(\omega)$ for a clamped-hinged beam.

41. Determine the boundary determinant $D(\omega)$ for a hinged-free beam.

42. A rotating beam, such as a helicopter blade, is sometimes considered as pinned at the hub. Establish the boundary determinant $D(\omega)$ for such a case.

43. Assume a helicopter blade to be represented by three lumped masses equally spaced, with the hub end clamped. On the basis of constant bending stiffness determine the natural frequencies for rotational speed Ω.

44. Determine the flexure-torsion vibration for the system of Fig. 7-21.

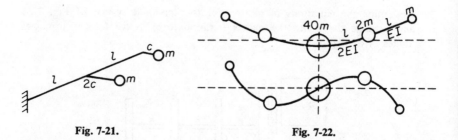

Fig. 7-21.   Fig. 7-22.

45. Using the matrix formulation, establish the boundary conditions for the symmetric and antisymmetric bending modes for the system shown in Fig. 7-22. Plot the boundary determinant against the frequency ω to establish the natural frequencies, and draw the first two mode shapes.

46. Set up the difference equations for the torsional system of Fig. 7-23. Determine the boundary equations and solve for the natural frequencies.

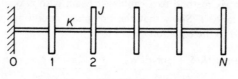

Fig. 7-23.

47. Set up the difference equations for N equal masses on a string with tension T, which is shown in Fig. 7-24. Determine the boundary equations and the natural frequencies.

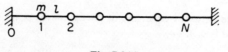

Fig. 7-24.

48. Write the difference equations for the spring-mass system of Fig. 7-25 and find the natural frequencies of the system.

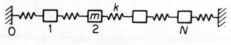

Fig. 7-25.

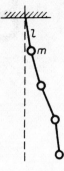

49. An N-mass pendulum is shown in Fig. 7-26. Determine the difference equations, boundary conditions, and the natural frequencies.

Fig. 7-26.

50. If the left end of the system of Fig. 7-23 is connected to a heavy flywheel, as shown in Fig. 7-27, show that the boundary conditions lead to the equation

$$(-\sin N\beta \cos \beta + \sin N\beta)\left(1 + 4\frac{K}{K_a}\frac{J_a}{J}\sin^2\frac{\beta}{2}\right) = -2\frac{J_a}{J}\sin^2\frac{\beta}{2}\sin \beta \cos N\beta$$

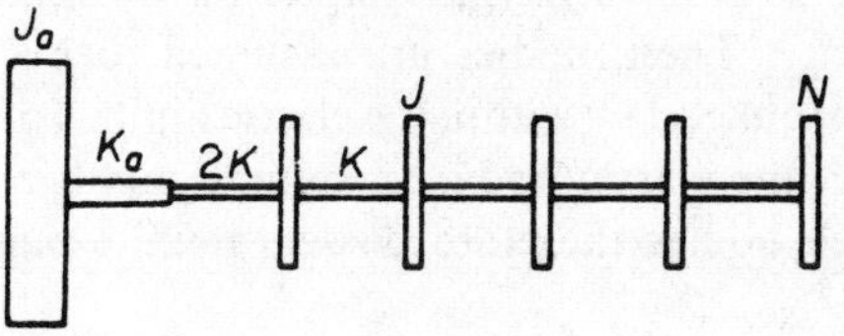

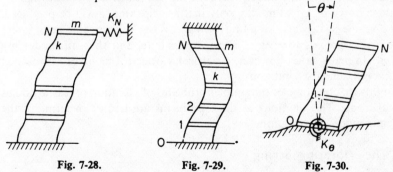

Fig. 7-27.

51. If the top story of a building is restrained by a spring of stiffness K_N, as shown in Fig. 7-28, determine the natural frequencies of the N-story building.

52. A ladder-type structure is fixed at both ends, as shown in Fig. 7-29. Determine the natural frequencies.

Fig. 7-28.   Fig. 7-29.   Fig. 7-30.

53. If the base of an N-story building is allowed to rotate against a resisting spring K_0, as shown in Fig. 7-30, determine the boundary equations and the natural frequencies.

8

Continuous Systems

8.1 Introductory

The systems to be studied in this chapter have continuously distributed masses and elasticity. These bodies are assumed to be homogeneous and isotropic, obeying Hooke's law within the elastic limit. To specify the position of every particle in the elastic body, an infinite number of coordinates are necessary, and such bodies therefore possess infinite number of degrees of freedom.

In general, the free vibration of these bodies is the sum of the principal modes as previously stated in Chapter 6. For the principal mode of vibration every particle of the body performs simple harmonic motion at the frequency corresponding to the particular root of the frequency equation, each particle passing simultaneously through its respective equilibrium position. If the elastic curve of the body under which the motion is started coincides exactly with one of the principal modes, only that principal mode will be produced. However, the elastic curve resulting from a blow or a sudden removal of forces seldom corresponds to that of a principal mode, and thus all modes are excited. In many cases, however, a particular principal mode can be excited by proper boundary conditions.

In this chapter some of the simpler problems of vibration of elastic bodies are taken up. The solutions to these problems are treated in terms of the principal modes of vibration.

8.2 The Vibrating String

A flexible string of mass ρ per unit length is stretched under tension T. By assuming the lateral deflection y of the string to be small, the change in tension with deflection is negligible and can be ignored.

In Fig. 8.2-1, a free-body diagram of an elementary length dx of the string is shown. Assuming small deflections and slopes, the equation of motion in the y-direction is

$$T\left(\theta + \frac{\partial\theta}{\partial x}\,dx\right) - T\theta = \rho\,dx\,\frac{\partial^2 y}{\partial t^2}$$

or

$$\frac{\partial\theta}{\partial x} = \frac{\rho}{T}\frac{\partial^2 y}{\partial t^2} \tag{8.2-1}$$

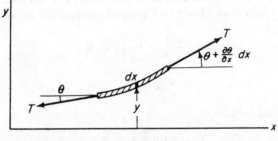

Fig. 8.2-1. String element in lateral vibration.

Since the slope of the string is $\theta = \partial y/\partial x$, the above equation reduces to

$$\frac{\partial^2 y}{\partial x^2} = \frac{1}{c^2}\frac{\partial^2 y}{\partial t^2} \tag{8.2-2}$$

where $c = \sqrt{T/\rho}$ can be shown to be the velocity of wave propagation along the string.

The general solution of Eq. (8.2-2) can be expressed in the form

$$y = F_1(ct - x) + F_2(ct + x) \tag{8.2-3}$$

where F_1 and F_2 are arbitrary functions. Regardless of the type of function F, the argument $(ct \pm x)$ upon differentiation leads to the equation

$$\frac{\partial^2 F}{\partial x^2} = \frac{1}{c^2}\frac{\partial^2 F}{\partial t^2} \tag{8.2-4}$$

and hence the differential equation is satisfied.

Considering the component $y = F_1(ct - x)$, its value is determined by the argument $(ct - x)$ and hence by a range of values of t and x. For example, if $c = 10$, the equation for $y = F_1(100)$ is satisfied by $t = 0$, $x = -100$; $t = 1, x = -90; t = 2, x = -80$, etc. Therefore, the wave profile moves in the positive x-direction with speed c. In a similar manner we can show that

$F_2(ct + x)$ represents a wave moving toward the negative x-direction with speed c. We therefore refer to c as the velocity of wave propagation.

We can next examine a solution of the form

$$y(x, t) = Y(x)G(t) \tag{8.2-5}$$

By substitution into Eq. (8.2-2) we obtain

$$\frac{1}{Y}\frac{d^2Y}{dx^2} = \frac{1}{c^2}\frac{1}{G}\frac{d^2G}{dt^2} \tag{8.2-6}$$

Since the left side of this equation is independent of t, whereas the right side is independent of x, it follows that each side must be a constant. Letting this constant be $-(\omega/c)^2$ we obtain two ordinary differential equations:

$$\frac{d^2Y}{dx^2} + \left(\frac{\omega}{c}\right)^2 Y = 0 \tag{8.2-7}$$

$$\frac{d^2G}{dt^2} + \omega^2 G = 0 \tag{8.2-8}$$

with the general solutions:

$$Y = A \sin\frac{\omega}{c}x + B \cos\frac{\omega}{c}x \tag{8.2-9}$$

$$G = C \sin \omega t + D \cos \omega t \tag{8.2-10}$$

The arbitrary constants A, B, C, D depend on the boundary conditions and the initial conditions. For example, if the string is stretched between two fixed points with distance l between them, the boundary conditions are $y(0, t) = y(l, t) = 0$. The condition that $y(0, t) = 0$ will require that $B = 0$ so that the solution will appear as

$$y = (C \sin \omega t + D \cos \omega t) \sin\frac{\omega}{c}x \tag{8.2-11}$$

The condition $y(l, t) = 0$ then leads to the equation

$$\sin\frac{\omega l}{c} = 0$$

or

$$\frac{\omega_n l}{c} = \frac{2\pi l}{\lambda} = n\pi, \qquad n = 1, 2, 3 \cdots$$

and $\lambda = c/f$ is the wavelength and f is the frequency of oscillation. Each n represents a normal-mode vibration with natural frequency determined from the equation

$$f_n = \frac{n}{2l}c = \frac{n}{2l}\sqrt{\frac{T}{\rho}}, \qquad n = 1, 2, 3, \cdots \tag{8.2-12}$$

The mode shape is sinusoidal with the distribution

$$Y = \sin n\pi \frac{x}{l} \qquad (8.2\text{-}13)$$

In the more general case of free vibration initiated in any manner, the solution will contain many of the normal modes and the equation for the displacement may be written as

$$y(x, t) = \sum_{n=1}^{\infty} (C_n \sin \omega_n t + D_n \cos \omega_n t) \sin \frac{n\pi x}{l} \qquad (8.2\text{-}14)$$

$$\omega_n = \frac{n\pi c}{l}$$

Fitting this equation to the initial conditions of $y(x, 0)$ and $\dot{y}(x, 0)$, the C_n and D_n can be evaluated.

EXAMPLE 8.2-1. A uniform string of length l is fixed at the ends and stretched under tension T. If the string is displaced into an arbitrary shape $y(x, 0)$ and released, determine C_n and D_n of Eq. (8.2-14).

At $t = 0$, the displacement and velocity are

$$y(x, 0) = \sum_{n=1}^{\infty} D_n \sin \frac{n\pi x}{l}$$

$$\dot{y}(x, 0) = \sum_{n=1}^{\infty} \omega_n C_n \sin \frac{n\pi x}{l} = 0$$

Multiplying each equation by $\sin k\pi x/l$ and integrating from $x = 0$ to $x = l$ all of the terms on the right side will be zero, except the term $n = k$. Thus we arrive at the result

$$D_k = \frac{2}{l} \int_0^l y(x, 0) \sin \frac{k\pi x}{l} \, dx$$

$$C_k = 0 \qquad\qquad k = 1, 2, 3, \cdots$$

8.3 Longitudinal Vibration of Rods

The rod considered in this section is assumed to be thin and uniform along its length. Due to axial forces there will be displacements u along the rod which will be a function of both the position x and the time t. Since the rod has an infinite number of natural modes of vibration, the distribution of the displacement will differ with each mode.

Let us consider an element of this rod of length dx (Fig. 8.3-1). If u is the displacement at x, the displacement at $x + dx$ will be $u + (\partial u/\partial x)\, dx$. It is evident then that the element dx in the new position has changed in length by an amount $(\partial u/\partial x)\, dx$, and thus the unit strain is $\partial u/\partial x$. Since from Hooke's

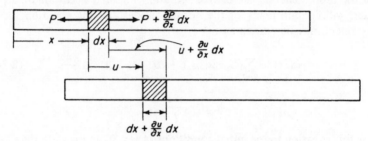

Fig. 8.3-1. Displacement of rod element.

law the ratio of unit stress to unit strain is equal to the modulus of elasticity E, we can write

$$\frac{\partial u}{\partial x} = \frac{P}{AE} \tag{8.3-1}$$

where A is the cross-sectional area of the rod. Differentiating with respect to x,

$$AE \frac{\partial^2 u}{\partial x^2} = \frac{\partial P}{\partial x} \tag{8.3-2}$$

We now apply Newton's law of motion for the element and equate the unbalanced force to the product of the mass and acceleration of the element

$$\frac{\partial P}{\partial x}\, dx = \rho \frac{A\, dx}{g} \frac{\partial^2 u}{\partial t^2} \tag{8.3-3}$$

where ρ is the density of the rod in pounds per unit volume. Eliminating $\partial P/\partial x$ between Eqs. (8.3-2) and (8.3-3), we obtain the partial differential equation

$$\frac{\partial^2 u}{\partial t^2} = \left(\frac{Eg}{\rho}\right) \frac{\partial^2 u}{\partial x^2} \tag{8.3-4}$$

or

$$\frac{\partial^2 u}{\partial x^2} = \frac{1}{c^2} \frac{\partial^2 u}{\partial t^2} \tag{8.3-5}$$

which is similar to that of Eq. (8.2-2) for the string. The velocity of propagation of the displacement or stress wave in the rod is then equal to

$$c = \sqrt{Eg/\rho} \tag{8.3-6}$$

and a solution of the form

$$u(x, t) = U(x)G(t) \tag{8.3-7}$$

will result in two ordinary differential equations similar to Eqs. (8.2-7) and (8.2-8), with

$$U(x) = A \sin \frac{\omega}{c} x + B \cos \frac{\omega}{c} x \tag{8.3-8}$$

$$G(t) = C \sin \omega t + D \cos \omega t \tag{8.3-9}$$

EXAMPLE 8.3-1. Determine the natural frequencies and mode shapes of a free-free rod (a rod with both ends free).

Solution. For such a bar, the stress at the ends must be zero. Since the stress is given by the equation $E \, \partial u / \partial x$, the unit strain at the ends must also be zero; that is,

$$\frac{\partial u}{\partial x} = 0 \quad \text{at} \quad x = 0, \quad \text{and} \quad x = l$$

The two equations corresponding to the above boundary conditions, are therefore,

$$\left(\frac{\partial u}{\partial x} \right)_{x=0} = A \frac{\omega}{c} (C \sin \omega t + D \cos \omega t) = 0$$

$$\left(\frac{\partial u}{\partial x} \right)_{x=l} = \frac{\omega}{c} \left(A \cos \frac{\omega l}{c} - B \sin \frac{\omega l}{c} \right) (C \sin \omega t + D \cos \omega t) = 0$$

Since these equations must be true for any time t, A must be equal to zero from the first equation. Since B must be finite in order to have vibration, the second equation is satisfied when

$$\sin \frac{\omega l}{c} = 0$$

or

$$\frac{\omega_n l}{c} = \omega_n l \sqrt{\rho/Eg} = \pi, \quad 2\pi, \quad 3\pi, \quad \cdots, \quad n\pi$$

The frequency of vibration is thus given by

$$\omega_n = \frac{n\pi}{l} \sqrt{\frac{Eg}{\rho}}, \qquad f_n = \frac{n}{2l} \sqrt{\frac{Eg}{\rho}}$$

where n represents the order of the mode. The solution of the free-free rod with zero initial displacement can then be written as

$$u = u_0 \cos \frac{n\pi}{l} x \sin \frac{n\pi}{l} \sqrt{\frac{Eg}{\rho}} t$$

The amplitude of the longitudinal vibration along the rod is therefore a cosine wave having n nodes.

8.4 Torsional Vibration of Rods

The equation of motion of a rod in torsional vibration is similar to that of longitudinal vibration of rods discussed in the preceding section.

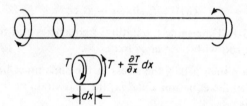

Fig. 8.4-1. Torque acting on an element dx.

Letting x be measured along the length of the rod, the angle of twist in any length dx of the rod due to a torque T is

$$d\theta = \frac{T\,dx}{I_P G} \qquad (8.4\text{-}1)$$

where $I_P G$ is the torsional stiffness given by the product of the polar moment of inertia I_P of the cross-sectional area and the shear modulus of elasticity G. The torque on the two faces of the element being T and $T + (\partial T/\partial x)\,dx$, as shown in Fig. 8.4-1, the net torque from Eq. (8.4-1) becomes

$$\frac{\partial T}{\partial x}\,dx = I_P G\,\frac{\partial^2 \theta}{\partial x^2}\,dx \qquad (8.4\text{-}2)$$

Equating this torque to the product of the mass moment of inertia $(\rho/g)I_P\,dx$ of the element and the angular acceleration $\partial^2\theta/\partial t^2$, where ρ is the density of the rod in pounds per unit volume, the differential equation of motion becomes

$$\frac{\rho}{g} I_P\,dx\,\frac{\partial^2 \theta}{\partial t^2} = I_P G\,\frac{\partial^2 \theta}{\partial x^2}\,dx, \qquad \frac{\partial^2 \theta}{\partial t^2} = \left(\frac{Gg}{\rho}\right)\frac{\partial^2 \theta}{\partial x^2} \qquad (8.4\text{-}3)$$

This equation is of the same form as that of longitudinal vibration of rods where θ and Gg/ρ replace u and Eg/ρ, respectively. The general solution may hence be written immediately by comparison as

$$\theta = \left(A \sin \omega \sqrt{\frac{\rho}{Gg}}\,x + B \cos \omega \sqrt{\frac{\rho}{Gg}}\,x\right)(C \sin \omega t + D \cos \omega t) \qquad (8.4\text{-}4)$$

EXAMPLE 8.4-1. Determine the equation for the natural frequencies of a uniform rod in torsional oscillation with one end fixed and the other end free, as in Fig. 8.4-2.

Solution. Starting with equation

$$\theta = (A \sin \omega \sqrt{\rho/Gg}\, x + B \cos \omega \sqrt{\rho/Gg}\, x) \sin \omega t$$

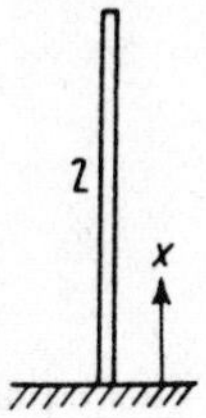

apply the boundary conditions, which are
(1) when $x = 0$, $\theta = 0$,
(2) when $x = l$, torque $= 0$, or

$$\frac{\partial \theta}{\partial x} = 0$$

Fig. 8.4-2.

Boundary condition (1) results in $B = 0$.
Boundary condition (2) results in the equation

$$\cos \omega \sqrt{\rho/Gg}\, l = 0$$

which is satisfied by the following angles:

$$\omega_n \sqrt{\frac{\rho}{Gg}}\, l = \frac{\pi}{2}, \frac{3\pi}{2}, \frac{5\pi}{2}, \cdots, \left(n + \frac{1}{2}\right)\pi$$

The natural frequencies of the bar are hence determined by the equation

$$\omega_n = \left(n + \frac{1}{2}\right)\frac{\pi}{l} \sqrt{\frac{Gg}{\rho}}$$

where $n = 0, 1, 2, 3, \cdots$

EXAMPLE 8.4-2. The drill pipe of an oil well terminates at the lower end to a rod containing a cutting bit. Derive the expression for the natural frequencies, assuming the drill pipe to be uniform and fixed at the upper end and the rod and cutter to be represented by an end mass of moment of inertia J_0, as shown in Fig. 8.4-3.

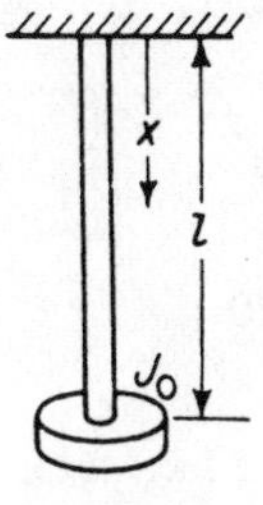

Fig. 8.4-3.

Solution. The boundary condition at the upper end is $x = 0$, $\theta = 0$, which requires B to be zero in Eq. (8.4-4).

For the lower end, the boundary condition requires the shaft torque to be equal to the inertia torque $-J_0(\partial^2\theta/\partial t^2)$; that is, $x = l$, torque $= J_0\omega^2\theta_{x=l}$. Since the torque is $GI_P(d\theta/dx)$, we obtain the equation

$$GI_P\left(\frac{d\theta}{dx}\right)_{x=l} = J_0\omega^2(\theta)_{x=l}$$

Substituting from Eq. (8.4-4) with $B = 0$.

$$GI_P\omega \sqrt{\frac{\rho}{Gg}} \cos \omega \sqrt{\frac{\rho}{Gg}} \, l = J_0\omega^2 \sin \omega \sqrt{\frac{\rho}{Gg}} \, l$$

$$\tan \omega l \sqrt{\frac{\rho}{Gg}} = \frac{I_P}{\omega J_0} \sqrt{\frac{G\rho}{g}} = \frac{I_P\rho l}{gJ_0\omega l} \sqrt{\frac{Gg}{\rho}}$$

$$= \frac{J_{\text{rod}}}{J_0\omega l} \sqrt{\frac{Gg}{\rho}}$$

This equation is of the form

$$\beta \tan \beta = \frac{J_{\text{rod}}}{J_0}, \qquad \beta = \omega l \sqrt{\frac{\rho}{Gg}}$$

which can be solved graphically or from tables.*

EXAMPLE 8.4-5. Using the frequency equation developed in the previous example, determine the first two natural frequencies of an oil-well drill pipe 5000 ft. long, fixed at the upper end and terminating at the lower end to a drill collar 120 ft. long. The average values for the drill pipe and drill collar are given as:

Drill pipe: outside diameter = $4\frac{1}{2}$ in.
        inside diameter   = 3.83 in.
        $I_P = 0.00094$ ft.^{4}
        $l = 5000$ ft.

$$J_{\text{rod}} = I_P \frac{\rho l}{g} = 0.00094 \times \frac{490}{32.2} \times 5000 = 71.4 \text{ lb. ft. sec.}^2$$

Drill collar: outside diameter = $7\frac{5}{8}$ in.
         inside diameter = 2.0 in.
         $J_0 = 0.244 \times 120$ ft. = 29.3 lb. ft. sec.^{2}

Solution. The equation to be solved is

$$\beta \tan \beta = \frac{J_{\text{rod}}}{J_0} = 2.44$$

From Table V, p. 32, Jahnke and Emde, $\beta = 1.135, 3.722, \cdots$

$$\beta = \omega l \sqrt{\frac{\rho}{Gg}} = 5000\omega \sqrt{\frac{490}{12 \times 10^6 \times 12^2 \times 32.2}} = 0.470\omega$$

* See Jahnke and Emde, Tables of Functions, 4th Ed. Dover Publications, Inc., 1945, Table V, p. 32.

Solving for ω, the first two natural frequencies are found to be

$$\omega_1 = \frac{1.135}{0.470} = 2.41 \text{ rad./sec.} = 0.384 \text{ c.p.s.}$$

$$\omega_2 = \frac{3.722}{0.470} = 7.93 \text{ rad./sec.} = 1.26 \text{ c.p.s.}$$

8.5 The Euler Equation for the Beam

To determine the differential equation for the lateral vibration of beams, consider the forces and moments acting on an element of the beam shown in Fig. 8.5-1.

V and M are shear and bending moments, respectively, and $p(x)$ represents the loading per unit length of the beam.

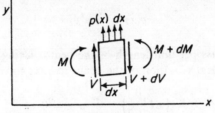

Fig. 8.5-1.

By summing forces in the vertical direction,

$$dV - p(x) \, dx = 0 \tag{8.5-1}$$

By summing moments about any point on the right face of the element

$$dM - V \, dx - \tfrac{1}{2}p(x)(dx)^2 = 0 \tag{8.5-2}$$

In the limiting process these equations result in the following important relationships:

$$\frac{dV}{dx} = p(x), \qquad \frac{dM}{dx} = V \tag{8.5-3}$$

The first part of Eq. (8.5-3) states that the rate of change of shear along the length of the beam is equal to the loading per unit length, and the second states that the rate of change of the moment along the beam is equal to the shear.

From Eq. (8.5-3) we obtain the following:

$$\frac{d^2M}{dx^2} = \frac{dV}{dx} = p(x) \tag{8.5-4}$$

The bending moment is related to the curvature by the flexure equation, which, for the coordinates indicated in Fig. 8.5-1, is

$$M = EI\frac{d^2y}{dx^2} \tag{8.5-5}$$

Substituting this relation into Eq. (8.5-4), we obtain

$$\frac{d^2}{dx^2}\left(EI\frac{d^2y}{dx^2}\right) = p(x) \tag{8.5-6}$$

For a beam vibrating under its own weight, the load per unit length is equal to the inertia load due to its mass and acceleration. Since the inertia force is in the same direction as $p(x)$ as shown in Fig. 8.5-1, we have, by assuming harmonic motion,

$$p(x) = \frac{w}{g}\omega^2 y \tag{8.5-7}$$

where w/g is the mass per unit length of the beam. Using this relation, the equation for the lateral vibration of the beam reduces to

$$\frac{d^2}{dx^2}\left(EI\frac{d^2y}{dx^2}\right) - \frac{w}{g}\omega^2 y = 0 \tag{8.5-8}$$

In the special case where the flexural rigidity EI is a constant, the above equation may be written as

$$EI\frac{d^4y}{dx^4} - \frac{w}{g}\omega^2 y = 0 \tag{8.5-9}$$

On substituting

$$n^4 = \frac{w}{g}\frac{\omega^2}{EI} \tag{8.5-10}$$

we obtain the fourth-order differential equation

$$\frac{d^4y}{dx^4} - n^4 y = 0 \tag{8.5-11}$$

for the motion of a uniform beam in lateral vibration.

The general solution of Eq. (8.5-11) can be shown to be

$$y = A\cosh nx + B\sinh nx + C\cos nx + D\sin nx \tag{8.5-12}$$

To arrive at this result, we assume a solution of the form

$$y = e^{ax}$$

which will satisfy the differential equation when

$$a = \pm n, \quad \text{and} \quad a = \pm in$$

Since

$$e^{\pm nx} = \cosh nx \pm \sinh nx$$

$$e^{\pm inx} = \cos nx \pm i \sin nx$$

the solution in the form of Eq. (8.5-12) is readily established.

The natural frequencies of vibration are found from Eq. (8.5-10) to be

$$\omega_n = n^2 \sqrt{gEI/w}$$

where the number n depends on the boundary conditions of the problem. The following table lists numerical values of $(nl)^2$ for typical end conditions.

Beam configuration	$(n_1l)^2$ Fundamental	$(n_2l)^2$ Second Mode	$(n_3l)^2$ Third Mode
Simply supported	9.87	39.5	88.9
Cantilever................	3.52	22.4	61.7
Free-free.................	22.4	61.7	121.0
Clamped-clamped	22.4	61.7	121.0
Clamped-hinged	15.4	50.0	104.0
Hinged-free	0	15.4	50.0

EXAMPLE 8.5-1. Determine the natural frequencies of vibration of a uniform beam clamped at one end and free at the other.

Solution: The boundary conditions are

$$\text{at } x = 0 \begin{cases} y = 0 \\ \dfrac{dy}{dx} = 0 \end{cases}$$

$$\text{at } x = l \begin{cases} M = 0 \quad \text{or} \quad \dfrac{d^2y}{dx^2} = 0 \\ V = 0 \quad \text{or} \quad \dfrac{d^3y}{dx^3} = 0 \end{cases}$$

Substituting these boundary conditions in the general solution, we obtain

$$(y)_{x=0} = A + C = 0, \quad \therefore A = -C$$

$$\left(\frac{dy}{dx}\right)_{x=0} = n[A \sinh nx + B \cosh nx - C \sin nx + D \cos nx]_{x=0} = 0$$

$$n[B + D] = 0, \quad \therefore B = -D$$

$$\left(\frac{d^2y}{dx^2}\right)_{x=l} = n^2[A \cosh nl + B \sinh nl - C \cos nl - D \sin nl] = 0$$

$$A(\cosh nl + \cos nl) + B(\sinh nl + \sin nl) = 0$$

$$\left(\frac{d^3y}{dx^3}\right)_{x=l} = n^3[A \sinh nl + B \cosh nl + C \sin nl - D \cos nl] = 0$$

$$A(\sinh nl - \sin nl) + B(\cosh nl + \cos nl) = 0$$

From the last two equations we obtain

$$\frac{\cosh nl + \cos nl}{\sinh nl - \sin nl} = \frac{\sinh nl + \sin nl}{\cosh nl + \cos nl}$$

which reduces to

$$\cosh nl \cos nl + 1 = 0$$

This last equation is satisfied by a number of values of nl, corresponding to each normal mode of oscillation, which for the first and second modes are 1.875 and 4.695, respectively. The natural frequency for the first mode is hence given as

$$\omega_1 = \frac{1.875^2}{l^2} \sqrt{\frac{gEI}{w}} = \frac{3.515}{l^2} \sqrt{\frac{gEI}{w}}$$

8.6 Effect of Rotary Inertia and Shear Deformation

The Timoshenko theory accounts for both the rotary inertia and shear deformation of the beam. The free-body diagram and the geometry for the beam element are shown in Fig. 8.6-1. If the shear deformation is zero, the center line of the beam element will coincide with the perpendicular to the face of the cross section. Due to shear, the rectangular element tends to go into a diamond shape without rotation of the face and the slope of the center

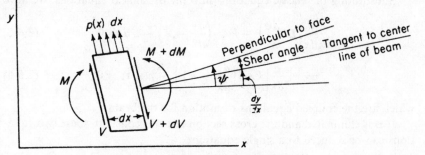

Fig. 8.6-1. Effect of shear deformation.

line is diminished by the shear angle $(\psi - dy/dx)$. The following quantities can then be defined:

$$y = \text{deflection of the center line of the beam}$$

$$\frac{dy}{dx} = \text{slope of the center line of the beam}$$

$$\psi = \text{slope due to bending}$$

$$\psi - \frac{dy}{dx} = \text{loss of slope, equal to the shear angle}$$

There are two elastic equations for the beam, which are

$$\psi - \frac{dy}{dx} = \frac{V}{kAG} \tag{8.6-1}$$

$$\frac{d\psi}{dx} = \frac{M}{EI} \tag{8.6-2}$$

where A is the cross-sectional area, G the shear modulus, k a factor depending on the shape of the cross section, and EI the bending stiffness. In addition, there are two dynamical equations:

$$\text{(moment)} \quad J\ddot{\psi} = \frac{dM}{dx} - V \tag{8.6-3}$$

$$\text{(force)} \quad m\ddot{y} = -\frac{dV}{dx} + p(x, t) \tag{8.6-4}$$

where J and m are the rotary inertia and mass of the beam per unit length.

Substituting the elastic equations into the dynamical equations, we have

$$\frac{d}{dx}\left(EI\frac{d\psi}{dx}\right) + kAG\left(\frac{dy}{dx} - \psi\right) - J\ddot{\psi} = 0 \qquad (8.6\text{-}5)$$

$$m\ddot{y} - \frac{d}{dx}\left[kAG\left(\frac{dy}{dx} - \psi\right)\right] - p(x, t) = 0 \qquad (8.6\text{-}6)$$

which are the coupled equations of motion for the beam.

If ψ is eliminated and the cross section remains constant, these two equations can be reduced to a single equation:

$$EI\frac{\partial^4 y}{\partial x^4} + m\frac{\partial^2 y}{\partial t^2} - \left(J + \frac{EIm}{kAG}\right)\frac{\partial^4 y}{\partial x^2\,\partial t^2} + \frac{Jm}{kAG}\frac{\partial^4 y}{\partial t^4} = p(x, t) \quad (8.6\text{-}7)$$

It is evident then that the Euler equation

$$EI\frac{\partial^4 y}{\partial x^4} + m\frac{\partial^2 y}{\partial t^2} = p(x, t)$$

is a special case of the general beam equation including the rotary inertia and the shear deformation.

8.7 Digital Computing Technique

Continuous systems with varying properties can be solved by means of the digital computer. The procedure is that of solving the several first order differential equations associated with the system, and satisfying the boundary conditions.

The computing technique may be illustrated in terms of the beam vibration problem of the previous section, where the effect of the shear deformation and rotary inertia are included. The four first-order equations for such a beam are:

$$\frac{d\psi}{dx} = \frac{M}{EI} \qquad\qquad = F(x, \psi, y, M, V)$$

$$\frac{dy}{dx} = \psi - \frac{V}{kAG} = G(x, \psi, y, M, V)$$

$$\frac{dM}{dx} = V - \omega^2 J\psi = H(x, \psi, y, M, V) \qquad (8.7\text{-}1)$$

$$\frac{dV}{dx} = \omega^2 my \qquad\quad = K(x, \psi, y, M, V)$$

where ψ, y, M, V, EI, kAG, m, and J are functions of x. The digital machine must proceed in computation from some point x_0 to a neighboring point x. For a point in the neighborhood of x_0, each of the dependent variables ψ, y, M, and V may be expanded in terms of the integration interval $h = (x - x_0)$ by the Taylor series:

$$\psi = \psi_0 + h\left(\frac{d\psi}{dx}\right)_0 + \tfrac{1}{2}h^2\left(\frac{d^2\psi}{dx^2}\right)_0 + \cdots$$

$$y = y_0 + h\left(\frac{dy}{dx}\right)_0 + \cdots \tag{8.7-2}$$

$$\cdot$$
$$\cdot$$
$$\cdot$$

It is also possible to write these equations in terms of a single average slope:

$$\psi = \psi_0 + h\left(\frac{d\psi}{dx}\right)_{av}$$

$$y = y_0 + h\left(\frac{dy}{dx}\right)_{av} \tag{8.7-3}$$

$$\cdot$$
$$\cdot$$
$$\cdot$$

where the average slope can be determined from Simpson's rule as

$$\left(\frac{d\psi}{dx}\right)_{av} = \frac{1}{6}\left\{\left(\frac{d\psi}{dx}\right)_{x_0} + 4\left(\frac{d\psi}{dx}\right)_{x_0+h/2} + \left(\frac{d\psi}{dx}\right)_{x_0+h}\right\} \tag{8.7-4}$$

In the Runga-Kutta procedure, the average slope is replaced by a somewhat different form where the middle term is split into two terms. The procedure is best illustrated in terms of a problem in two variables y and u, as follows.

Let the two first-order equations be

$$\frac{dy}{dx} = F(x, y, u)$$

$$\frac{du}{dx} = G(x, y, u) \tag{8.7-5}$$

With $y(x_0) = y_0$ and $u(x_0) = u_0$, y and u at $x_0 + h$ are expressed by the equations

$$y = y_0 + \tfrac{1}{6}(k_0 + 2k_1 + 2k_2 + k_3) + O(h^5)$$

$$u = u_0 + \tfrac{1}{6}(g_0 + 2g_1 + 2g_2 + g_3) + O(h^5)$$

$$(8.7\text{-}6)$$

where

$$k_0 = hF(x_0, y_0, u_0)$$

$$k_1 = hF\left(x_0 + \frac{h}{2}, y_0 + \tfrac{1}{2}k_0, u_0 + \tfrac{1}{2}g_0\right)$$

$$k_2 = hF\left(x_0 + \frac{h}{2}, y_0 + \tfrac{1}{2}k_1, u_0 + \tfrac{1}{2}g_1\right)$$

$$k_3 = hF(x_0 + h, y_0 + k_2, u_0 + g_2)$$

$$g_0 = hG(x_0, y_0, u_0)$$

$$(8.7\text{-}7)$$

$$g_1 = hG\left(x_0 + \frac{h}{2}, y_0 + \tfrac{1}{2}k_0, u_0 + \tfrac{1}{2}g_0\right)$$

$$g_2 = hG\left(x_0 + \frac{h}{2}, y_0 + \tfrac{1}{2}k_1, u_0 + \tfrac{1}{2}g_1\right)$$

$$g_3 = hG(x_0 + h, y_0 + k_2, u_0 + g_2)$$

Since $F(x_0, y_0, u_0)$ and $G(x_0, y_0, u_0)$ are slopes $(dy/dx)_0$ and $(du/dx)_0$ at x_0, the procedure is analogous to the average-slope method of Simpson.

Returning to the beam equations, the boundary conditions at the beginning end x_0 provide a starting point. For example, in the cantilever beam with origin at the fixed end, the boundary conditions at the starting end are

$$\psi_0 = 0, \qquad M_0 = M_0$$

$$y_0 = 0, \qquad V_0 = V_0$$

$$(8.7\text{-}8)$$

These can be considered to be the linear combination of two boundary vectors as follows,

$$\begin{Bmatrix} \psi_0 \\ y_0 \\ M_0 \\ V_0 \end{Bmatrix} = \begin{Bmatrix} 0 \\ 0 \\ 1 \\ 0 \end{Bmatrix} + \alpha \begin{Bmatrix} 0 \\ 0 \\ 0 \\ 1 \end{Bmatrix} = C_0 + \alpha D_0 \qquad (8.7\text{-}9)$$

Since the system is linear, we can start with each boundary vector separately. Starting with C_0, we obtain

$$C_N = \begin{Bmatrix} \psi_N \\ y_N \\ M_N \\ V_N \end{Bmatrix}_C$$

Starting with D_0, we obtain

$$\alpha D_N = \begin{Bmatrix} \psi_N \\ y_N \\ M_N \\ V_N \end{Bmatrix}_D$$

These must now add to satisfy the actual boundary conditions at the terminal end, which for a cantilever free end are:

$$\begin{Bmatrix} \psi \\ y \\ M \\ V \end{Bmatrix}_N = \begin{Bmatrix} \psi \\ y \\ 0 \\ 0 \end{Bmatrix} = C_N + \alpha D_N \tag{8.7-10}$$

If the frequency chosen is correct, the above boundary equations lead to

$$M_{NC} + \alpha M_{ND} = 0$$

$$V_{NC} + \alpha V_{ND} = 0$$

$$\alpha = -\frac{M_{NC}}{M_{ND}} = -\frac{V_{NC}}{V_{ND}}$$

which is satisfied by the determinant

$$\begin{vmatrix} M_{NC} & V_{NC} \\ M_{ND} & V_{ND} \end{vmatrix} = 0 \tag{8.7-11}$$

The iteration can be started with three different frequencies, which results in three values of the determinant. A parabola is passed through these three points and the zero of the curve is chosen for a new estimate of the frequency. When the frequency is close to the correct value, the new estimate may be made by a straight line between two values of the boundary determinant.

REFERENCES

1. Hildebrand, F., *Introduction to Numerical Analysis*, pp. 223–239. (New York: McGraw-Hill Book Company, 1956.)

2. Liebowitz, B. H., "Numerical Solutions to Differential Equations," *Electro-Technology* (April, 1962).

PROBLEMS

1. Find the wave velocity along a rope whose density is $\frac{1}{4}$ lb. per foot when stretched to a tension of 100 lb.

2. Derive the equation for the natural frequencies of a uniform cord of length l fixed at the two ends. The cord is stretched to a tension T and its mass per unit length is ρ.

3. A cord of length l and mass per unit length ρ is under tension T with the left end fixed and the right end attached to a spring-mass system as shown in Fig. 8-1. Determine the equation for the natural frequencies.

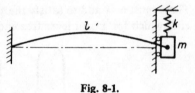

Fig. 8-1.

4. A harmonic vibration has an amplitude that varies as a cosine function along the x-direction such that

$$y = a \cos kx \cdot \sin \omega t$$

Show that if another harmonic vibration of same frequency and equal amplitude displaced in space phase and time phase by a quarter wave length is added to the first vibration, the resultant vibration will represent a traveling wave with a propagation velocity equal to $c = \omega/k$.

5. Find the velocity of longitudinal waves along a thin steel bar. The modulus of elasticity and weight per unit volume of steel are 29×10^6 lb./in.^{2} and 0.282 lb./in.^{3}

6. A uniform bar of length l is fixed at one end and free at the other end. Show that the frequencies of normal longitudinal vibrations are $f = (n + \frac{1}{2})c/2l$, where $c = \sqrt{Eg/\rho}$ is the velocity of longitudinal waves in the bar, and $n = 0, 1, 2, \cdots$.

7. A uniform rod of length l and cross-sectional area A is fixed at the upper end and is loaded with a weight W on the other end. Show that the natural frequencies are determined from the equation

$$\omega l \sqrt{\frac{\rho}{Eg}} \tan \omega l \sqrt{\frac{\rho}{Eg}} = \frac{A\rho l}{W}$$

8. Show that the fundamental frequency for the system of Prob. 7 can be expressed in the form

$$\omega_1 = \beta_1 \sqrt{k/rM}$$

where $\quad n_1 l = \beta_1, \qquad r = \dfrac{M_{rod}}{M},$

$$k = \frac{AE}{l}, \qquad M = \text{end mass}$$

Reducing the above system to a spring k and an end mass equal to $M + \frac{1}{3}M_{rod}$, determine an approximate equation for the fundamental frequency. Show that the ratio of the approximate to the exact frequency as found above is $\dfrac{1}{\beta_1}\sqrt{\dfrac{3r}{3+r}}$.

9. The frequency of magnetostriction oscillators is determined by the length of the nickel alloy rod which generates an alternating voltage in the surrounding coils (see Fig. 8-2) equal to the frequency of longitudinal vibration of the rod. Determine the proper length of the rod clamped at the middle for a frequency of 20 kc.p.s. if the modulus of elasticity and density are given as $E = 30 \times 10^6$ lb./in.^{2} and $\rho = 0.31$ lb./in.^{3}

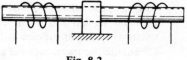

Fig. 8-2.

10. Show that $c = \sqrt{Gg/\rho}$ is the velocity of propagation of torsional strain along the rod. What is the numerical value of c for steel?

11. Determine the expression for the natural frequencies of torsional oscillations of a uniform rod of length l clamped at the middle and free at the two ends.

12. Determine the natural frequencies of a torsional system consisting of a uniform shaft of mass moment of inertia J_s with a disk of inertia J_0 attached to each end. Check the fundamental frequency by reducing the uniform shaft to a torsional spring with end masses.

13. Determine the expression for the natural frequencies of a free-free bar in lateral vibration.

14. Determine the node position for the fundamental mode of the free-free beam by Rayleigh's method, assuming the curve to be $y = \sin\dfrac{\pi x}{l} - b$. By equating the momentum to zero, determine b. Substitute this value of b to find ω_1.

15. A concrete test beam $2 \times 2 \times 12$ in., supported at two points $0.224l$ from the ends, was found to resonate at 1690 c.p.s. If the density of concrete is 153 lb./ft.^{3}, determine the modulus of elasticity, assuming the beam to be slender.

16. Determine the natural frequencies of a uniform beam of length l clamped at both ends.

17. Determine the natural frequencies of a uniform beam of length l, clamped at one end and pinned at the other end.

18. A uniform beam of length l and weight W_b is clamped at one end and carries a concentrated weight W_0 at the other end. State the boundary conditions and determine the frequency equation.

19. The pinned end of a pinned-free beam is given a harmonic motion of amplitude y_0 perpendicular to the beam. Show that the boundary conditions result in the equation

$$\frac{y_0}{y_l} = \frac{\sinh nl \cos nl - \cosh nl \sin nl}{\sinh nl - \sin nl}$$

which for $y_0 \rightarrow 0$, reduce to

$$\tanh nl = \tan nl$$

20. A uniform bar has these specifications: length l, mass density per unit volume ρ, and torsional stiffness $I_P G$ where I_P is the polar moment of inertia of the cross section and G the shear modulus. The end $x = 0$ is fastened to a torsional spring of stiffness K lb. in./rad., while the end l is fixed (see Fig. 8-3). Determine the transcendental equation from which the natural frequencies can be established. Verify the correctness of this equation by considering special cases for $K = 0$ and $K = \infty$.

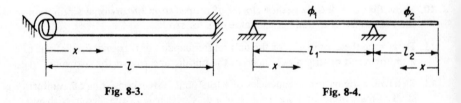

Fig. 8-3.   Fig. 8-4.

21. A simply supported beam has an overhang of length l_2, as shown in Fig. 8-4. If the end of the overhang is free, show that boundary conditions require the deflection equation for each span to be

$$\phi_1 = C\left(\sin \beta x - \frac{\sin \beta l_1}{\sinh \beta l_1} \sinh \beta x\right)$$

$$\phi_2 = A\left\{\cos \beta x + \cosh \beta x - \left(\frac{\cos \beta l_2 + \cosh \beta l_2}{\sin \beta l_2 + \sinh \beta l_2}\right)(\sin \beta x + \sinh \beta x)\right\}$$

where x is measured from the left and right ends.

22. A particular satellite consists of two equal masses m each, connected by a cable of length $2l$ and mass density ρ, as shown in Fig. 8-5. The assembly rotates in space with angular speed ω_0. Show that if the variation in the cable tension is

neglected, the differential equation of lateral motion of the cable is

$$\frac{\partial^2 y}{\partial x^2} = \frac{\rho}{m\omega_0^2 l}\left(\frac{\partial^2 y}{\partial t^2} - \omega_0^2 y\right)$$

and that its fundamental frequency of oscillation is

$$\omega^2 = \left(\frac{\pi}{2l}\right)^2\left(\frac{m\omega_0 l}{\rho}\right) - \omega_0^2.$$

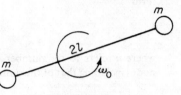

Fig. 8-5.

23. Figure 8-6 shows a flexible cable supported at the upper end and free to oscillate under the influence of gravity. Show that the equation of lateral motion is

$$\frac{\partial^2 y}{\partial t^2} = g\left(x\frac{\partial^2 y}{\partial x^2} + \frac{\partial y}{\partial x}\right)$$

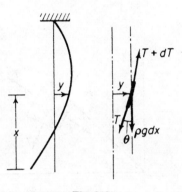

Fig. 8-6.

24. In Prob. 23, assume a solution in the form $y = Y(x)\cos \omega t$ and show that $Y(x)$ can be reduced to a Bessel's differential equation:

$$\frac{d^2 Y(z)}{dz^2} + \frac{1}{z}\frac{dY(z)}{dz} + Y(z) = 0$$

with solution

$$Y(z) = J_0(z) \quad \text{or} \quad Y(x) = J_0\left(2\omega\sqrt{\frac{x}{g}}\right)$$

by a change in variable $z^2 = 4\omega^2 x/g$.

25. A membrane is stretched with large tension T lb./in., so that its lateral deflection y does not increase T appreciably. Using polar coordinates shown in Fig. 8-7, show that the differential equation of lateral vibration is

$$\frac{\partial^2 y}{\partial t^2} = \frac{T}{\rho}\left(\frac{\partial^2 y}{\partial r^2} + \frac{1}{r}\frac{\partial y}{\partial r} + \frac{1}{r^2}\frac{\partial^2 y}{\partial \theta^2}\right)$$

26. Apply the results of Prob. 25 to a circular membrane of radius a with the boundary conditions $y(a) = 0$. The deflection of the symmetric modes without radial node lines can be shown to be given by $J_0(r\sqrt{\rho\omega^2/T})$. For the general case of radial and circumferential nodes, the natural frequencies are evaluated from

the boundary conditions at $r = a$ and $r = 0$, which result in an equation of the form

$$\omega = \frac{\alpha_{n,m}}{a}\sqrt{\frac{T}{\rho}}$$

where n refers to the number of radial nodes, and m the number of circular nodes including that of the outer boundary. A few shapes are shown in Fig. 8-8.

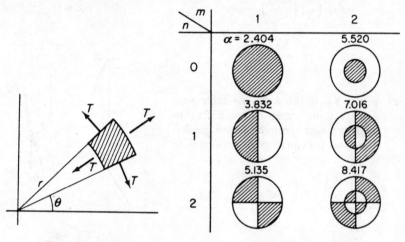

Fig. 8-7.   Fig. 8-8. Deflection of membranes.

27. The equation for the longitudinal oscillations of a slender rod with viscous damping is

$$m\,\frac{\partial^2 u}{\partial t^2} = AE\,\frac{\partial^2 u}{\partial x^2} - \alpha\,\frac{\partial u}{\partial t} + \frac{p_0}{l}p(x)f(x)$$

where the loading per unit length is assumed to be separable. Letting $u = \sum_i \phi_i(x)q_i(t)$ and $p(x) = \sum_i b_i\phi_i(x)$ show that

$$u = \frac{p_0}{ml\sqrt{1-\zeta^2}}\sum_j \frac{b_j\phi_j}{\omega_j}\int_0^t f(t-\tau)e^{-\zeta\omega_j\tau}\sin\omega_j\sqrt{1-\zeta^2}\,\tau\,d\tau$$

$$b_j = \frac{1}{l}\int_0^l p(x)\phi_j(x)\,dx$$

Derive the equation for the stress at any point x.

9 | Lagrange's Equation

9.1 Generalized Coordinates

We have already shown that the equations of motion of a system can be formulated in a number of different coordinate systems. However, to describe completely the motion of a system of n degrees of freedom, n independent coordinates were found to be necessary. For example, in Sec. 6.3 we used the coordinates x and θ to describe the vibrational motion of a rigid bar in translation and rotation. Such independent coordinates are called generalized coordinates, and for the development of Lagrange's equation it is convenient to distinguish them from the constrained or nonindependent coordinates by the letters $q_1, q_2 \cdots q_n$.

9.2 Principle of Virtual Work

Lagrange's method enables us to write the equations of motion in terms of the generalized coordinates when the kinetic energy and the work done are expressed in these coordinates. In its development the concepts of virtual displacement and the work done in such displacement are of importance.

A virtual displacement δx, $\delta \theta$, δq, etc., is an infinitesimal change in the coordinate which may be conceived irrespective of the time, and which must be compatible with the constraints of the system.

The principle of virtual work, enounced by Jean Bernoulli (1717) states that if a system is in equilibrium, the work done by the applied forces in a virtual displacement is zero.

To illustrate these two concepts, we will consider the problem of establishing the equilibrium position of the rigid bar constrained in its movement as shown in Fig. 9.2-1.

The position of the bar is completely established by the coordinate θ which can serve as the generalized coordinate. If the bar is given a virtual displacement, $\delta\mathbf{r}_1$, $\delta\mathbf{r}_2$, $\delta\mathbf{r}_G$, etc., they must all be compatible with the constraints of the system. They can all be expressed in terms of $\delta\theta$ which is

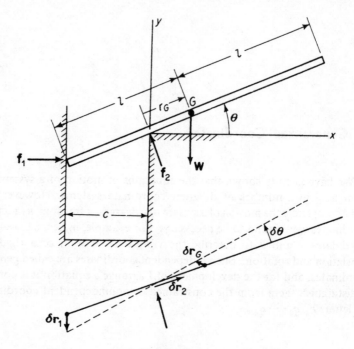

Fig. 9.2-1.

the only independent quantity and which can be assigned any arbitrary value.

There are two types of forces acting on the bar. $\mathbf{f}_1$ and $\mathbf{f}_2$ are constraint forces, whereas $\mathbf{W}$ is an applied force. Assuming frictionless contacts, the constraint forces $\mathbf{f}_1$ and $\mathbf{f}_2$ are normal to the virtual displacements $\delta\mathbf{r}_1$ and $\delta\mathbf{r}_2$, respectively, and hence do no work on the bar when it undergoes a virtual displacement $\delta\theta$. Thus the virtual work of the system is due to the applied force only.

$$\delta W = \mathbf{f}_1 \cdot \delta\mathbf{r}_1 + \mathbf{f}_2 \cdot \delta\mathbf{r}_2 + \mathbf{w} \cdot \delta\mathbf{r}_G$$
$$= \mathbf{w} \cdot \delta\mathbf{r}_G$$

Since $\mathbf{r}_G$ is some function of θ,

$$\delta\mathbf{r}_G = \frac{\partial \mathbf{r}_G}{\partial \theta}\,\delta\theta$$

and the above equation for the virtual work becomes

$$\delta W = \mathbf{w} \cdot \frac{\partial \mathbf{r}_G}{\partial \theta}\, \delta\theta = Q_\theta\, \delta\theta$$

where $Q_\theta = \mathbf{w}(\partial \mathbf{r}_G/\partial\theta)$ is called the generalized force.

Using unit vectors i and j along the x and y axes with origin at the corner where $\mathbf{f}_2$ is acting, the equation for $\mathbf{r}_G$ is

$$\mathbf{r}_G = r_G(\mathbf{i}\cos\theta + \mathbf{j}\sin\theta)$$

$$= \left(l - \frac{c}{\cos\theta}\right)(\mathbf{i}\cos\theta + \mathbf{j}\sin\theta)$$

$$= (l\cos\theta - c)\mathbf{i} + (l\sin\theta - c\tan\theta)\mathbf{j}$$

Differentiating with respect to θ,

$$\delta\mathbf{r}_G = (-l\sin\theta)\,\delta\theta\mathbf{i} + (l\cos\theta - c\sec^2\theta)\,\delta\theta\mathbf{j}$$

and taking the dot product with $\mathbf{w} = -w\mathbf{j}$ results in the equation*

$$\delta W = -w(l\cos\theta - c\sec^2\theta)\,\delta\theta = 0$$

The equation is satisfied by $(l\cos\theta - c\sec^2\theta) = 0$ or

$$\cos\theta = \sqrt[3]{c/l}$$

which defines the equilibrium position of the bar.

9.3 Development of Lagrange's Equation

The principle of virtual work, established for the case of static equilibrium, can be extended to dynamics by a reasoning advanced by D'Alembert (1743). D'Alembert reasoned that since the sum of the forces acting on a particle results in its acceleration $m_i\ddot{\mathbf{r}}_i$, the application of a force equal to $-m_i\ddot{\mathbf{r}}_i$ would produce a condition of equilibrium. The equation for the particle can then be written as

$$\mathbf{F}_i + \mathbf{f}_i - m_i\ddot{\mathbf{r}}_i = 0 \qquad (9.3\text{-}1)$$

where $\mathbf{F}_i$ and $\mathbf{f}_i$ are the applied and constraint forces, respectively. It then follows from the principle of virtual work that for a system of particles

$$\sum_i (\mathbf{F}_i - m_i\ddot{\mathbf{r}}_i) \cdot \delta\mathbf{r}_i = 0 \qquad (9.3\text{-}2)$$

where the work done by the constraint forces $\mathbf{f}_i$ is again zero. Thus, for a

---

* This requires G to move horizontally.

dynamical system, the principle of virtual work requires that the applied forces $\mathbf{F}_i$ be replaced by $(\mathbf{F}_i - m_i\ddot{\mathbf{r}}_i)$ which introduces a new term $\sum_i m_i\ddot{\mathbf{r}}_i \cdot \delta\mathbf{r}_i$.
We will now show that this new term is related to the kinetic energy T by the equation

$$\sum_{k=1}^{n}\left[\frac{d}{dt}\left(\frac{\partial T}{\partial \dot{q}_k}\right) - \frac{\partial T}{\partial q_k}\right]\delta q_k$$

Considering a body to be representable by a system of particles, its kinetic energy is equal to

$$T = \sum_i \tfrac{1}{2}m_i\dot{r}_i^2 = \sum_i \tfrac{1}{2}m_i\dot{\mathbf{r}}_i \cdot \dot{\mathbf{r}}_i \tag{9.3-3}$$

For a system of n degrees of freedom the position of any particle can be expressed in terms of the n generalized coordinates $q_1, q_2, \cdots, q_n$ and in some cases time t,

$$\mathbf{r}_i = \mathbf{r}_i(q_1, q_2, \cdots, q_n, t) \tag{9.3-4}$$

and its velocity is

$$\dot{\mathbf{r}}_i = \frac{\partial \mathbf{r}_i}{\partial q_1}\dot{q}_1 + \frac{\partial \mathbf{r}_i}{\partial q_2}\dot{q}_2 + \cdots \frac{\partial \mathbf{r}_i}{\partial q_n}\dot{q}_n + \frac{\partial \mathbf{r}_i}{\partial t} \tag{9.3-5}$$

From these we form two important relationships. First, if we take the partial derivative of $\dot{\mathbf{r}}_i$ with respect to $\dot{q}_k$, it will be equal to the coefficient of $\dot{q}_k$

$$\frac{\partial \dot{\mathbf{r}}_i}{\partial \dot{q}_k} = \frac{\partial \mathbf{r}_i}{\partial q_k} \tag{9.3-6}$$

Second, the virtual displacement of $\mathbf{r}_i$ from Eq. (9.3-4) is

$$\delta\mathbf{r}_i = \frac{\partial \mathbf{r}_i}{\partial q_1}\delta q_1 + \frac{\partial \mathbf{r}_i}{\partial q_2}\delta q_2 + \cdots \frac{\partial \mathbf{r}_i}{\partial q_n}\delta q_n = \sum_{k=1}^{n}\frac{\partial \mathbf{r}_i}{\partial q_k}\delta q_k \tag{9.3-7}$$

where it should be noted that the time t does not enter into the equation (definition of virtual displacement, irrespective of time).

Making use of the above equation for $\delta\mathbf{r}_i$ we have

$$m_i\ddot{\mathbf{r}}_i \cdot \delta\mathbf{r}_i = \sum_{k=1}^{n}m_i\ddot{\mathbf{r}}_i \cdot \frac{\partial \mathbf{r}_i}{\partial q_k}\delta q_k \tag{9.3-8}$$

Next examine one of the terms of this summation

$$m_i\ddot{\mathbf{r}}_i \cdot \frac{\partial \mathbf{r}_i}{\partial q_k} = \frac{d}{dt}\left(m_i\dot{\mathbf{r}}_i \cdot \frac{\partial \mathbf{r}_i}{\partial q_k}\right) - m_i\dot{\mathbf{r}}_i \cdot \frac{d}{dt}\left(\frac{\partial \mathbf{r}_i}{\partial q_k}\right) \tag{9.3-9}$$

From Eq. (9.3-6), $\partial \mathbf{r}_i/\partial q_k$ in the first term can be replaced by $\partial \dot{\mathbf{r}}_i/\partial \dot{q}_k$, and the order of differentiation in the second term can be reversed so that

$$\frac{d}{dt}\left(\frac{\partial \mathbf{r}_i}{\partial q_k}\right) = \frac{\partial \dot{\mathbf{r}}_i}{\partial q_k}$$

The result is

$$m_i\ddot{\mathbf{r}}_i \cdot \frac{\partial \dot{\mathbf{r}}_i}{\partial q_k} = \frac{d}{dt}\left(m_i\dot{\mathbf{r}}_i \cdot \frac{\partial \dot{\mathbf{r}}_i}{\partial \dot{q}_k}\right) - m_i\dot{\mathbf{r}}_i \cdot \frac{\partial \dot{\mathbf{r}}_i}{\partial q_k}$$

$$= \left[\frac{d}{dt}\frac{\partial}{\partial \dot{q}_k} - \frac{\partial}{\partial q_k}\right](\tfrac{1}{2}m_i\dot{\mathbf{r}}_i \cdot \dot{\mathbf{r}}_i) \qquad (9.3\text{-}10)$$

and

$$m_i\ddot{\mathbf{r}}_i \cdot \delta\mathbf{r}_i = \sum_{k=1}^{n}\left[\frac{d}{dt}\frac{\partial}{\partial \dot{q}_k} - \frac{\partial}{\partial q_k}\right](\tfrac{1}{2}m_i\dot{r}_i^2)\,\delta q_k \qquad (9.3\text{-}11)$$

Summing over the i particles, we arrive at the result:

$$\sum_i m_i\ddot{\mathbf{r}}_i \cdot \delta\mathbf{r}_i = \sum_{k=1}^{n}\left[\frac{d}{dt}\frac{\partial T}{\partial \dot{q}_k} - \frac{\partial T}{\partial q_k}\right]\delta q_k \qquad (9.3\text{-}12)$$

where $T = \tfrac{1}{2}\sum_i m_i\dot{r}_i^2$ is the kinetic energy of the system.

To complete the development, the work done by the applied forces in the virtual displacement is written as

$$\delta W = \sum_i \mathbf{F}_i \cdot \delta\mathbf{r}_i = \sum_i \mathbf{F}_i \cdot \sum_{k=1}^{n} \frac{\partial \mathbf{r}_i}{\partial q_k}\,\delta q_k$$

$$= \sum_{k=1}^{n}\left(\sum_i \mathbf{F}_i \cdot \frac{\partial \mathbf{r}_i}{\partial q_k}\right)\delta q_k$$

$$= \sum_{k=1}^{n} Q_k\,\delta q_k \qquad (9.3\text{-}13)$$

where

$$Q_k = \sum_i \mathbf{F}_i \cdot \frac{\partial \mathbf{r}_i}{\partial q_k} \qquad (9.3\text{-}14)$$

is called the generalized force associated with the coordinate q_k. The dimensions of Q_k will depend on the dimensions of q_k, so that if q_k is an angle θ, the generalized force will be a moment.

We now put Eqs. (9.3-12) and (9.3-13) back into the original Eq. (9.3-2):

$$\sum_{k=1}^{n}\left(\frac{d}{dt}\frac{\partial T}{\partial \dot{q}_k} - \frac{\partial T}{\partial q_k} - Q_k\right)\delta q_k = 0 \qquad (9.3\text{-}15)$$

Since the n δq_k corresponding to the n degrees of freedom are independent quantities, we can choose them in any manner we please. By singling out one of the $\delta q_j \neq 0$ and letting the remaining δq_s be zero, we obtain Lagrange's equation for the coordinate q_j:

$$\frac{d}{dt}\frac{\partial T}{\partial \dot{q}_j} - \frac{\partial T}{\partial q_j} - Q_j = 0 \qquad (9.3\text{-}16)$$

By repeating the procedure with other coordinates, a similar equation can be established for the n coordinates of the system.

There are a few variations of Lagrange's equation which can now be mentioned. If we have a conservative system, the work done is equal to the negative of the potential energy

$$W = -U(q_1, q_2, \cdots, q_n) \qquad (9.3\text{-}17)$$

and the virtual work of Eq. (9.3-13) can be replaced by

$$\delta W = -\sum \frac{\partial U}{\partial q_k} \delta q_k \qquad (9.3\text{-}18)$$

Thus in place of Q_k we use $-(\partial U/\partial q_k)$ and rewrite Lagrange's equation as

$$\frac{d}{dt}\frac{\partial T}{\partial \dot{q}_k} - \frac{\partial T}{\partial q_k} + \frac{\partial U}{\partial q_k} = 0 \qquad (9.3\text{-}19)$$

The second variant results from recognizing that U is not a function of $\dot{q}$ so that if we define a Lagrangian L as

$$L = T - U \qquad (9.3\text{-}20)$$

we can write Eq. (9.3-16) as

$$\frac{d}{dt}\frac{\partial L}{\partial \dot{q}_k} - \frac{\partial L}{\partial q_k} = 0 \qquad (9.3\text{-}21)$$

When nonconservative forces exist in the system, the work done by them can be separated out in the form

$$\delta W = \sum_{k=1}^{n} \bar{Q}_k \delta q_k \qquad (9.3\text{-}22)$$

in which case it is possible to present Lagrange's equation for a nonconservative system as

$$\frac{d}{dt}\frac{\partial L}{\partial \dot{q}_k} - \frac{\partial L}{\partial q_k} = \bar{Q}_k$$

$$\frac{d}{dt}\frac{\partial T}{\partial \dot{q}_k} - \frac{\partial T}{\partial q_k} + \frac{\partial U}{\partial q_k} = \bar{Q}_k$$

$$(9.3\text{-}23)$$

These last forms enable us to extend the use of Lagrange's method to nonconservative systems and hence the method of Lagrange is applicable to all dynamical systems including damped vibrations.

Although all of the equations of motion so far have been written down without the use of Lagrange's equation, there are complex problems which would be difficult to set up in the usual way. For some problems the forces and accelerations may be difficult to determine, whereas the kinetic and potential energy equations can be written down with ease. We need only to mention the vibrational equations for the flexible missile with gimballed engines and fluid propellants as an example where the Lagrangian approach has decided advantage. Also in interconnected structures the energy formulation avoids the necessity of calculating the internal or interconnecting forces which are required in the non-Lagrangian procedure. There are also dynamical concepts which can be more easily presented by the Lagrangian procedure.

EXAMPLE 9.3-1. The much discussed double pendulum is an excellent example for the demonstration of Lagrange's method. In spite of the simplicity of this system, for large angles it is much simpler to determine the

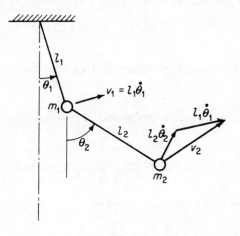

Fig. 9.3-1.

velocities of the masses than to attempt a force and acceleration analysis. Choosing θ_1 and θ_2 as generalized coordinates, the square of the velocities of the two masses are:

$$v_1^2 = l_1^2 \dot{\theta}_1^2$$

$$v_2^2 = (l_1 \dot{\theta}_1 \sin \theta_1 + l_2 \dot{\theta}_2 \sin \theta_2)^2 + (l_1 \dot{\theta}_1 \cos \theta_1 + l_2 \dot{\theta}_2 \cos \theta_2)^2$$

and the kinetic energy of the system is

$$T = \tfrac{1}{2} m_1 l_1^2 \dot{\theta}_1^2 + \tfrac{1}{2} m_2 [(l_1 \dot{\theta}_1 \sin \theta_1 + l_2 \dot{\theta}_2 \sin \theta_2)^2 + (l_1 \dot{\theta}_1 \cos \theta_1 + l_2 \dot{\theta}_2 \cos \theta_2)^2]$$

Letting the potential energy be zero at the equilibrium position $\theta_1 = \theta_2 = 0$, it can be written as

$$U = m_1 g l_1 (1 - \cos \theta_1) + m_2 g [l_1 (1 - \cos \theta_1) + l_2 (1 - \cos \theta_2)]$$

Substituting into the Lagrange's equations,

$$\frac{d}{dt}\frac{\partial T}{\partial \dot\theta_1} - \frac{\partial T}{\partial \theta_1} + \frac{\partial U}{\partial \theta_1} = 0$$

$$\frac{d}{dt}\frac{\partial T}{\partial \dot\theta_2} - \frac{\partial T}{\partial \theta_2} + \frac{\partial U}{\partial \theta_2} = 0$$

the two equations of motion are found to be lengthy nonlinear expressions which could not be solved without simplification. When the equations are linearized by limiting θ to small angles, the result is the dynamically coupled equations

$$(m_1 + m_2)l_1^2\ddot\theta_1 + m_2 l_1 l_2 \ddot\theta_2 + (m_1 + m_2)gl_1\theta_1 = 0$$

$$m_2 l_2^2 \ddot\theta_2 + m_2 l_1 l_2 \ddot\theta_1 + m_2 g l_2 \theta_2 = 0$$

9.4 Solution in Terms of Normal Coordinates

If we start with the kinetic and potential energies equal to

$$T = \tfrac{1}{2}\sum_i \sum_j m_{ij}\dot q_i \dot q_j \tag{9.4-1}$$

$$U = \tfrac{1}{2}\sum_i \sum_j k_{ij} q_i q_j \tag{9.4-2}$$

their substitution into Lagrange's equation

$$\frac{d}{dt}\frac{\partial T}{\partial \dot q_i} - \frac{\partial T}{\partial q_i} + \frac{\partial U}{\partial q_i} = 0 \tag{9.4-3}$$

leads to the matrix equation

$$[m_{ij}]\{\ddot q_i\} + [k_{ij}]\{q_i\} = 0 \tag{9.4-4}$$

When q_i are principal coordinates, the off-diagonal terms in the matrix equation are zero and the equations of motion uncouple to

$$m_i \ddot q_i + k_i q_i = 0 \tag{9.4-5}$$

The solution is then the normal-mode vibration $q_i = A_i \sin \omega_i t$ whose substitution into the differential equation results in $k_i = -\omega_i^2 m_i$. Thus, when q_i are principal (normal) coordinates, the kinetic and potential energy expressions simplify to the form

$$T = \tfrac{1}{2}\sum_i m_i \dot q_i^2 \tag{9.4-6}$$

$$U = \tfrac{1}{2}\sum_i k_i q_i^2 = \tfrac{1}{2}\sum_i \omega_i^2 m_i q_i^2 \tag{9.4-7}$$

Extending this notion to a continuous structure, we can represent its displacement in terms of its normal modes. Consider for example the general motion of a beam loaded by a distributed force $p(x, t)$ whose equation of motion is

$$[EIy''(x, t)]'' + m(x)\ddot{y}(x, t) = p(x, t) \qquad (9.4\text{-}8)$$

The normal modes $\phi_i(x)$ of such a beam must satisfy the equation

$$(EI\phi_i'')'' - \omega_i^2 m(x)\phi_i = 0 \qquad (9.4\text{-}9)$$

and its boundary conditions. They are also orthogonal functions satisfying the relation

$$\int_0^l m(x)\phi_i\phi_j\, dx = \begin{cases} 0 & \text{for} \quad j \neq i \\ M_i & \text{for} \quad j = i \end{cases} \qquad (9.4\text{-}10)$$

If we represent the solution to the general problem defined by Eq. (9.4-3) in terms of the normal modes $\phi_i(x)$

$$y(x, t) = \sum_i \phi_i(x)q_i(t) \qquad (9.4\text{-}11)$$

the $q_i(t)$ can be determined from Lagrange's equation. The kinetic and potential energies are then

$$
\begin{aligned}
T &= \tfrac{1}{2}\int_0^l \dot{y}^2(x, t)m(x)\, dx = \tfrac{1}{2}\sum_i \sum_j \dot{q}_i\dot{q}_j \int_0^l \phi_i\phi_j m(x)\, dx \\
&= \tfrac{1}{2}\sum_i M_i\dot{q}_i^2
\end{aligned}
\qquad (9.4\text{-}12)
$$

$$
\begin{aligned}
U &= \tfrac{1}{2}\int_0^l EIy''^2(x, t)\, dx = \tfrac{1}{2}\sum_i \sum_j q_iq_j \int_0^l EI\phi_i''\phi_j''\, dx \\
&= \tfrac{1}{2}\sum_i K_iq_i^2 = \tfrac{1}{2}\sum_i \omega_i^2 M_iq_i^2
\end{aligned}
\qquad (9.4\text{-}13)^*
$$

* The orthogonality relation for $i \neq j$ can be rewritten in terms of the differential equation, Eq. (9.4-9), as

$$\int_0^l m(x)\phi_i\phi_j\, dx = \frac{1}{\omega_i^2}\int_0^l (EI\phi_i'')''\phi_j\, dx = 0 \qquad j \neq i$$

Integrating by parts, the last integral becomes

$$(EI\phi_i'')'\phi_j\Big]_0^l - (EI\phi_i'')\phi_j'\Big]_0^l + \int_0^l EI\phi_i''\phi_j''\, dx = 0 \qquad j \neq i$$

and since for any boundary conditions the first two terms can be shown to be zero, we have proved that

$$K_{ij} = \int_0^l EI\phi_i''\phi_j''\, dx = 0 \qquad j \neq i$$

In addition to T and U we need the expression for the generalized force Q_i, which is determined from the work done by the applied force $p(x, t)\, dx$ in the virtual displacement δq_i

$$\delta W = \int_0^l p(x, t)\left(\sum_i \phi_i\, \delta q_i \right) dx$$

$$= \sum_i \delta q_i \int_0^l p(x, t)\phi_i(x)\, dx = \sum_i Q_i\, \delta q_i \tag{9.4-14}$$

Thus, substituting into Lagrange's equation,

$$\frac{d}{dt}\left(\frac{\partial T}{\partial \dot{q}_i}\right) - \frac{\partial T}{\partial q_i} + \frac{\partial U}{\partial q_i} = Q_i$$

the equation for $q_i(t)$ is found as

$$\ddot{q}_i + \omega_i^2 q_i = \frac{1}{M_i} \int_0^l p(x, t)\phi_i(x)\, dx \tag{9.4-15}$$

and the solution $y(x, t) = \sum \phi_i q_i$ is completed.

It is convenient at this point to consider the case where the loading per unit length $p(x, t)$ is separable in the form

$$p(x, t) = \frac{P_0}{l}\, p(x)f(t) \tag{9.4-16}$$

Equation (9.4-15) then reduces to

$$\ddot{q}_i(t) + \omega_i^2 q_i(t) = \frac{P_0}{M_i} \left\{ \frac{1}{l} \int_0^l p(x)\phi(x)\, dx \right\} f(t) \tag{9.4-17}$$

where

$$K_i = \frac{1}{l} \int_0^l p(x)\phi_i(x)\, dx \tag{9.4-18}$$

can be defined as the mode participation factor for mode i. The solution of the above equation is then

$$q_i(t) = q_i(0) \cos \omega_i t + \frac{1}{\omega_i} \dot{q}_i(0) \sin \omega_i t$$

$$+ \frac{P_0 K_i}{\omega_i^2 M_i} \omega_i \int_0^t f(\xi) \sin \omega_i(t - \xi)\, d\xi \tag{9.4-19}$$

and since the statical deflection [with $\ddot{q}_i(t) = 0$] expanded in terms of $\phi_i(x)$ is $P_0 K_i/\omega_i^2 M_i$, the quantity

$$D_i(t) = \omega_i \int_0^t f(\xi) \sin \omega_i(t - \xi)\, d\xi \tag{9.4-20}$$

can be called the dynamic load factor for the ith mode.

EXAMPLE 9.4-1. A simply supported uniform beam of mass M_0 is suddenly loaded by the force shown in Fig. 9.4-1. Determine the equation of motion.

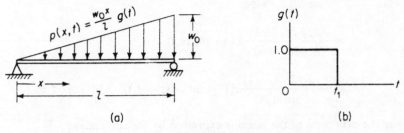

(a) (b)

Fig. 9.4-1.

Solution. The normal modes of the beam are

$$\phi_n(x) = \sqrt{2} \sin \frac{n\pi x}{l}$$

$$\omega_n = (n\pi)^2 \sqrt{EI/M_0 l^3}$$

and the generalized mass is

$$M_n = \frac{M_0}{l} \int_0^l 2 \sin^2 \frac{n\pi x}{l} \, dx = M_0$$

The generalized force is

$$\int_0^l p(x,t)\phi_n \, dx = g(t) \int_0^l \frac{w_0 x}{l} \sqrt{2} \sin \frac{n\pi x}{\cdot l} \, dx$$

$$= g(t) \frac{w_0\sqrt{2}}{l} \left[\frac{\sin(n\pi x/l)}{(n\pi/l)^2} - \frac{x \cos(n\pi x/l)}{(n\pi/l)} \right]_0^l$$

$$= -g(t) \frac{w_0\sqrt{2}\, l}{n\pi} \cos n\pi$$

$$= - \frac{\sqrt{2}\, l w_0}{n\pi} g(t)(-1)^n$$

where $g(t)$ is the time history of the load. The equation for q_n is then

$$\ddot{q}_n + \omega_n^2 q_n = - \frac{\sqrt{2}\, l w_0}{n\pi M_0} (-1)^n g(t)$$

which has the solution

$$q_n(t) = \frac{-\sqrt{2}\, lw_0}{n\pi M_0} \frac{(-1)^n}{\omega_n^2}\, (1 - \cos \omega_n t) \qquad\qquad 0 \le t \le t_1$$

$$= \frac{-\sqrt{2}\, lw_0}{n\pi M_0} \frac{(-1)^n}{\omega_n^2}\, (1 - \cos \omega_n t)$$

$$+ \frac{2\sqrt{2}\, lw_0 (-1)^n}{n\pi M_0 \omega_n^2}\, [1 - \cos \omega_n (t - t_1)] \qquad t_1 \le t \le \infty$$

Thus the deflection of the beam is expressed by the summation

$$y(x, t) = \sum_{n=1}^{\infty} q_n(t) \sqrt{2} \sin \frac{\pi n x}{l}$$

EXAMPLE 9.4-2. A missile in flight is excited longitudinally by the thrust $F(t)$ of its rocket engine at the end $x = 0$. Determine the equation for the displacement $u(x, t)$ and the acceleration $\ddot{u}(x, t)$.

Solution. We assume the solution for the displacement to be

$$u(x, t) = \sum_i q_i(t)\varphi_i(x)$$

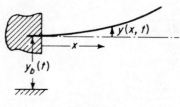

where $\varphi_i(x)$ and ω_i are normal modes of the missile in longitudinal oscillation. The generalized coordinate q_i satisfies the differential equation

Fig. 9.4-2.

$$\ddot{q}_i + \omega_i^2 q_i = \frac{F(t)\varphi_i(0)}{M_i}$$

If, instead of $F(t)$, a unit impulse acted at $x = 0$, the above equation would have the solution $(\varphi_i(0)/M_i\omega_i) \sin \omega_i t$ for initial conditions $q_i(0) = \dot{q}_i(0) = 0$. Thus the response to the arbitrary force $F(t)$ is

$$q_i(t) = \frac{\varphi_i(0)}{M_i\omega_i} \int_0^t F(\xi) \sin \omega_i (t - \xi)\, d\xi$$

and the displacement at any point x is

$$u(x, t) = \sum_i \frac{\varphi_i(x)\varphi_i(0)}{M_i\omega_i} \int_0^t F(\xi) \sin \omega_i (t - \xi)\, d\xi$$

The acceleration $\ddot{q}_i(t)$ of mode i can be determined by rewriting the differential equation and substituting the former solution for $q_i(t)$:

$$\ddot{q}_i(t) = \frac{F(t)\varphi_i(0)}{M_i} - \omega_i^2 q_i$$

$$= \frac{F(t)\varphi_i(0)}{M_i} - \frac{\varphi_i(0)\omega_i}{M_i} \int_0^t F(\xi) \sin \omega_i(t - \xi) \, d\xi$$

Thus the equation for the acceleration of any point x is found as

$$\ddot{u}(x, t) = \sum_i \ddot{q}_i(t)\varphi_i(x)$$

$$= \sum_i \left\{ \frac{F(t)\varphi_i(0)\varphi_i(x)}{M_i} - \frac{\varphi_i(0)\varphi_i(x)\omega_i}{M_i} \int_0^t F(\xi) \sin \omega_i(t - \xi) \, d\xi \right\}$$

EXAMPLE 9.4-3. Determine the response of a cantilever beam when its base is given a motion $y_b(t)$ normal to the beam axis as shown in Fig. 9.4-2.

Solution. The differential equation for the beam with base motion is

$$[EIy''(x, t)]'' + m(x)[\ddot{y}_b(t) + \ddot{y}(x, t)] = 0$$

which can be rearranged to

$$[EIy''(x, t)]'' + m(x)\ddot{y}(x, t) = -m(x)\ddot{y}_b(t)$$

Thus, instead of the force per unit length $F(x, t)$ we have the inertial force per unit length $-m(x)\ddot{y}_b(t)$. Assuming the solution in the form

$$y(x, t) = \sum_i q_i(t)\varphi_i(x)$$

the equation for the generalized coordinate q_i becomes

$$\ddot{q}_i + \omega_i^2 q_i = -\ddot{y}_b(t) \frac{1}{M_i} \int_0^l \varphi_i(x) \, dx$$

The solution for q_i then differs from that of a simple oscillator only by the factor $-\dfrac{1}{M_i} \int_0^l \varphi_i(x) \, dx$ so that for the initial conditions $y(0) = \dot{y}(0) = 0$:

$$q_i(t) = \left\{ -\frac{1}{M_i} \int_0^l \varphi_i(x) \, dx \right\} \frac{1}{\omega_i} \int_0^t \ddot{y}_b(\xi) \sin \omega_i(t - \xi) \, d\xi$$

9.5 Normal Modes of Constrained Structures

A question frequently encountered is: How will the addition of a mass or a spring, at certain points on the structure, influence the natural frequencies and mode shapes of the structure? Such problems can be formulated in terms of generalized coordinates and the mode-summation techniques.

Consider the forced vibration of any one-dimensional structure (i.e., the points on the structure defined by one coordinate x) excited by a force per unit length $f(x, t)$ and moment per unit length $M(x, t)$. If we know the normal modes of the structure, ω_i and $\varphi_i(x)$, its deflection at any point x can be represented by

$$y(x, t) = \sum_i q_i(t)\varphi_i(x) \tag{9.5-1}$$

where the generalized coordinate q_i must satisfy the equation

$$\ddot{q}_i(t) + \omega_i^2 q_i(t) = \frac{1}{M_i}\left[\int f(x, t)\varphi_i(x)\, dx + \int M(x, t)\varphi_i'(x)\, dx\right] \tag{9.5-2}$$

The right side of this equation is $1/M_i$ times the generalized force Q_i, which can be determined from the virtual work of the applied loads as $Q_i = \delta W/\delta q_i$.

If, instead of distributed loads, we have a concentrated force $F(a, t)$ and a concentrated moment $M(a, t)$ at some point $x = a$, the generalized force for such loads is found from

$$\delta W = F(a, t)\, \delta y(a, t) + M(a, t)\, \delta y'(a, t)$$

$$= F(a, t) \sum_i \varphi_i(a)\, \delta q_i + M(a, t) \sum_i \varphi_i'(a)\, \delta q_i$$

$$Q_i = \frac{\delta W}{\delta q_i} = F(a, t)\varphi_i(a) + M(a, t)\varphi_i'(a) \tag{9.5-3}$$

Then, instead of Eq. (9.5-2), we obtain the equation

$$\ddot{q}_i(t) + \omega_i^2 q_i(t) = \frac{1}{M_i}[F(a, t)\varphi_i(a) + M(a, t)\varphi_i'(a)] \tag{9.5-4}$$

These equations form the starting point for the analysis of constrained structures, provided the constraints are expressible as external loads on the structure.

As an example, let us consider attaching a linear and torsional spring to the simply supported beam of Fig. 9.5-1. The linear spring exerts a force on

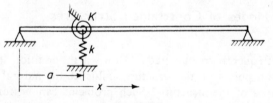

Fig. 9.5-1.

the beam equal to

$$F(a, t) = -ky(a, t) = -k \sum_j q_j(t)\varphi_j(a) \qquad (9.5\text{-}5)$$

whereas the torsional spring exerts a moment

$$M(a, t) = -Ky'(a, t) = -K \sum_j q_j(t)\varphi_j'(a) \qquad (9.5\text{-}6)$$

Substituting these equations into Eq. (9.5-4), we obtain

$$\ddot{q}_i + \omega_i^2 q_i = \frac{1}{M_i}\left[-k\varphi_i(a) \sum_j q_j\varphi_j(a) - K\varphi_i'(a) \sum_j q_j\varphi_j'(a)\right] \qquad (9.5\text{-}7)$$

The normal modes of the constrained modes are also harmonic and so we can write

$$q_i = \bar{q}_i e^{i\omega t}$$

The solution to the ith equation is then

$$\bar{q}_i = \frac{1}{M_i(\omega_i^2 - \omega^2)}\left[-k\varphi_i(a) \sum_j \bar{q}_j\varphi_j(a) - K\varphi_i'(a) \sum_j \bar{q}_j\varphi_j'(a)\right] \qquad (9.5\text{-}8)$$

If we use n modes, there will be n values of $\bar{q}_j$ and n equations such as the one above. The determinant formed by the coefficients of the $\bar{q}_j$ will then lead to

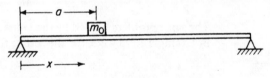

Fig. 9.5-2.

the natural frequencies of the constrained modes, and the mode shapes of the constrained structure are found by substituting the $\bar{q}_j$ into Eq. (9.5-1).

If, instead of springs, a mass m_0 is placed at a point $x = a$, as shown in Fig. 9.5-2, the force exerted by m_0 on the beam is

$$F(a, t) = -m_0\ddot{y}(a, t) = -m_0 \sum_j \ddot{q}_j\varphi_j(a) \qquad (9.5\text{-}9)$$

Thus, in place of Eq. (9.5-8), we would obtain the equation

$$\bar{q}_i = \frac{1}{M_i(\omega_i^2 - \omega^2)}\left[\omega^2 m_0\varphi_i(a) \sum_j \bar{q}_j\varphi_j(a)\right] \qquad (9.5\text{-}10)$$

EXAMPLE 9.5-1. Give a single mode approximation for the natural frequency of a simply supported beam when a mass m_0 is attached to it at $x = l/3$.

Solution. When only a single mode is used, Eq. (9.5-10) reduces to

$$M_1(\omega_1^2 - \omega^2) = \omega^2 m_0 \varphi_1^2(a)$$

Solving for ω^2, we obtain

$$\left(\frac{\omega}{\omega_1}\right)^2 = \frac{1}{1 + \dfrac{m_0}{M_1}\varphi_1^2(a)}$$

For the first mode of the unconstrained beam, we have

$$\omega_1 = \pi^2\sqrt{\frac{EI}{Ml^3}}, \qquad \varphi_1(x) = \sqrt{2}\sin\frac{\pi x}{l}$$

$$\varphi_1\left(\frac{l}{3}\right) = \sqrt{2}\sin\frac{\pi}{3} = \sqrt{2}\times 0.866$$

$$M_1 = M = \text{mass of the beam}$$

Thus its substitution into the above equation gives the one-mode approximation for the constrained beam the value

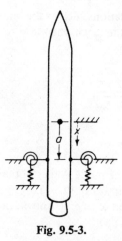

Fig. 9.5-3.

$$\left(\frac{\omega}{\omega_1}\right)^2 = \frac{1}{1 + 1.5\dfrac{m_0}{M}}$$

The same problem treated by the Dunkerley equation in Example 7.3-5 gave, for this ratio, the result

$$\frac{1}{1 + 1.6\dfrac{m_0}{M}}$$

EXAMPLE 9.5-2. A missile is constrained in a test stand by linear and torsional springs, as shown in Fig. 9.5-3. Formulate the inverse problem of determining its free-free modes from the normal modes of the constrained missile, which are designated as Φ_i and Ω_i.

Solution. The problem is approached in a manner similar to that of the direct problem where, in place of φ_i and ω_i, we use Φ_i and Ω_i. We now relieve the constraints at the supports by introducing opposing forces $-F(a)$ and $-M(a)$ equal to $ky(a)$ and $Ky'(a)$.

To carry out this problem in greater detail, we start with the equation

$$\bar{q}_i = \frac{-F(a)\Phi_i(a) - M(a)\Phi_i'(a)}{M_i\Omega_i^2[1 - (\omega/\Omega_i)^2]}$$

which replaces Eq. (9.5-8). Letting $D_i(\omega) = M_i\Omega_i^2[1 - (\omega/\Omega_i)^2]$, the displacement at $x = a$ is

$$y(a) = \sum_i \Phi_i(a)\bar{q}_i = \sum_i \frac{-F(a)\Phi_i^2(a) - M(a)\Phi_i'(a)\Phi_i(a)}{D_i(\omega)}$$

We now replace $-F(a)$ and $-M(a)$ with $ky(a)$ and $Ky'(a)$ and write

$$y(a) = \sum_i \frac{ky(a)\Phi_i^2(a) + Ky'(a)\Phi_i'(a)\Phi_i(a)}{D_i(\omega)}$$

$$y'(a) = \sum_i \frac{ky(a)\Phi_i'(a)\Phi_i(a) + Ky'(a)\Phi_i'^2(a)}{D_i(\omega)}$$

These equations may now be rearranged as

$$y(a)\left[1 - k\sum_i \frac{\Phi_i^2(a)}{D_i(\omega)}\right] = y'(a)K\sum_i \frac{\Phi_i'(a)\Phi_i(a)}{D_i(\omega)}$$

$$y(a)k\sum_i \frac{\Phi_i'(a)\Phi_i(a)}{D_i(\omega)} = y'(a)\left[1 - K\sum_i \frac{\Phi_i'^2(a)}{D_i(\omega)}\right]$$

The frequency equation then becomes

$$\left[1 - k\sum_i \frac{\Phi_i^2(a)}{D_i(\omega)}\right]\left[1 - K\sum_i \frac{\Phi_i'^2(a)}{D_i(\omega)}\right] - kK\left[\sum_i \frac{\Phi_i'(a)\Phi_i(a)}{D_i(\omega)}\right]^2 = 0$$

The slope to deflection ratio at $x = a$ is

$$\frac{y'(a)}{y(a)} = \frac{1 - k\sum_i \dfrac{\Phi_i^2(a)}{D_i(\omega)}}{K\sum_i \dfrac{\Phi_i'(a)\Phi_i(a)}{D_i(\omega)}}$$

The free-free mode shape is then given by

$$\frac{y(x)}{y(a)} = \sum_i \frac{k\Phi_i(a)\Phi_i(x) + K\dfrac{y'(a)}{y(a)}\Phi_i'(a)\Phi_i(x)}{D_i(\omega)}$$

EXAMPLE 9.5-3. Determine the constrained modes of the missile of Fig. 9.5-3, using only the first free-free mode $\varphi_1(x)$, ω_1, together with translation $\varphi_T = 1$, $\Omega_T = 0$ and rotation $\varphi_R = x$, $\Omega_R = 0$, where x is measured positively toward the tail of the missile.

Solution. The generalized mass for each of the three modes is

$$M_T = \int dm = M$$

$$M_R = \int x^2 \, dm = I = M\rho^2$$

$$M_1 = \int \varphi_1^2(x) \, dm = M$$

where the $\varphi_1(x)$ mode was normalized such that $M_1 = M =$ actual mass.
The frequency dependent factors D_i are

$$D_T = -M_T\omega^2 = -M\omega^2 = -M\omega_1^2\lambda$$

$$D_R = -M\rho^2\omega^2 = -M\rho^2\omega_1^2\lambda$$

$$D_1 = M\omega_1^2\left[1 - \left(\frac{\omega}{\omega_1}\right)^2\right] = M\omega_1^2(1 - \lambda)$$

$$\left(\frac{\omega}{\omega_1}\right)^2 = \lambda$$

The frequency equation for this problem is the same as that of Example 9.5-2, except that the minus k's are replaced by positive k's and $\varphi(x)$ and ω replace $\Phi(x)$ and Ω. Substituting the above quantities into the frequency equation, we have

$$\left\{1 - \frac{k}{M\omega_1^2}\left[\frac{1}{\lambda} + \frac{a^2}{\rho^2\lambda} - \frac{\varphi_1^2(a)}{(1-\lambda)}\right]\right\}\left\{1 - \frac{K}{M\omega_1^2}\left[\frac{1}{\rho^2\lambda} - \frac{\varphi_1'^2(a)}{(1-\lambda)}\right]\right\}$$
$$- \frac{kK}{M^2\omega_1^4}\left\{\frac{-a}{\rho^2\lambda} + \frac{\varphi_1'(a)\varphi_1(a)}{(1-\lambda)}\right\}^2 = 0$$

which can be simplified to:

$$\lambda^2(1 - \lambda) + \left(\frac{k}{M\omega_1^2}\right)\left[\varphi_1^2(a) + \frac{K}{k}\varphi_1'^2(a)\right]\lambda^2 - \left(\frac{k}{M\omega_1^2}\right)\left[1 + \frac{a^2}{\rho^2} + \frac{K}{k\rho^2}\right]\lambda(1 - \lambda)$$
$$+ \left(\frac{k}{M\omega_1^2}\right)^2\frac{K}{k\rho^2}(1 - \lambda) - \left(\frac{k}{M\omega_1^2}\right)^2\frac{K}{k}\lambda\left\{\varphi_1'^2(a) + \frac{1}{\rho^2}[\varphi_1(a) - a\varphi_1'(a)]^2\right\} = 0$$

A number of special cases of the above equation are of interest, and we mention one of these. If $K = 0$, the frequency equation simplifies to:

$$\lambda^2 - \left\{1 + \left(\frac{k}{M\omega_1^2}\right)\left[1 + \frac{a^2}{\rho^2} + \varphi_1^2(a)\right]\right\}\lambda + \left(\frac{k}{M\omega_1^2}\right)\left(1 + \frac{a^2}{\rho^2}\right) = 0$$

Here $x = a$ might be taken negatively so that the missile is hanging by a spring.

9.6 Mode-Acceleration Method

One of the difficulties encountered in any mode summation method has to do with the convergence of the procedure. If this convergence is poor, a large number of modes must be used, thereby increasing the order of the frequency determinant. The mode-acceleration method tends to overcome this difficulty by improving the convergence so that a fewer number of normal modes are needed.

The mode-acceleration method starts with the same differential equation for the generalized coordinate q_i, but rearranged in order. For example, we can start with Eq. (9.5-4) and write it in the order:

$$q_i(t) = \frac{F(a, t)\varphi_i(a)}{M_i\omega_i^2} + \frac{M(a, t)\varphi_i'(a)}{M_i\omega_i^2} - \frac{\ddot{q}_i(t)}{\omega_i^2} \qquad (9.6\text{-}1)$$

Substituting this into Eq. (9.5-1), we obtain

$$y(x, t) = \sum_i q_i(t)\varphi_i(x)$$

$$= F(a, t) \sum_i \frac{\varphi_i(a)\varphi_i(x)}{M_i\omega_i^2} + M(a, t) \sum_i \frac{\varphi_i'(a)\varphi_i(x)}{M_i\omega_i^2} - \sum_i \frac{\ddot{q}_i(t)\varphi_i(x)}{\omega_i^2} \qquad (9.6\text{-}2)$$

We note here that, if $F(a, t)$ and $M(a, t)$ were static loads, the last term containing the acceleration would be zero. Thus the terms

$$\sum_i \frac{\varphi_i(a)\varphi_i(x)}{M_i\omega_i^2} = \alpha(a, x)$$

$$\sum_i \frac{\varphi_i'(a)\varphi_i(x)}{M_i\omega_i^2} = \beta(a, x) \qquad (9.6\text{-}3)$$

must represent influence functions, where $\alpha(a, x)$ and $\beta(a, x)$ are the deflections at x due to a unit load and unit moment at a, respectively. We can therefore rewrite Eq. (9.6-2) as

$$y(x, t) = F(a, t)\alpha(a, x) + M(a,\cdot t)\beta(a, x) - \sum \frac{\ddot{q}_i(t)\varphi_i(x)}{\omega_i^2} \qquad (9.6\text{-}4)$$

Because of ω_i^2 in the denominator of the terms summed, the convergence is improved over the mode-summation method.

In the forced vibration problem where $F(a, t)$ and $M(a, t)$ are excitations, Eq. (9.5-4) is first solved for $q_i(t)$ in the conventional manner, and then substituted into Eq. (9.6-4) for the deflection. For the normal modes of constrained structures, $F(a, t)$ and $M(a, t)$ are again the forces and moments exerted by the constraints, and the problem is treated in a manner similar to those of Sec. 9.5. However, because of the improved convergence, fewer number of modes will be found to be necessary.

EXAMPLE 9.6-1. Using the mode-acceleration method, solve the problem of Fig. 9.5-2 of a concentrated mass m_0 attached to the structure.

Solution. Assuming harmonic oscillations,

$$F(a, t) = F(a)e^{i\omega t}$$

$$q_i(t) = \bar{q}_i e^{i\omega t}$$

$$y(x, t) = \bar{y}(x)e^{i\omega t}$$

Substituting these equations into Eq. (9.6-4) and letting $x = a$,

$$\bar{y}(a) = F(a)\alpha(a, a) + \omega^2 \sum_j \frac{\bar{q}_j \varphi_j(a)}{\omega_j^2}$$

Since the force exerted by m_0 on the structure is

$$F(a) = m_0 \omega^2 \bar{y}(a)$$

we can eliminate $\bar{y}(a)$ between the above two equations, obtaining

$$\frac{F(a)}{m_0 \omega^2} = F(a)\alpha(a, a) + \omega^2 \sum_j \frac{\bar{q}_j \varphi_j(a)}{\omega_j^2}$$

or

$$F(a) = \frac{\omega^2 \sum_j \dfrac{\bar{q}_j \varphi_j(a)}{\omega_j^2}}{\dfrac{1}{m_0 \omega^2} - \alpha(a, a)}$$

If we now substitute this equation into Eq. (9.5-4) and assume harmonic motion, we obtain the equation

$$(\omega_i^2 - \omega^2)\bar{q}_i = \frac{F(a)\varphi_i(a)}{M_i} = \frac{\omega^2 \varphi_i(a) \sum_j \bar{q}_j \dfrac{\varphi_j(a)}{\omega_j^2}}{M_i \left[\dfrac{1}{m_0 \omega^2} - \alpha(a, a) \right]}$$

Rearranging, we have

$$[1 - m_0 \omega^2 \alpha(a, a)](\omega_i^2 - \omega^2)\bar{q}_i = \frac{\omega^4 m_0 \varphi_i(a)}{M_i} \sum_j \frac{\bar{q}_j \varphi_j(a)}{\omega_j^2}$$

which represents a set of linear equations in $\bar{q}_k$. The series represented by the summation will, however, converge rapidly because of ω_j^2 in the denominator. Offsetting this advantage of smaller number of modes is the disadvantage that these equations are now quartic rather than quadratic in ω.

9.7 Beam Orthogonality, Including Rotary Inertia and Shear Deformation

The equations for the beam, including rotary inertia and shear deformation, were derived in Sec. 8.6. For such beams the orthogonality is no longer expressed by the equations of Sec. 7.4, but by the equation

$$\int [m(x)\varphi_j\varphi_i + J(x)\psi_j\psi_i]\, dx = \begin{cases} 0 & \text{if } j \neq i \\ M_i & \text{if } j = i \end{cases} \tag{9.7-1}$$

which can be proved in the following manner.

For convenience we will rewrite Eqs. (8.6-5) and (8.6-6), including a distributed moment per unit length $\mathcal{M}(x, t)$:

$$\frac{d}{dx}\left(EI\,\frac{d\psi}{dx}\right) + kAG\left(\frac{dy}{dx} - \psi\right) - J\ddot{\psi} + \mathcal{M}(x, t) = 0 \tag{8.6-5}$$

$$m\ddot{y} - \frac{d}{dx}\left[kAG\left(\frac{dy}{dx} - \psi\right)\right] - p(x, t) = 0 \tag{8.6-6}$$

For the forced oscillation with excitation $p(x, t)$ and $\mathcal{M}(x, t)$ per unit length of beam, the deflection $y(x, t)$ and the bending slope $\psi(x, t)$ can be expressed in terms of the generalized coordinates:

$$\begin{aligned} y &= \sum_j q_j(t)\varphi_j(x) \\ \psi &= \sum_j q_j(t)\psi_j(x) \end{aligned} \tag{9.7-2}$$

With these summations substituted into the two beam equations, we obtain

$$J\sum_j \ddot{q}_j\psi_j = \sum_j q_j\left\{\frac{d}{dx}(EI\psi_j') + kAG(\varphi_j' - \psi_j)\right\} + \mathcal{M}(x, t)$$

$$m\sum_j \ddot{q}_j\varphi_j = \sum_j q_j\frac{d}{dx}\{kAG(\varphi_j' - \psi_j)\} + p(x, t) \tag{9.7-3}$$

However, normal-mode vibrations are of the form

$$\begin{aligned} y &= \varphi_j(x)e^{i\omega_j t} \\ \psi &= \psi_j(x)e^{i\omega_j t} \end{aligned} \tag{9.7-4}$$

which, when substituted into the beam equations with zero excitation, lead to

$$-\omega_j^2 J\psi_j = \frac{d}{dx}(EI\psi_j') + kAG(\varphi_j' - \psi_j)$$

$$-\omega_j^2 m\varphi_j = \frac{d}{dx}\{kAG(\varphi_j' - \psi_j)\}$$

(9.7-5)

The right sides of this set of equations are the coefficients of the generalized coordinates q_j in the forced vibration equations, so that we can write Eqs. (9.7-3) as

$$J\sum_j \ddot{q}_j\psi_j = -\sum_j q_j\omega_j^2 J\psi_j + \mathcal{M}(x, t)$$

$$m\sum_j \ddot{q}_j\varphi_j = -\sum_j q_j\omega_j^2 m\varphi_j + p(x, t)$$

(9.7-6)

Multiplying these two equations by $\varphi_i\, dx$ and $\psi_i\, dx$, adding, and integrating, we obtain

$$\sum_j \ddot{q}_j \int_0^l (m\varphi_j\varphi_i + J\psi_j\psi_i)\, dx + \sum_j q_j\omega_j^2 \int_0^l (m\varphi_j\varphi_i + J\psi_j\psi_i)\, dx$$

$$= \int_0^l p(x, t)\varphi_i\, dx + \int_0^l \mathcal{M}(x, t)\psi_i\, dx \quad (9.7\text{-}7)$$

If the q's in these equations are generalized coordinates, they must be independent coordinates which satisfy the equation

$$\ddot{q}_i + \omega_i^2 q_i = \frac{1}{M_i}\left\{\int_0^l p(x, t)\varphi_i\, dx + \int_0^l \mathcal{M}(x, t)\psi_i\, dx\right\}$$

(9.7-8)

We see then that this requirement is satisfied only if

$$\int_0^l (m\varphi_j\varphi_i + J\psi_j\psi_i)\, dx = \begin{cases} 0 & \text{if } j \neq i \\ M_i & \text{if } j = i \end{cases}$$

(9.7-9)

which defines the orthogonality for the beam, including rotary inertia and shear deformation.

PROBLEMS

1. Show that the dynamic load factor for a suddenly applied constant force reaches a maximum value of 2.0.

2. If a suddenly applied constant force is applied to a system for which the damping factor of the ith mode is $\zeta = c/c_{cr}$, show that the dynamic load factor is given approximately by the equation

$$D_i = 1 - e^{-\zeta\omega_i t}\cos\omega_i t$$

3. Determine the mode participation factor for a uniformly distributed force.

4. If a concentrated force acts at $x = a$, the loading per unit length corresponding to it can be represented by a delta function $l\,\delta(x - a)$. Show that the mode-participation factor then becomes $K_i = \varphi_i(a)$ and the deflection is expressible as

$$y(x, t) = \frac{P_0 l^3}{EI} \sum_i \frac{\varphi_i(a)\varphi_i(x)}{(\beta_i l)^4}\, D_i(t)$$

where $\omega_i^2 = (\beta_i l)^4 (EI/Ml^3)$ and $(\beta_i l)$ is the eigenvalue of the normal-mode equation.

5. For a couple of moment M_0 acting at $x = a$, show that the loading $p(x)$ is the limiting case of two delta functions shown in Fig. 9-1 as $\epsilon \to 0$. Show also that the mode-participation factor for this case is

$$K_i = l \frac{d\varphi_i(x)}{dx} = (\beta_i l)\varphi_i'(x)$$

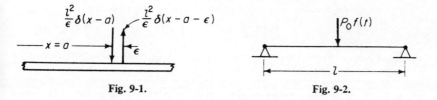

Fig. 9-1.   **Fig. 9-2.**

6. A concentrated force $P_0 f(t)$ is applied to the center of a simply supported uniform beam, as shown in Fig. 9-2. Show that the deflection is given by

$$y(x, t) = \frac{P_0 l^3}{EI} \sum_i \frac{K_i \varphi_i(x)}{(\beta_i l)^4}\, D_i$$

$$= \frac{2P_0 l^3}{EI} \left\{ \frac{\sin \pi \dfrac{x}{l}}{\pi^4} D_1(t) - \frac{\sin 3\pi \dfrac{x}{l}}{(3\pi)^4} D_3(t) + \frac{\sin 5\pi \dfrac{x}{l}}{(5\pi)^4} D_5(t) \cdots \right\}$$

7. A couple of moment M_0 is applied at the center of the beam of Prob. 6, as shown in Fig. 9-3. Show that the deflection at any point is given by the equation:

$$y(x, t) = \frac{M_0 l^2}{EI} \sum_i \frac{\varphi_i(a)\varphi_i(x)}{(\beta_i l)^3}\, D_i(t)$$

$$= \frac{2M_0 l^2}{EI} \left\{ -\frac{\sin 2\pi \dfrac{x}{l}}{(2\pi)^3} D_2(t) + \frac{\sin 4\pi \dfrac{x}{l}}{(4\pi)^3} D_4(t) - \frac{\sin 6\pi \dfrac{x}{l}}{(6\pi)^3} D_6(t) + \cdots \right\}$$

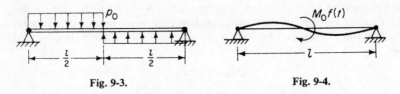

Fig. 9-3.   Fig. 9-4.

8. A simply supported uniform beam has suddenly applied to it the load distribution shown in Fig. 9-4, where the time variation is a step function. Determine the response $y(x, t)$ in terms of the normal modes of the beam. Indicate what modes are absent and write down the first two existing modes.

9. A slender rod of length l, free at $x = 0$, is struck longitudinally by a time-varying force concentrated at the end $x = 0$. Show that all modes are equally excited (i.e., that the mode-participation factor is independent of the mode number), the complete solution being

$$u(x, t) = \frac{2F_0 l}{AE} \left\{ \frac{\cos \dfrac{\pi x}{2 l}}{\left(\dfrac{\pi}{2}\right)^2} D_1(t) + \frac{\cos \dfrac{3\pi x}{2 l}}{\left(\dfrac{3\pi}{2}\right)^2} D_2(t) + \cdots \right\}$$

10. If the force of Prob. 9 is concentrated at $x = l/3$, determine which modes will be absent in the solution.

11. In Prob. 10, determine the participation factor of the modes present and obtain a complete solution for an arbitrary time variation of the applied force.

12. Consider a uniform beam of mass M and length l supported on equal springs of total stiffness k, as shown in Fig. 9-5. Assume the deflection to be

$$y(x, t) = \varphi_1(x) q_1(t) + \varphi_2(x) q_2(t)$$

and choose $\varphi_1 = \sin \dfrac{\pi x}{l}$ and $\varphi_2 = 1.0$.

Using Lagrange's equation, show that

$$\ddot{q}_1 + \frac{4}{\pi} \ddot{q}_2 + \omega_{11}^2 q_1 = 0$$

$$\frac{2}{\pi} \ddot{q}_1 + \ddot{q}_2 + \omega_{22}^2 q_2 = 0$$

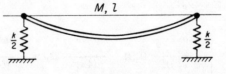

M, l

Fig. 9-5.

where $\omega_{11}^2 = \pi^4(EI/Ml^3) :=$ natural frequency of beam on rigid supports
$\qquad \omega_{22}^2 = k_M =$ natural frequency of rigid beam on springs
Solve these equations and show that

$$\omega^2 = \omega_{22}^2 \frac{\pi}{2} \left\{ \frac{(R + 1) \pm \sqrt{(R - 1)^2 + \dfrac{32}{\pi^2} R}}{\pi^2 - 8} \right\}$$

$$\frac{q_2}{q_1} = b = \frac{\pi}{8} \left\{ (R - 1) \mp \sqrt{(R - 1)^2 + \frac{32}{\pi^2} R} \right\}$$

$$R = \left(\frac{\omega_{11}}{\omega_{22}} \right)^2$$

A plot of the natural frequencies of the system is shown in Fig. 9-6.

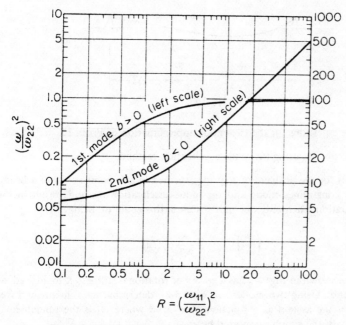

$$R = \left(\frac{\omega_{11}}{\omega_{22}} \right)^2$$

Fig. 9-6. First two natural frequencies of the system of Fig. 9-5.

13. A uniform beam of Fig. 9-7, clamped at both ends, is excited by a concentrated

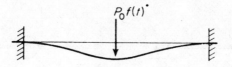

$P_0 f(t)^{\cdot}$

Fig. 9-7.

force $P_0 f(t)$ at midspan. Determine the deflection under the load and the resulting bending moment at the clamped ends.

14. If a uniformly distributed load of arbitrary time variation is applied to a uniform cantilever beam, determine the participation factor for the first three modes.

15. A spring of stiffness k is attached to a uniform beam, as shown in Fig. 9-8. Show that the one-mode approximation results in the frequency equation

$$\left(\frac{\omega}{\omega_1}\right)^2 = 1 + 1.5\left(\frac{k}{M}\right)\left(\frac{Ml^3}{\pi^4 EI}\right)$$

where

$$\omega_1^2 = \frac{\pi^4 EI}{Ml^3}$$

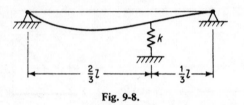

Fig. 9-8.

16. Write the equations for the two-mode approximation of Prob. 15.

17. Repeat Prob. 16, using the mode-acceleration method.

18. Show that for the problem of a spring attached to any point $x = a$ of a beam, both the constrained-mode and the mode-acceleration methods result in the same equation when only one mode is used, this equation being:

$$\left(\frac{\omega}{\omega_1}\right)^2 = 1 + \frac{k}{M\omega_1^2}\varphi_1^2(a)$$

19. The beam shown in Fig. 9-9 has a spring of rotational stiffness K lb. in./rad. at the left end. Using two modes in Eq. (9.5-8), determine the fundamental frequency of the system as a function of $K/M\omega_1^2$ where ω_1 is the fundamental frequency of the simply supported beam.

20. If both ends of the beam of Fig. 9-9 are restrained by springs of stiffness K, determine the fundamental frequency. As K approaches infinity, the result should approach that of the clamped-ended beam.

Fig. 9-9.   **Fig. 9-10.**

21. An airplane is idealized to a simplified model of a uniform beam of length l and mass per unit length m with a lumped mass M_0 at its center, as shown in Fig. 9-10. Using the translation of M_0 as one of the generalized coordinates, write the equations of motion and establish the natural frequency of the symmetric mode.

22. For the system of Prob. 21, determine the antisymmetric mode by using the rotation of the fuselage as one of the generalized coordinates.

23. If wing tip tanks of mass M_1 are added to the system of Prob. 21, determine the new frequency.

24. Using the method of constrained modes, show that the effect of adding a mass m_1 with moment of inertia J_1 to a point x_1 on the structure changes the first natural frequency ω_1 to

$$\omega_1' = \frac{\omega_1}{\sqrt{1 + \dfrac{m_1}{M_1}\,\varphi_1^2(x_1) + \dfrac{J_1}{M_1}\,\varphi_1'^2(x_1)}}$$

and the generalized mass and damping to

$$M_1' = M_1\left\{1 + \frac{m_1\varphi_1^2(x_1)}{M_1} + \frac{J_1}{M_1}\,\varphi_1'^2(x_1)\right\}$$

$$\zeta_1' = \frac{\zeta_1}{\sqrt{1 + \dfrac{m_1}{M_1}\,\varphi_1^2(x_1) + \dfrac{J_1}{M_1}\,\varphi_1'^2(x_1)}}$$

where a one mode approximation is used for the inertia forces.

10 | Random Vibrations

10.1 Introduction

With the development of the jet and rocket engines, engineers have become aware of a new aspect of vibrations. These engines generate intense noise and vibration energy which are random over a wide frequency range. The resulting vibrations are also random and require for their interpretation an approach which differs from those discussed earlier in the text.

Fig. 10.1-1 A record of random vibration.

An example of a random vibration is shown in Fig. 10.1-1. By random vibration we mean that the instantaneous value is not predictable as a time function. However, there are certain properties of random functions which can be described statistically. For example, it is possible to predict the probability of finding the instantaneous value within a certain range x to $x + \Delta x$. Other quantities, such as the mean or the mean square values, can be established by averaging, and the frequency content of the record can be determined by various methods based on Fourier analysis.

314

10.2 The Frequency Response Function

If a linear spring-mass system with viscous damping is excited by a force $F(t)$, its equation of motion is

$$\ddot{x} + 2\zeta\omega\dot{x} + \omega^2 x = \frac{1}{m} F(t) \tag{10.2-1}$$

When the excitation is in the form of a motion $y(t)$ of the base, an equation similar in form is obtained:

$$\ddot{z} + 2\zeta\omega\dot{z} + \omega^2 z = -\ddot{y}(t) \tag{10.2-2}$$

where $z = (x - y)$ is the relative motion between the mass m and the base. Thus either equation may be treated similarly by noting that $(1/m)F(t)$ and $-\ddot{y}(t)$ are equivalent excitations.

The general solution of these equations consists of a homogeneous solution, which depends on the initial conditions and which diminishes with time due to damping, and a particular solution which depends on the excitation $(1/m)F(t)$ or $-\ddot{y}(t)$.

To establish certain fundamental concepts necessary in understanding random vibrations, we will first consider the particular solution of Eq. (10.2-1) when the excitation is harmonic and described by the equation

$$F(t) = \mathcal{R}(F_0 e^{i\omega t}) \tag{10.2-3}$$

where $\mathcal{R}$ denotes the real part of the complex quantity. Using the method of complex algebra discussed in Sec. 3.2, the response x and the impressed force are vectors of fixed magnitude differing by a phase ϕ and rotating together with common angular speed ω according to the equation

$$x = \mathcal{R}(X e^{i\omega t}) \tag{10.2-4}$$

where

$$X = \left(\frac{F_0}{k}\right) H(\omega) \tag{10.2-5}$$

and

$$H(\omega) = \frac{1}{\left[1 - \left(\dfrac{\omega}{\omega_n}\right)^2\right] + i\left[2\zeta\dfrac{\omega}{\omega_n}\right]} \tag{10.2-6}$$

Here the quantity $H(\omega)$ is referred to as the admittance or the *frequency response function*, which we have already encountered in Chapter 3. For small damping, its peak value occurs at $\omega/\omega_n = 1.0$ and the sharpness of the resonance curve is indicated by $Q = 1/2\zeta$.

In random vibrations, the phase ϕ has little meaning and is therefore ignored. We are concerned mainly with the average energy which we can associate with the mean square value of x. The mean square value, designated by the notation $\overline{x^2}$ is found by integrating x^2 over a time interval T and taking its average value

$$\overline{x^2} = \frac{1}{T} \int_0^T x^2 \, dt \qquad (10.2\text{-}7)$$

This equation can, of course, be applied to the exciting force or the response. For example, if we have a harmonic force $F = F_0 \sin \omega t$, its mean square value is

$$\overline{F^2} = \frac{1}{T} \int_0^T \frac{F_0^2}{2} (1 - \cos 2\omega t) \, dt = \frac{F_0^2}{2} \qquad (10.2\text{-}8)$$

where T is a multiple of π/ω.

In applying this to the response given by Eq. (10.2-4), we note that the equation can be written as

$$x = \tfrac{1}{2}(Xe^{i\omega t} + X^* e^{-i\omega t}) \qquad (10.2\text{-}9)$$

where X^* is the complex conjugate of X.

Thus by squaring and substituting into Eq. (10.2-7), the mean square value of x is

$$\overline{x^2} = \frac{1}{T} \int_0^T \tfrac{1}{4}(X^2 e^{i2\omega t} + 2XX^* + X^{*2} e^{-i2\omega t}) \, dt$$

$$= \tfrac{1}{2} XX^* = \frac{|X|^2}{2} \qquad (10.2\text{-}10)$$

where an integer multiple of a period has been chosen for T to make the exponential terms integrate to zero. Due to the factor $1/T$ in front of the integral, the above result is accurate even when integrated over a time interval not a multiple of one period, provided T is large in comparison to the period.

Actually, for a random record the averaging time must be large (a long record is required for accurate evaluation of $\overline{x^2}$) and the mean square value is redefined as

$$\overline{x^2} = \lim_{T \to \infty} \frac{1}{T} \int_0^T x^2 \, dt \qquad (10.2\text{-}11)$$

If we substitute Eq. (10.2-5) for X into Eq. (10.2-10), the result is

$$\overline{x^2} = \frac{F_0^2}{2k^2} H(\omega) H^*(\omega) = \frac{\overline{F^2}}{k^2} |H(\omega)|^2 \qquad (10.2\text{-}12)$$

Thus the mean square response is equal to the mean square excitation multiplied by the square of the absolute value of the system response function.

10.3 Spectral Density

If the excitation $F(t)$ is a periodic time function, it can be regarded as a superposition of steady simple harmonic functions of different amplitudes, frequencies, and phases. It can then be represented by the equation

$$F(t) = \mathscr{R} \sum_n F_n e^{in\omega_0 t} \tag{10.3-1}$$

where F_n is a complex number and $\mathscr{R}$ stands for the real part of the series. Since each term of the above equation is a vector in the complex plane, the real part of $F(t)$ is equal to

$$F(t) = \tfrac{1}{2}\left\{\sum_n F_n e^{in\omega_0 t} + \sum_n F_n^* e^{-in\omega_0 t}\right\} \tag{10.3-2}$$

and the mean-square value of the excitation becomes

$$\overline{F^2} = \frac{1}{T}\int_0^T \frac{1}{4}\left\{\sum_n F_n e^{in\omega_0 t} + \sum_n F_n^* e^{-in\omega_0 t}\right\}^2 dt$$

$$= \sum_n \frac{F_n F_n^*}{2} = \sum_n \tfrac{1}{2}|F_n|^2 = \sum_n \overline{F_n^2} \tag{10.3-3}$$

Thus the mean-square value of the multi-frequency wave is simply the sum of the mean-square values of each harmonic component present, the result being a discrete frequency spectrum shown in Fig. 10.3-1.

We are, in general, interested in the mean-square contribution in each frequency interval $\Delta\omega$. Thus, by letting $S(n\omega_0)$ be the density of the

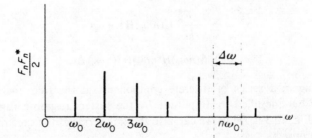

Fig. 10.3-1. Discrete frequency spectrum of a periodic function.

mean-square value in the interval $\Delta\omega$ at the frequency $n\omega_0$, we obtain

$$\sum_n S(n\omega_0)\,\Delta\omega = \sum_n \frac{F_n F_n^*}{2} \qquad (10.3\text{-}4)$$

and the quantity $S(n\omega_0) = F_n F_n^*/2\Delta\omega$ is the discrete spectral density function.

It is evident that, when $F(t)$ contains a large number of frequency components, the discrete spectrum approaches a continuous spectrum and the

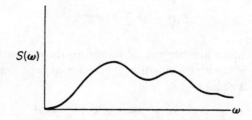

Fig. 10.3-2. Continuous spectral density curve.

discrete spectral density function $S(n\omega_0)$ becomes more nearly a continuous spectral density function $S(\omega)$, such as the one shown in Fig. 10.3-2. Random time functions generally contain a large number of frequency components approaching a continuous spectrum.

We are already aware of the fact that the response to a harmonic time function $F_n e^{in\omega_0}$ is

$$x = \tfrac{1}{2}(Xe^{in\omega_0 t} + X^* e^{-in\omega_0 t})$$

where $X = (F_n/k)H(n\omega_0)$ and its mean·square value is equal to

$$\overline{x^2} = \frac{F_n F_n^*}{2k^2}\,H(n\omega_0)H^*(n\omega_0) \qquad (10.2\text{-}10)$$

Thus for a multi-frequency input the mean-square response is the super-position of all such values or

$$\overline{x^2} = \sum_n \frac{F_n F_n^*}{2k^2}\,H(n\omega_0)H^*(n\omega_0)$$

$$= \sum_n S(n\omega_0)H(n\omega_0)H^*(n\omega_0)\,\Delta\omega \qquad (10.3\text{-}5)$$

which is again a series of discrete components at the frequencies of the excitation but modified in amplitude by the system response function, as shown in Fig. 10.3-3. Thus we can also think in terms of the spectral density of the response as

$$S_x(n\omega_0) = S_F(n\omega_0)H(n\omega_0)H^*(n\omega_0)$$

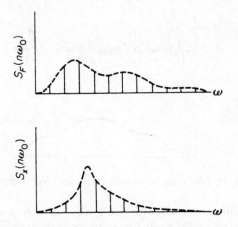

Fig. 10.3-3. Spectral density of excitation and response.

For a continuous spectrum the summation is replaced by an integral and the mean-square response is given by the equation:

$$\overline{x^2} = \int_0^\infty S(\omega)H(\omega)H^*(\omega)\, d\omega \qquad (10.3\text{-}6)$$

In practice, the spectral density function is generally given in terms of the frequency $f = \omega/2\pi$ c.p.s. and hence the equation becomes

$$\overline{x^2} = \int_0^\infty S(f)H(f)H^*(f)\, df \qquad (10.3\text{-}7)$$

where

$$S(f) = 2\pi S(\omega)$$

$$H(f) = \frac{1}{[1 - (f/f_n)^2] + i[2\zeta(f/f_n)]}$$

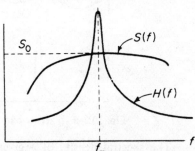

Fig. 10.3-4. $S(f)$ and $H(f)$ leading to $\bar{x}^2$ of Eq. 10.3-8.

For a lightly damped system, the response function $H(f)$ is peaked steeply at resonance, and if the spectral density of the excitation is broad, as in Fig. 10.3-4, the mean-square response can be approximated by the equation:

$$\overline{x^2} \cong f_n S(f_n)\frac{\pi}{4\zeta} \qquad (10.3\text{-}8)$$

Typical spectral density functions for two common types of random records are shown in Figs. 10.3-5 and 10.3-6. The first is a wide-band noise-type of record which has a broad spectral density function. The second is a

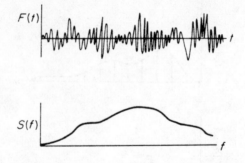

Fig. 10.3-5. Wide band record and its spectral density.

narrow-band random record which is typical of a response of a sharply resonant system to a wide-band input. Its spectral density function is concentrated around the frequency of the instantaneous variation within the envelope.

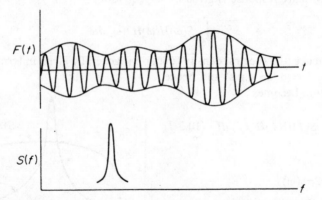

Fig. 10.3-6. Narrow band record and its spectral density.

The spectral density of a given record can be measured electronically by the circuit of Fig. 10.3-7. Here the spectral density is noted as the contribution of the mean-square value in the frequency interval $\Delta\omega$ divided by $\Delta\omega$.

$$S(\omega) = \lim_{\Delta\omega \to 0} \frac{\Delta(\overline{x^2})}{\Delta\omega} \qquad (10.3-9)$$

The band-pass filter of pass band $B = \Delta\omega$ passes $x(t)$ in the frequency interval ω to $\omega + \Delta x$, and the output is squared, averaged, and divided by $\Delta\omega$.

For high resolution, $\Delta\omega$ should be made as narrow as possible; however, the pass band of the filter cannot be reduced indefinitely without losing the

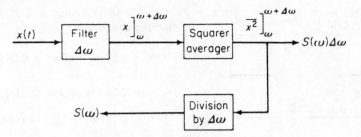

Fig. 10.3-7. Power spectral density analyzer.

reliability of the measurement. Also, a long record is required for the true estimate of the mean-square value [see Eq. (10.2-11)], but actual records are always of finite length. It is evident now that a parameter of importance is the product of the record length and the band width, $2BT$, which must be sufficiently large.

EXAMPLE 10.3-1. A single-degree-of-freedom system with natural frequency $\omega_n = \sqrt{k/m}$ and damping $\zeta = 0.20$ is excited by the force

$$F(t) = F \cos \tfrac{1}{2}\omega_n t + F \cos \omega_n t + F \cos \tfrac{3}{2}\omega_n t = \sum_{m=1/2,1,3/2} F \cos m\omega_n t$$

Determine the mean-square response and compare the output spectrum with that of the input.

Solution. The response of the system is simply the sum of the response of the single-degree-of-freedom system to each of the harmonic components of the exciting force.

$$x(t) = \sum_{m=1/2,1,3/2} \frac{1}{k} |H(n\omega)| F \cos (m\omega_n t - \phi_n)$$

where

$$|H(\tfrac{1}{2}\omega_n)| = \frac{1}{\sqrt{\frac{9}{16} + (0.20)^2}} = 1.29$$

$$|H(\omega_n)| = \frac{1}{\sqrt{4(0.20)^2}} = 2.50$$

$$|H(\tfrac{3}{2}\omega_n)| = \frac{1}{\sqrt{\frac{25}{16} + 9(0.02)^2}} = 0.72$$

$$\phi_{1/2} = \tan^{-1} \frac{4\zeta}{3} = 0.083\pi$$

$$\phi_1 = \tan^{-1} \infty = 0.50\pi$$

$$\phi_{3/2} = \tan^{-1} \frac{-12\zeta}{5} = -0.142\pi$$

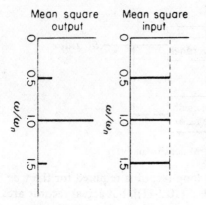

Mean square output	Mean square input

Substituting these values into $x(t)$, we obtain the equation

$$x(t) = \frac{F}{k} 1.29 \cos (0.5\omega_n - 0.083\pi)$$

$$+ 2.50 \cos (\omega_n t - 0.50\pi)$$

$$+ 0.72 \cos (1.5\omega_n t + 0.142\pi)$$

The mean-square response is then

$$\overline{x^2} = \frac{F^2}{2k^2} [(1.29)^2 + (2.50)^2 + (0.72)^2]$$

Fig. 10.3-8. Input and output spectra with discrete frequencies.

Figure 10.3-8 shows the input and output spectra for the problem. The components of the mean-square input are the same for each frequency and equal to $F^2/2$. The output spectra is modified by the system frequency response function.

10.4 Probability Distribution

For a random time function $x(t)$, shown in Fig. 10.4-1, we might ask: What is the probability at any instant of $x(t)$ lying between x_1 and $x_1 + \Delta x$? To answer this question, we could draw two horizontal lines at x_1 and $x_1 + \Delta x$, and proceed to sum the time intervals Δt during which $x(t)$ occupied this interval. This sum divided by the total time then represents the fraction of the time that $x(t)$ remained in the amplitude interval x_1 to $x_1 + \Delta x$, which is the probability that $x(t)$ will be found within this interval.

We can establish this information in still another way by starting at a very large negative value of x_1 and counting the time intervals for which $x(t)$

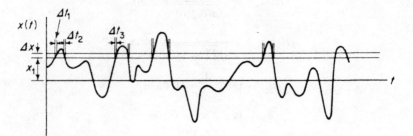

Fig. 10.4-1. Probability determination for a random function.

exceeded this amplitude; i.e., for which $x(t) < x_1$. If x_1 is chosen large enough, none of the curve will lie beyond it and hence $(1/t)\Sigma \Delta t$ will be zero. As x_1 is reduced $x(t)$ will exceed $(x(t) < x_1)$ the specified value of x more frequently and $(1/t)\Sigma \Delta t$ will tend to increase, as shown in Fig. 10.4-2. Such a curve at any x then gives the probability that $x(t)$ will lie on the negative side of x. As $x \rightarrow \infty$, all of the curve will lie in the region less than $x = \infty$, and hence the probability of $x(t)$ being less than

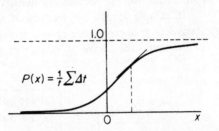

$$P(x) = \frac{1}{t}\sum \Delta t$$

Fig. 10.4-2. Cumulative probability.

$x = \infty$ is certain or $(1/t)\Sigma \Delta t = 1$. The curve of Fig. 10.4-2 must monotonically increase from zero at $x = -\infty$ to 1 at $x = \infty$, and is called the *cumulative probability* curve. Stated in mathematical terms, we have

$$P(-\infty) = 0, \qquad 0 \leq P(x) \leq 1, \qquad P(\infty) = 1 \qquad (10.4\text{-}1)$$

For a smooth curve, it is possible to define another function $p(x)$, representing the slope of the cumulative probability curve,

$$\frac{d}{dx} P(x) = p(x) \qquad (10.4\text{-}2)$$

which is plotted in Fig. 10.4-3, and called the *probability density* curve. It is evident that

$$P(x_1 + dx) - P(x_1) = p(x_1)\, dx \qquad (10.4\text{-}3)$$

which is the probability of $x(t)$ having a value in the interval x_1 and $x_1 + dx$. It also follows that

$$P(x_1) = \int_{-\infty}^{x_1} p(x)\, dx$$

and $(10.4\text{-}4)$

$$P(\infty) = \int_{-\infty}^{\infty} p(x)\, dx = 1.0$$

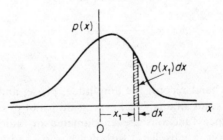

Fig. 10.4-3. Probability density.

Figure 10.4-4 shows a block diagram of a circuit which will perform the calculation for the probability density electronically.

If sample lengths of records of sufficient length all result in the same statistical properties, the record is said to be *stationary*. For nonstationary phenomena, the above procedure of time averaging cannot be used and the

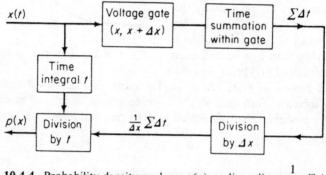

Fig. 10.4-4. Probability density analyzer $p(x) = \lim_{t \to \infty} \lim_{\Delta x \to 0} \frac{1}{t \Delta x} \Sigma \, \Delta t.$

statistical properties must be established at any time t_1 by examining a large ensemble of random records belonging to the same phenomenon. In this chapter we will discuss only stationary random phenomena with *ergodic* property, which implies that time averaging and ensemble averaging lead to the same statistical results.

Three important examples of time records frequently encountered in practice are shown in Fig. 10.4-5 where the mean value is arbitrarily chosen to be zero. The cumulative probability distribution for the sine wave is easily shown to be

$$P(x) = \tfrac{1}{2} + \frac{1}{\pi} \sin^{-1} \frac{x}{A} \qquad (10.4\text{-}5)$$

and its probability density, by differentiation, is

$$p(x) = \frac{1}{\pi \sqrt{A^2 - x^2}} \qquad |x| < A$$

$$= 0 \qquad\qquad\quad |x| > A \qquad (10.4\text{-}6)$$

For the wide band record, the amplitude, phase, and frequency are all varying randomly and an analytical expression is not possible for its instantaneous value. Such functions are encountered in radio noise, jet engine pressure fluctuation, atmospheric turbulence, etc., and a most likely probability distribution for such records is the *Gaussian distribution*, the normalized forms of which are given by the equations

$$p(x) = \frac{1}{\sqrt{2\pi}} e^{-(1/2)x^2}$$

$$P(x) = \frac{1}{\sqrt{2\pi}} \int_{-\infty}^{x} e^{-\frac{1}{2}\xi^2} d\xi \tag{10.4-7}$$

When a wide band record is put through a narrow band filter, or a resonance system where the filter bandwidth is small compared to its central

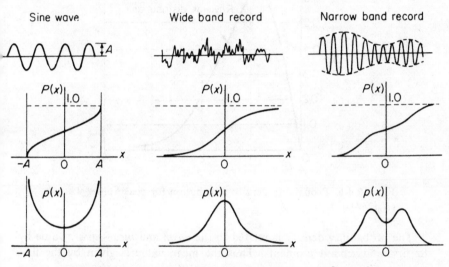

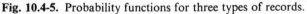

Fig. 10.4-5. Probability functions for three types of records.

frequency f_0, we obtain the third type of wave which is essentially a constant frequency oscillation with slowly varying amplitude and phase. Such waves can be considered to be a combination of a sine wave of frequency f_0 and noise in a neighboring frequency band. The typical probability density tends to drop off sharply outside some maximum amplitude and has a concave dip near zero, as in the sine wave case. The relative amounts of the sine wave and wide band fluctuations present in a narrow band record can often be estimated by a careful analysis of the probability density distribution.

Another quantity of great interest is the distribution of the peak values. Rice shows that the distribution of the peak values depends on a quantity $N_0/2M$ where N_0 is the number of zero crossings and $2M$ is the number of positive and negative peaks. For a sine wave or a narrow band, N_0 is equal to $2M$ so that the ratio $N_0/2M = 1$. For a wide band random record, the number of peaks will greatly exceed the number of zero crossings so that

$N_0/2M$ tends to approach zero. When $N_0/2M = 0$, the probability density distribution of peak values turns out to be Gaussian, whereas when $N_0/2M = 1$, as in the narrow band case, the probability density distribution of the peak values tends to a *Rayleigh distribution* given by the equation

$$p(x_p) = x_p e^{-(1/2)x_p^2} \tag{10.4-8}$$

and shown in Fig. 10.4-6.

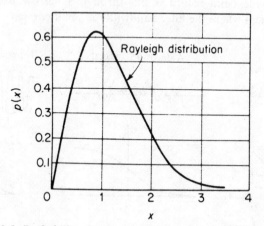

Fig. 10.4-6. Probability density distribution for peak values when $(N_0/2M) = 1$.

The probability density is related to the *mean* and *mean-square* value by the first and second moments. Thus the mean value is given by the first moment

$$\bar{x} = \int_{-\infty}^{\infty} x p(x) \, dx \tag{10.4-9}$$

which for a stationary random function becomes

$$\bar{x} = \lim_{T \to \infty} \frac{1}{2T} \int_{-T}^{T} x(t) \, dt \tag{10.4-10}$$

The mean-square value is determined from the second moment

$$\overline{x^2} = \int_{-\infty}^{\infty} x^2 p(x) \, dx \tag{10.4-11}$$

which again for the stationary random function is given by the equation

$$\overline{x^2} = \lim_{T \to \infty} \frac{1}{2T} \int_{-T}^{T} x^2(t) \, dt \tag{10.4-12}$$

Similarly, higher moments may be found.

The *variance* is defined as the mean-square value about the mean, or

$$\int_{-\infty}^{\infty} (x - \bar{x})^2 p(x)\, dx = \overline{x^2} - (\bar{x})^2 = \sigma^2 \tag{10.4-13}$$

and the *standard deviation* σ is the positive square root of the variance. When the mean value is zero, the standard deviation is equal to the positive square root of the mean-square value which is called the *root mean-square* (r.m.s.) value. Thus for a sine wave, $\bar{x} = 0$ and the standard deviation σ is equal to its r.m.s. value which is $A/\sqrt{2}$. In terms of the standard deviation, the Gaussian and the Rayleigh distributions are given by the equations

$$\text{Gaussian,} \quad p(x) = \frac{1}{\sigma\sqrt{2\pi}}\, e^{-x^2/2\sigma^2} \tag{10.4-14}$$

$$\text{Rayleigh,} \quad p(X) = \frac{X}{\sigma^2}\, e^{-X^2/2\sigma^2} \quad X > 0 \tag{10.4-15}$$

The standard deviation is a measure of the spread of the distribution about its mean value, and as σ decreases the Gaussian curve becomes narrower and taller, the area under the curve being always unity.

Many random phenomena are Gaussian (or normal) in distribution. In fact, when the random phenomenon is the result of the cumulative effects of a large number of independent causes, the distribution tends to become Gaussian, a fact which can be shown by the central limit theorem. The probability of $x(t)$ having a specified value is then completely established by the mean value and the standard deviation. It is generally convenient to let $\bar{x} = 0$, in which case $\sigma = \sqrt{\overline{x^2}}$.

When the distribution is Gaussian, the probability of $x(t)$ being between $\pm\lambda\sigma$ where λ is any positive number, is found from the equation

$$P[-\lambda\sigma \leq x(t) \leq \lambda\sigma] = \frac{1}{\sigma\sqrt{2\pi}} \int_{-\lambda\sigma}^{\lambda\sigma} e^{-x^2/2\sigma^2}\, dx \tag{10.4-16}$$

The following numerical values for $\lambda = 1$, 2, and 3 are convenient to remember

λ	$P[-\lambda\sigma \leq x(t) \leq \lambda\sigma]$	$P[|x| > \lambda\sigma]$
1	68.3%	31.7%
2	95.4%	4.6%
3	99.7%	0.3%

The probability of $x(t)$ lying outside $\pm\lambda\sigma$ is the probability of $|x|$ exceeding $\lambda\sigma$ which is 1 minus the above values, or expressible by the equation

$$P[|x| > \lambda\sigma] = \frac{2}{\sigma\sqrt{2\pi}} \int_{\lambda\sigma}^{\infty} e^{-x^2/2\sigma^2}\, dx = \text{erfc}\left(\frac{\lambda}{\sqrt{2}}\right) \tag{10.4-17}$$

For the Rayleigh function which is often encountered for the distribution of a positive quantity X, its mean value and the mean-square value are given as

$$\bar{X} = \int_0^\infty XP(X)\,dX = \int_0^\infty \frac{X^2}{\sigma^2} e^{-X^2/2\sigma^2}\,dX = \sqrt{\frac{\pi}{2}}\,\sigma \qquad (10.4\text{-}18)$$

$$\bar{X}^2 = \int_0^\infty X^2 P(X)\,dX = \int \frac{X^3}{\sigma^2} e^{-X^2/2\sigma^2}\,dX = 2\sigma^2 \qquad (10.4\text{-}19)$$

The variance associated with the Rayleigh distribution is

$$\sigma_X^2 = \bar{X}^2 - (\bar{X})^2 = \left(\frac{4-\pi}{2}\right)\sigma^2 \quad \text{or} \quad \sigma_X \cong \tfrac{2}{3}\sigma$$

Also the probability of X exceeding a specified value $\lambda\sigma$ is

$$P[X > \lambda\sigma] = \int_{\lambda\sigma}^\infty \frac{X}{\sigma^2} e^{-X^2/2\sigma^2}\,dX \qquad (10.4\text{-}20)$$

which has the following numerical values

λ	$P[X > \lambda\sigma]$
0	100%
1	60.7%
2	13.5%
3	1.2%

EXAMPLE 10.4-1. A random vibration test specification calls for:

Mean value of acceleration = 0
Acceleration density, 0.025 g^2/c.p.s.
Frequency range, 20 to 2000 c.p.s.

Determine the r.m.s. acceleration.

Solution. The r.m.s. acceleration is found by multiplying the acceleration density by the bandwidth and taking the square root of this number.

$$\text{rms accel} = \sqrt{0.025 \times (2000 - 20)} = \sqrt{49.5} = 7.03g$$

EXAMPLE 10.4-2. A random signal has a spectral density which is constant,

$$S(f) = 0.004 \text{ in.}^2/\text{c.p.s.}$$

between 20 to 1200 c.p.s., and zero outside this frequency range. Its mean value is 2.0 in. Determine its standard deviation and its rms value.

Solution. If the mean value is not zero, we must use Eq. (10.4-13):

$$\overline{x^2} = (\bar{x})^2 + \sigma^2$$

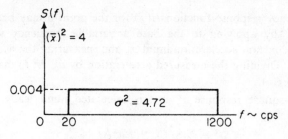

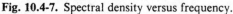

Fig. 10.4-7. Spectral density versus frequency.

The standard deviation is found from

$$\sigma^2 = \int_0^\infty S(f)\,df = \int_{20}^{1200} 0.004\,df = 4.72$$

and the mean-square value is

$$\overline{x^2} = 2^2 + 4.72 = 8.72$$

The rms value is then

$$\mathrm{rms} = \sqrt{\overline{x^2}} = \sqrt{8.72} = 2.95 \text{ in.}$$

The problem is graphically displayed by Fig. 10.4-7.

EXAMPLE 10.4-3. The response of any structure to a single point random excitation can be computed by a simple numerical procedure, provided the spectral density of the excitation and the frequency response curve of the structure are known. For example, consider the structure of Fig. 10.4-8(a) whose base is subjected to a random acceleration input with the power spectral density function shown in Fig. 10.4-8(b). It is desired to compute the response of the point p and establish the probability of exceeding any specified acceleration.

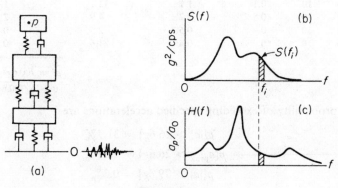

Fig. 10.4-8.

The frequency response function $H(f)$ for the point p may be obtained experimentally by applying to the base a variable frequency sinusoidal shaker with a constant acceleration input a_0, and measuring the acceleration response at p. Dividing the measured acceleration by a_0, $H(f)$ may appear as in Fig. 10.4-8(c).

The mean-square response $\overline{a_p^2}$ at p is calculated numerically from the equation

$$\overline{a_p^2} = \sum_i S(f_i) \, |H(f_i)|^2 \, \Delta f_i$$

The following numerical table illustrates the computational procedure.

Numerical Example

f c.p.s.	Δf c.p.s.	$S(f_i)$ g^2/c.p.s.	$\|H(f_i)\|$ Nondimensional	$\|H(f_i)\|^2 \Delta f$ c.p.s.	$S(f_i)\|H(f_i)\|^2 \Delta f$ g^2 units
0	10	0	1.0	10	0
10	10	0	1.0	10	0
20	10	0.2	1.1	12.1	2.4
30	10	0.6	1.4	19.6	11.8
40	10	1.2	2.0	40	48.0
50	10	1.8	1.3	16.9	30.5
60	10	1.8	1.3	16.9	30.5
70	10	1.1	2.0	40	44.0
80	10	0.9	3.7	137	123
90	10	1.1	5.4	291	320
100	10	1.2	2.2	48.4	57.7
110	10	1.1	1.3	16.9	18.6
120	10	0.8	0.8	6.4	5.1
130	10	0.6	0.6	3.6	2.2
140	10	0.3	0.5	2.5	0.8
150	10	0.2	0.6	3.6	0.7
160	10	0.2	0.7	4.9	0.1
170	10	0.1	1.3	16.9	1.7
180	10	0.1	1.1	12.1	1.2
190	10	0.5	0.7	4.9	2.3
200	10	0	0.5	2.5	0
210	10	0	0.4	1.6	0

$$\overline{a^2} = 700.6g^2$$
$$\sigma = \sqrt{700.6g^2} = 26.6g$$

The probability of exceeding specified accelerations are

$$p[|a| > 26.6g] = 31.7\%$$
$$p[a_{\text{peak}} > 26.6g] = 60.7\%$$
$$p[|a| > 79.8g] = 0.3\%$$
$$p[a_{\text{peak}} > 79.8g] = 1.2\%$$

10.5 Autocorrelation

For stationary ergodic phenomena, it is useful to introduce the concept of autocorrelation, defined by the equation

$$\psi(\tau) = \lim_{T \to \infty} \frac{1}{2T} \int_{-T}^{T} x(t)x(t + \tau)\, dt \qquad (10.5\text{-}1)$$

where $\psi(\tau)$ is a function only of the difference in time $t_2 - t_1 = \tau$.

Figure 10.5-1 shows a circuit diagram for the autocorrelation analyzer. The time delay τ is fixed during any run, and either changed for the next run or given a slow sweep in value.

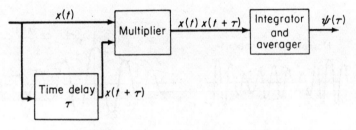

Fig. 10.5-1. Analyzer for autocorrelation.

It is of interest to examine again the three types of records shown in Fig. 10.4-5 to see what distinguishing features are revealed by their autocorrelation. These are shown in Fig. 10.5-2, and it is evident that the three types of records can be distinguished clearly by the autocorrelation function.

For the sine wave, the autocorrelation retains the same frequency ω_0 possessed by the original wave. Its value at $\tau = 0$ is equal to the mean square $\psi(0) = A^2/2$, and $\psi(\tau)$ is an even function independent of the phase. It should also be noted that $\psi(\tau)$ for the sine wave does not approach zero as in the other two cases.

The autocorrelation for the wide band noise is a peaked curve at $\tau = 0$, which tends to approach zero very quickly on either side. It implies that wide band random records have little or no correlation except near $\tau = 0$.

For the narrow band record, the autocorrelation has some of the characteristics found for the sine wave in that it is again an even function with a maximum at $\tau = 0$ and frequency ω_0 corresponding to the dominant or central frequency. The difference appears in the fact that $\psi(\tau)$ approaches zero for large τ for the narrow band record. It is evident then that hidden

Type of record	Autocorrelation

Sine wave $x(t) = A \sin (\omega_0 t + \theta)$

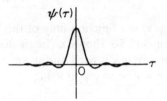

$$\psi(\tau) = \frac{A^2}{2} \cos \omega_0 \tau$$

Wide band noise $x(t)$

$\psi(\tau)$

Narrow band response

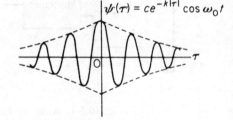

$$\psi(\tau) = ce^{-k|\tau|} \cos \omega_0 t$$

Fig. 10.5-2. Autocorrelation for three types of records.

periodicities in a random record can be detected by determining $\psi(\tau)$ for large values of τ.

10.6 Method of Fourier Transform

In Sec. 10.3 we examined the frequency content of periodic time functions which resulted in discrete frequency spectra. The concept of the spectral density was then introduced as the ratio of the mean-square value in the frequency interval divided by the frequency interval, this quantity approaching a continuous variation as the discrete spectrum tended to a continuous spectrum. Thus, when the spectral density function is known, the task of determining the mean-square value of a time function reduces to that of summing the spectral density over the frequency.

To determine the spectral density function, the Fourier series or the

Fourier transform formulation may be used. For the periodic function, $x(t)$ can be represented by the real part of the Fourier series in complex form

$$x(t) = \mathscr{R} \sum_{n=-\infty}^{\infty} c_n e^{in\omega_0 t} = \tfrac{1}{2} \sum_{n=-\infty}^{\infty} \{C_n e^{in\omega_0 t} + C_n^* e^{-in\omega_0 t}\} \quad (10.6\text{-}1)$$

which compared to Eqs. (10.2-4) and (10.2-9) is simply the sum of the single frequency components. Thus the Fourier series leads to the discrete spectrum which also results from the principle of superposition of linear systems.

When the record is aperiodic (nonperiodic), the Fourier integral must be used in place of the Fourier series. The Fourier integral may be viewed as a limiting form of the Fourier series as the period $2T$ is extended to infinity. To visualize this intuitive approach, we first note that the coefficient of the Fourier series is equal to

$$C_n = \frac{1}{2T} \int_{-T}^{T} x(\xi) e^{-in\omega_0 \xi} \, d\xi \quad (10.6\text{-}2)$$

so that $x(t)$ can be written as

$$x(t) = \sum_{n=\infty}^{\infty} \frac{1}{2T} \int_{-T}^{T} x(\xi) e^{-in\omega_0 \xi} e^{in\omega_0 t} \, d\xi \quad (10.6\text{-}3)$$

Since the increment of the frequency is

$$\Delta\omega = (n+1)\omega_0 - n\omega_0 = \omega_0 = \frac{2\pi}{2T}$$

we can replace $1/T$ with $\Delta\omega/\pi$. Now as $T \to \infty$, $\Delta\omega \to d\omega$, and $n\omega_0 \to \omega$, the above equation becomes

$$x(t) = \frac{1}{2\pi} \int_{-\infty}^{\infty} \left\{ \int_{-\infty}^{\infty} x(\xi) e^{-i\omega\xi} \, d\xi \right\} e^{i\omega t} \, d\omega \quad (10.6\text{-}4)$$

which is the Fourier integral.

Since the quantity within the inner braces is a function only of $i\omega$, we can rewrite this equation in two parts as:

$$X(i\omega) = \int_{-\infty}^{\infty} x(\xi) e^{-i\omega\xi} \, d\xi \quad (10.6\text{-}5)$$

$$x(t) = \frac{1}{2\pi} \int_{-\infty}^{\infty} X(i\omega) e^{i\omega t} \, d\omega \quad (10.6\text{-}6)$$

The quantity $X(i\omega)$ is the Fourier transform of $x(t)$, and the two equations above are referred to as the Fourier transform pair.

We will now show that the autocorrelation function is related to the spectral density function (often called power spectral density) by the Fourier cosine transform. From Eq. (10.6-6),

$$x(t + \tau) = \frac{1}{2\pi} \int_{-\infty}^{\infty} X(i\omega)e^{i\omega(t+\tau)} \, d\omega \qquad (10.6\text{-}7)$$

Substituting this into Eq. (10.5-1),

$$\psi(\tau) = \lim_{T \to \infty} \frac{1}{2T} \int_{-T}^{T} x(t)x(t + \tau) \, dt$$

$$= \lim_{T \to \infty} \frac{1}{2T} \int_{-T}^{T} x(t) \frac{1}{2\pi} \int_{-\infty}^{\infty} X(i\omega)e^{i\omega t}e^{i\omega \tau} \, d\omega \, dt$$

$$= \frac{1}{2\pi} \int_{-\infty}^{\infty} \lim_{T \to \infty} \frac{1}{2T} \left\{ \int_{-\infty}^{\infty} x(t)e^{i\omega t} \, dt \right\} X(i\omega)e^{i\omega \tau} \, d\omega$$

$$= \frac{1}{2} \int_{-\infty}^{\infty} \lim_{T \to \infty} \frac{1}{2\pi T} X(-i\omega)X(i\omega)e^{i\omega \tau} \, d\omega \qquad (10.6\text{-}8)^*$$

If τ is allowed to go to zero, the autocorrelation as given by Eq. (10.5-1) reduces to the mean-square value as given by Eq. (10.2-7). The mean-square value is also the frequency integral of the spectral density function. Thus Eq. (10.6-8) with $\tau = 0$ is related to the mean-square value as follows:

$$\overline{x^2} = \lim \frac{1}{2T} \int_{-T}^{T} x^2(t) \, dt = \int_{0}^{\infty} S_x(\omega) \, d\omega = \frac{1}{2} \int_{-\infty}^{\infty} \lim_{T \to \infty} \frac{1}{2\pi T} X(-i\omega)X(i\omega) \, d\omega$$

$$= \int_{0}^{\infty} \lim_{T \to \infty} \frac{1}{2\pi T} X(i\omega)X(-i\omega) \, d\omega \qquad (10.6\text{-}9)$$

By comparing the integrands in the above equation, we arrive at the relationship

$$S_x(\omega) = \lim_{T \to \infty} \frac{1}{2\pi T} X(i\omega)X(-i\omega) = \lim_{T \to \infty} \frac{1}{2\pi T} X(i\omega)X^*(i\omega)$$

$$= \lim_{T \to \infty} \frac{1}{2\pi T} |X(i\omega)|^2 \qquad (10.6\text{-}10)$$

Substituting this result back into Eq. (10.6-8), the autocorrelation is expressed in terms of the power spectral density $S_x(\omega)$ as:

$$\psi(\tau) = \tfrac{1}{2} \int_{-\infty}^{\infty} S_x(\omega)e^{-i\omega \tau} \, d\tau \qquad (10.6\text{-}11)$$

* This result is directly available from Parseval's theorem:

$$\int_{-\infty}^{\infty} x(t)y(t + \tau) \, dt = \frac{1}{2\pi} \int_{-\infty}^{\infty} X(-i\omega) \, Y(i\omega)e^{i\omega \tau} \, d\omega$$

Furthermore, since $S_x(\omega)$ from Eq. (10.6-10) is seen to be an even function, the above equation can be rewritten as:

$$\psi(\tau) = \int_0^\infty S_x(\omega) \cos \omega\tau \, d\tau \qquad (10.6\text{-}12)$$

Its inverse is then

$$S_x(\omega) = \frac{1}{\pi} \int_{-\infty}^\infty \psi(\tau) e^{-i\omega\tau} \, d\tau = \frac{2}{\pi} \int_0^\infty \psi(\tau) \cos \omega\tau \, d\tau \qquad (10.6\text{-}13)$$

The Fourier cosine transform pair, Eqs. (10.6-12) and (10.6-13), relating the autocorrelation to the power spectral density, are the Wiener-Khinchin equations.

As a parallel to the Wiener-Khinchin equations, we can define the cross correlation between two quantities $x(t)$ and $y(t)$ as

$$\overline{x(t)y(t + \tau)} = \lim_{T \to \infty} \frac{1}{2T} \int_{-T}^T x(t)y(t + \tau) \, dt$$

$$= \frac{1}{2} \int_{-\infty}^\infty \lim_{T \to \infty} \frac{1}{2\pi T} X^*(i\omega)Y(i\omega)e^{i\omega\tau} \, d\omega$$

$$= \tfrac{1}{2} \int_{-\infty}^\infty S_{xy}(\omega) e^{-i\omega\tau} \, d\omega \qquad (10.6\text{-}14)$$

where the cross power density is defined by

$$S_{xy}(\omega) = \lim_{T \to \infty} \frac{1}{2\pi T} X^*(i\omega)Y(i\omega)$$

$$= \lim_{T \to \infty} \frac{1}{2\pi T} X(i\omega)Y^*(i\omega)$$

$$= S_{xy}^*(\omega) = S_{yx}(-\omega) \qquad (10.6\text{-}15)$$

Its inverse is then

$$S_{xy}(\omega) = \frac{1}{\pi} \int_{-\infty}^\infty x(t)y(t + \tau)e^{-i\omega\tau} \, d\tau \qquad (10.6\text{-}16)$$

which is the parallel to Eq. (10.6-13). Unlike the autocorrelation, the cross correlation and the cross power density functions are in general not even functions, and hence the limits $-\infty$ to $+\infty$ are retained.

EXAMPLE 10.6-1. Determine the Fourier coefficients C_n and the power spectral density of the periodic function shown in Fig. 10.6-1.

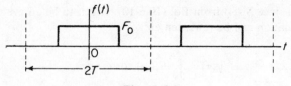

Fig. 10.6-1.

Solution. The period is $2T$ and C_n are found from Eq. (10.6-2):

$$C_0 = \frac{1}{2T} \int_{-T/2}^{T/2} F_0 \, d\xi = \frac{F_0}{2}$$

$$C_n = \frac{1}{2T} \int_{-T/2}^{T/2} F_0 e^{-in\omega_0 \xi} \, d\xi = \frac{F_0}{2}\left(\frac{\sin(n\pi/2)}{n\pi/2}\right)$$

Numerical values of C_n are computed as follows and plotted in Fig. 10.6-2.

n	$\dfrac{n\pi}{2}$	$\sin\dfrac{n\pi}{2}$	C_n
0	0	0	$\dfrac{F_0}{2} = 1.0\dfrac{F_0}{2}$
1	$\dfrac{\pi}{2}$	1	$\left(\dfrac{2}{\pi}\right)\dfrac{F_0}{2} = 0.636\dfrac{F_0}{2}$
2	π	0	0
3	$3\dfrac{\pi}{2}$	-1	$\left(-\dfrac{2}{3\pi}\right)\dfrac{F_0}{2} = -0.212\dfrac{F_0}{2}$
4	2π	0	0
5	$5\dfrac{\pi}{2}$	1	$\left(\dfrac{2}{5\pi}\right)\dfrac{F_0}{2} = 0.127\dfrac{F_0}{2}$

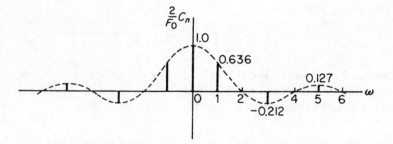

Fig. 10.6-2. Fourier coefficients versus n.

The mean-square value is determined from the equation

$$\bar{f}^2 = \frac{1}{2T} \int_{-T}^{T} f^2(t) \, dt = \frac{1}{2T} \int_{-T}^{T} \frac{1}{4} \left\{ \sum_n (C_n e^{in\omega_0 t} + C_n^* e^{-in\omega_0 t}) \right\}^2 dt$$

$$= \sum_{n=1}^{\infty} \frac{C_n C_n^*}{2}$$

and since $\bar{f}^2 = \int_0^{\infty} S_f(\omega) \, d\omega$, the spectral density function can be represented by a series of delta functions as:

$$S_f(\omega) = \sum_{n=1}^{\infty} \frac{C_n C_n^*}{2} \delta(\omega - n\omega_0)$$

10.7 Response of Continuous Structures to Distributed Random Forces

We consider here any elastic structure whose normal modes are defined by $\phi_j(x)$ and ω_j. Its deflection at any point x can be expressed in terms of the generalized coordinates $q_j(t)$ as

$$y(x, t) = \sum_j \phi_j(x) q_j(t) \qquad (10.7\text{-}1)$$

Generally, damping is an unknown quantity, however, we will assume it to be small and approximate it as viscous. In order not to introduce coupling between modes, its distribution over the structure will be assumed to give

$$\int_0^l c(x) \phi_j(x) \phi_k(x) \, dx = 0 \qquad \text{for } j \neq k$$

Under these assumptions, the equation of motion of the jth mode is

$$\ddot{q}_j(t) + 2\zeta_j \omega_j \dot{q}_j(t) + \omega_j^2 q_j(t) = \frac{1}{M_j} \int_0^l f(x, t) \phi_j(x) \, dx \qquad (10.7\text{-}2)$$

where

$$M_j = \int_0^l \phi_j^2(x) \, dm = \text{generalized mass}$$

$$\int_0^l f(x, t) \phi_j(x) \, dx = \text{generalized force}$$

When $f(x, t)$ is random, the above set of equations must be solved in a statistical sense. For this we now have at our disposal the method of Fourier transforms, F.T., the concepts of spectral density, and correlation between quantities.

Letting the capital letters stand for the F.T. of the corresponding quantities in lower case letters, and taking the F.T. of Eq. (10.7-2), we have

$$(-\omega^2 + i2\zeta_j\omega_j\omega + \omega_j^2)Q_j(i\omega) = \frac{1}{M_j} \int_0^l F(x, i\omega)\phi_j(x)\,dx$$

or

$$Q_j(i\omega) = \frac{H_j(i\omega)}{M_j\omega_j^2} \int_0^l F(x, i\omega)\phi_j(x)\,dx \qquad (10.7\text{-}3)$$

Substituting this into the F.T. of Eq. (10.7-1), we obtain

$$Y(x, i\omega) = \sum_j \phi_j(x) \frac{H_j(i\omega)}{M_j\omega_j^2} \int_0^l F(x, i\omega)\phi_j(x)\,dx \qquad (10.7\text{-}4)$$

Following the discussion of Sections 10.5 and 10.6, we next consider the correlation between the response at x and x'.

$$\overline{y(x, t)y(x', t)} = \lim_{T\to\infty} \frac{1}{2T} \int_{-T}^{T} y(x, t)y(x', t)\,dt \qquad (10.7\text{-}5)$$

By Parseval's theorem (See also Eq. 10.6-8 with $\tau = 0$), the above equation may be rewritten as

$$\overline{y(x, t)y(x', t)} = \frac{1}{2} \int_{-\infty}^{\infty} \lim_{T\to\infty} \frac{1}{2\pi T} Y^*(x, i\omega)Y(x', i\omega)\,d\omega \qquad (10.7\text{-}6)$$

Substituting Eq. (10.7-4) into Eq. (10.7-6), we obtain the result

$$\overline{y(x, t)y(x', t)} = \frac{1}{2} \sum_j \sum_k \phi_j(x)\phi_k(x') \int_{-\infty}^{\infty} \frac{H_j^*(i\omega)H_k(i\omega)}{M_j M_k \omega_j^2 \omega_k^2}$$
$$\times \int_0^l \int_0^l \lim_{T\to\infty} \frac{1}{2\pi T} F^*(x, i\omega)F(x', i\omega)\phi_j(x)\phi_k(x')\,dx\,dx'\,d\omega$$

$$(10.7\text{-}7)$$

To interpret this equation, let us examine the applied force without consideration of the structure, and define the *spacial correlation* of the applied forces at x and x' as

$$\overline{f(x, t)f(x', t)} = \lim_{T\to\infty} \frac{1}{2T} \int_{-T}^{T} f(x, t)f(x', t)\,dt \qquad (10.7\text{-}8)$$

Again from Parseval's theorem, Eq. (10.7-8) may be rewritten as

$$\overline{f(x, t)f(x', t)} = \frac{1}{2} \int_{-\infty}^{\infty} \lim_{T\to\infty} \frac{1}{2\pi T} F^*(x, i\omega)F(x', i\omega)\,d\omega$$
$$= \frac{1}{2} \int_{-\infty}^{\infty} R_f(x, x', \omega)\,d\omega \qquad (10.7\text{-}9)$$

where

$$R_f(x, x', \omega) = \lim_{T \to \infty} \frac{1}{2\pi T} F^*(x, i\omega(F)x', i\omega) \qquad (10.7\text{-}10)$$

is the *spacial correlation density* of the applied force at frequency ω. The quantity $R_f(x, x', \omega)$ can of course be obtained electronically by multiplying $f(x, t)$ and $f(x', t)$ and passing it through a narrow band filter whose central frequency is varied slowly through the desired frequency range.

Returning to Eq. (10.7-7) and substituting Eq. (10.7-10) into it, we obtain

$$\overline{y(x, t)y(x', t)} = \frac{1}{2} \sum_j \sum_k \phi_j(x)\phi_k(x') \int_{-\infty}^{\infty} \frac{H_j^*(i\omega)H_k(i\omega)}{M_j M_k \omega_j^2 \omega_k^2}$$

$$\times \int_0^l \int_0^l R_f(x, x', \omega)\phi_j(x)\phi_k(x')\, dx\, dx'\, d\omega \qquad (10.7\text{-}11)$$

It is convenient at this point to let

$$L_{jk}(i\omega) = \frac{1}{M_j M_k \omega_j^2 \omega_k^2} \int_0^l \int_0^l R_f(x, x', \omega)\phi_j(x)\phi_k(x')\, dx\, dx' \qquad (10.7\text{-}12)$$

and rewrite Eq. 10.7-11 as

$$\overline{y(x, t)y(x', t)} = \frac{1}{2} \sum_j \sum_k \phi_j(x)\phi_k(x') \int_{-\infty}^{\infty} L_{jk}(i\omega)H_j^*(i\omega)H_k(i\omega)\, d\omega \qquad (10.7\text{-}13)$$

Since the integrand of Eq. 10.7-13 is an even function of ω, the $\frac{1}{2}$ may be deleted and the lower limit of the integral changed from $-\infty$ to 0. The mean-square response at any point x is then found by letting $x' = x$ in Eq. 10.7-13.

$$\overline{y^2(x, t)} = \sum_j \sum_k \phi_j(x)\phi_k(x) \int_0^{\infty} L_{jk}(i\omega)H_j^*(i\omega)H_k(i\omega)\, d\omega \qquad (10.7\text{-}14)$$

REFERENCES

1. Bendat, J. S., *Principles and Applications of Random Noise Theory*. (New York: John Wiley & Sons, Inc., 1958.)

2. Blackman, R. B., and J. W. Tukey, *The Measurement of Power Spectra*. (New York: Dover Publications, Inc., 1958.).

3. Clarkson, B. L., "The Effect of Jet Noise on Aircraft Structures," *Aeronautical Quarterly*, Vol. 10, Part 2 (May 1959).

4. Cramer, H., *The Elements of Probability Theory*. (New York: John Wiley & Sons, Inc., 1955.)

5. Crandall, S. H., *Random Vibration*. (Cambridge, Mass.: The Technology Press of M.I.T., 1948.)

6. Crandall, S. H., *Random Vibration*, Vol. 2. (Cambridge, Mass.: The Technology Press of M.I.T., 1963.)

7. Eringen, A. C., "Response of Beams and Plates to Random Loads," *J. Appld. Mech.*, pp. 46–52 (March, 1957).

8. Fung, Y. C., "Statistical Aspects of Dynamic Loads," *J. Aero. Sci.*, pp. 317–330 (May, 1953).

9. Lyon, R. H., "Response of Strings to Random Noise Fields," *J. Acoustical Soc. of Amer.*, Vol. 28 No. 1, pp. 391–398.

10. Miles, J. W., "On Structural Fatigue Under Random Loading," *J. Aero. Sci.*, Vol. 21, pp. 753–762 (1954).

11. Powell, A., "On the Approximation to the Infinite Solution by the Method of Normal Modes for Random Vibrations," *J. Acoustical Soc. of Amer.*. Vol. 30, No. 12, pp. 1136–1139 (Dec., 1958).

12. Rice, S. O., *Mathematical Analysis of Random Noise*. (New York: Dover Publications, Inc., 1954.)

13. Thomson, W. T., and M. V. Barton, "The Response of Mechanical Systems to Random Excitation," *J. Appld. Mech.*, pp. 248–251 (June, 1957).

14. Thomson, W. T., "Continuous Structures Excited by Correlated Random Forces," *Int. Jour. Mech. Sci.*, Vol. 4 pp. 109–114 (1962).

PROBLEMS

1. Show that

$$\int_0^\infty \frac{d\eta}{[1 - \eta^2]^2 + [2\zeta\eta]^2} = \frac{\pi}{4\zeta} \quad \text{for} \quad \zeta \ll 1$$

2. The mean-square response of a single-degree-of-freedom system subjected to an excitation with a spectral density $S(f)$ is

$$\overline{x^2} = \int_0^\infty \frac{S(f)\,df}{[1 - (f/f_n)^2]^2 + [2\zeta(f/f_n)]^2}$$

If $S(f)$ is a constant and equal to S_0, show that

$$\overline{x^2} = \frac{S_0 \pi f_n}{4\zeta}$$

3. A single-degree-of-freedom system with natural frequency ω_n and damping factor $\zeta = 0.10$ is excited by the force

$$F(t) = F\cos(0.5\omega_n t - \theta_1) + F\cos(\omega_n t - \theta_2) + F\cos(2\omega_n t - \theta_3)$$

Show that the mean-square response is

$$\overline{y^2} = (1.575 + 25.0 + 0.110)\frac{1}{2}\left(\frac{F}{k}\right)^2 = 13.35\left(\frac{F}{k}\right)^2$$

4. The sharpness of the frequency response curve near resonance is often expressed in terms of $Q = \frac{1}{2}\zeta$. Points on either side of resonance where the response falls to a value $1/\sqrt{2}$ are called half-power points. Determine the respective frequencies of the half-power points in terms of ω_n and Q.

5. Determine the mean-square value of a periodic function represented by the real part of the series

$$f(t) = \sum_{n=-\infty}^{\infty} C_n e^{in\omega_0 t}$$

where $\omega_0 = 2\pi/\tau$ and τ = period.

6. Determine the mean-square response of a single-degree-of-freedom system in terms of $H(\omega)$, when excited by the periodic force of Prob. 5.

7. Throw a coin 20 times, recording 1 for heads and 0 for tails. Determine the probability of obtaining heads by dividing the cumulative heads by the number of throws and plot this number as a function of the number of throws. The curve should approach 0.50.

8. Derive the cumulative probability and the probability density equations for the sine wave; i.e., Eqs. (10.4-5) and (10.4-6).

9. What would the cumulative probability and the probability density curves look like for the rectangular wave shown in Fig. 10-1?

10. In Example 10.4-3, what is the probability of the instantaneous acceleration exceeding a value $53.2g$? Of the peak value exceeding this value?

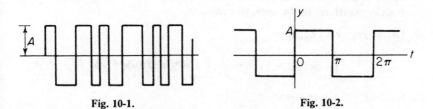

Fig. 10-1.   Fig. 10-2.

11. A random signal is found to have a constant spectral density of $S(f) = 0.002$ in.^{2}/c.p.s. between 20 c.p.s. and 2000 c.p.s. Outside this range, the spectral density is zero. Determine the standard deviation and the rms value if the mean value is 3.0 in. Plot this result.

12. Determine the spectral density of the periodic square wave with zero mean value, and compare this result with that of a harmonic wave of the same period.

13. Determine the complex form of the Fourier series for the rectangular wave shown in Fig. 10-2.

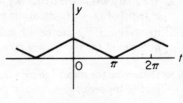

14. Determine the complex form of the Fourier series for the wave shown in Fig. 10-3.

Fig. 10-3.

15. Prove that the autocorrelation of a sine wave is another sine wave.

16. Determine the autocorrelation of the rectangular wave shown in Fig. 10-2.

17. Determine the autocorrelation of a rectangular pulse and plot it against τ.

18. Determine the mean-square value of a rectangular pulse between $-T$ and $+T$ by means of the Fourier integral.

19. Duhamel's integral can be written as

$$y(t) = \int_{-\infty}^{t} F(\tau)g(t - \tau)\, dt = \int_{0}^{\infty} F(t - \xi)g(\xi)\, d\xi$$

where $g(t)$ is the response of the system to a unit impulse $\delta(t)$, and $F(t)$ is the input which may have existed from time $t = -\infty$. If $F(t) = e^{i\omega t}$, show that

$$y(t) = H(i\omega)e^{i\omega t} = e^{i\omega t} \int_{0}^{\infty} e^{-i\omega \tau}g(\tau)\, d\tau$$

where $H(i\omega)$ is the frequency response function. Since $g(t) = 0$ for $t < 0$, the above equation indicates that the frequency response function is equal to the Fourier transform of the impulsive response.

20. Derive Parseval's equation

$$\int_{-\infty}^{\infty} x(t)y(t + \tau)\, dt = \frac{1}{2\pi} \int_{-\infty}^{\infty} X^*(i\omega)\, Y(i\omega)e^{i\omega \tau}\, d\omega$$

21. Starting with the relationship

$$x(t) = \int_{0}^{\infty} f(t - \xi)g(\xi)\, d\xi$$

and using the Fourier transform technique, show that

$$X(i\omega) = F(i\omega)H(i\omega)$$

and

$$\overline{x^2} = \int_{0}^{\infty} S_F(\omega)\, |H(i\omega)|^2\, d\omega$$

where

$$S_F(\omega) = \lim_{T \to \infty} \frac{1}{2\pi T} F(i\omega)F^*(i\omega)$$

22. Starting with the relationship

$$H(i\omega) = |H(i\omega)|\, e^{i\phi(\omega)}$$

show that

$$\frac{H(i\omega)}{H^*(i\omega)} = e^{i2\phi(\omega)}$$

23. Starting with the equations

$$S_{FX}(\omega) = \lim_{T\to\infty} \frac{1}{2\pi T} F^*(i\omega)X(i\omega) = \lim_{T\to\infty} \frac{1}{2\pi T} F^*(FH) = S_F H$$

and

$$S_{XF}(\omega) = \lim_{T\to\infty} \frac{1}{2\pi T} X^*F = \lim_{T\to\infty} \frac{1}{2\pi T}(F^*H^*)F = S_F H^*$$

show that

$$\frac{S_{FX}(\omega)}{S_{XF}(\omega)} = e^{i2\phi(x)}$$

and

$$\frac{S_F(\omega)}{S_{XF}(\omega)} = \frac{S_{FX}(\omega)}{S_F(\omega)} = H(i\omega)$$

24. The differential equation for the longitudinal motion of a uniform slender rod is

$$\frac{\partial^2 u}{\partial t^2} = c^2 \frac{\partial^2 u}{\partial x^2}$$

Show that for an arbitrary axial force at the end $x = 0$, with the other end $x = l$ free, the Laplace transform of the response is

$$\bar{u}(x, s) = \frac{-c\bar{F}(s)e^{-s(l/c)}}{sAE(1 - e^{-2s(l/c)})}\left\{e^{(s/c)(x-l)} + e^{-(s/c)(x-l)}\right\}$$

25. If the force in Prob. 24 is harmonic and equal to $F(t) = F_0 e^{i\omega t}$, show that

$$u(x, t) = \frac{cF_0 e^{i\omega t}\cos(\omega l/c)(x/l - 1)}{\omega AE \sin(\omega l/c)}$$

and

$$\sigma(x, t) = \frac{-\sin(\omega l/c)(x/l - 1)}{\sin(\omega l/c)}\frac{F}{A}e^{i\omega t}$$

where σ is the stress.

26. For structural damping γ [see Eq. (3.9-8)] in place of viscous damping, show that the frequency response function for the jth mode becomes

$$H_j(i\omega) = \frac{1}{1 - (\omega/\omega_j)^2 + i\gamma}$$

27. With $S(\omega)$ as the spectral density of the excitation stress at $x = 0$, show that the mean-square stress in Prob. 24 is

$$\overline{\sigma^2} \cong \frac{2\pi}{\gamma}\sum_n \frac{c}{n\pi l} S(\omega_n)\sin^2 n\pi \frac{x}{l}$$

where structural damping is assumed. The normal modes of the problem are

$$\varphi_n(x) = \sqrt{2} \cos n\pi(x/l - 1), \quad \omega_n = n\pi(c/l) \quad c = \sqrt{AE/m}.$$

28. If a structure is subjected to two concentrated forces at x_1 and x_2 which vary randomly, we have

$$f(x, t) = p_1(t)\delta(x - x_1) + p_2(t)\delta(x - x_2)$$

$$f(x', t) = p_1(t)\delta(x' - x_1) + p_2(t)\delta(x' - x_2)$$

Show that

$$\int_0^l \int_0^l R_f(x, x', \omega)\phi_j(x)\phi_k(x')\,dx\,dx' = \lim_{T \to \infty} \frac{1}{2\pi T} \{P_1^* P_1 \phi_j(x_1)\phi_k(x_1)$$

$$+ P_1^* P_2 \phi_j(x_1)\phi_k(x_2) + P_2^* P_1 \phi_j(x_2)\phi_k(x_1) + P_2^* P_2 \phi_j(x_2)\phi_k(x_2)\}$$

Interpret the terms $\lim\limits_{T \to \infty} \dfrac{1}{2\pi T} P_1^* P_1$, $\quad \lim\limits_{T \to \infty} \dfrac{1}{2\pi T} P_1^* P_2$ etc, and indicate how they can be obtained. Hint: examine Eqs. (10.7-8), (10.7-9), and (10.7-10).

Appendix A | Matrices

1. Definition of a Matrix

A rectangular array of m rows and n columns is called a matrix. For example

$$A = \begin{bmatrix} a_{11} & a_{12} & a_{13} & a_{14} \\ a_{21} & a_{22} & a_{23} & a_{24} \\ a_{31} & a_{32} & a_{33} & a_{34} \end{bmatrix} = [a_{ij}]$$

is a matrix of three rows and four columns. When $m = n$, the result is a square matrix. A unit matrix is a square matrix whose diagonal elements, from the top left to the bottom right, are equal to unity with all other elements equal to zero. For example

$$I = \begin{bmatrix} 1 & 0 & 0 \\ 0 & 1 & 0 \\ 0 & 0 & 1 \end{bmatrix}$$

is a 3×3 unit matrix.

A matrix consisting of elements in a single column is called a column matrix. Likewise, a matrix consisting of elements in a single row is called a row matrix.

The transpose of a matrix is one where the rows and columns are interchanged:

$$[a_{ij}]' = [a_{ji}]$$

345

2. Matrix Multiplication

The product of two matrices A and B is another matrix C.

$$AB = C$$

The element c_{ij} of C is determined by multiplying the elements of the ith row in A by the elements of the jth column in B according to the rule

$$c_{ij} = \sum_k a_{ik}b_{kj}$$

Example
   Let

$$A = \begin{bmatrix} 1 & 1 & 1 \\ 1 & 2 & 2 \\ 1 & 2 & 3 \end{bmatrix} \quad \text{and} \quad B = \begin{bmatrix} 2 & 0 \\ 0 & 1 \\ 3 & -1 \end{bmatrix}$$

Then

$$AB = \begin{bmatrix} 1 & 1 & 1 \\ 1 & 2 & 2 \\ 1 & 2 & 3 \end{bmatrix} \begin{bmatrix} 2 & 0 \\ 0 & 1 \\ 3 & -1 \end{bmatrix} = \begin{bmatrix} 5 & 0 \\ 8 & 0 \\ 11 & -1 \end{bmatrix} = C$$

i.e., $c_{32} = a_{31}b_{12} + a_{32}b_{22} + a_{33}b_{32} = 1 \times 0 + 2 \times 1 + 3 \times (-1) = -1$

It is evident that the number of columns in A must equal the number of rows in B, and that $AB \neq BA$.

3. Matrix Solution of Equations

A set of simultaneous algebraic equations

$$a_{11}x_1 + a_{12}x_2 + a_{13}x_3 = F_1$$
$$a_{21}x_1 + a_{22}x_2 + a_{23}x_3 = F_2$$
$$a_{31}x_1 + a_{32}x_2 + a_{33}x_3 = F_3$$

may be expressed in matrix form

$$\begin{bmatrix} a_{11} & a_{12} & a_{13} \\ a_{21} & a_{22} & a_{23} \\ a_{31} & a_{32} & a_{33} \end{bmatrix} \begin{Bmatrix} x_1 \\ x_2 \\ x_3 \end{Bmatrix} = \begin{Bmatrix} F_1 \\ F_2 \\ F_3 \end{Bmatrix}$$

or in abbreviated notation

$$AX = F$$

If the above equation is premultiplied by the inverse of A

$$A^{-1}AX = A^{-1}F$$

the solution

$$X = A^{-1}F$$

is obtained, provided

$$A^{-1}A = I$$

If the solution of only one of the x_i is desired, say x_1, Cramer's rule gives

$$x_1 = \frac{\begin{vmatrix} F_1 & a_{12} & a_{13} \\ F_2 & a_{22} & a_{23} \\ F_3 & a_{32} & a_{33} \end{vmatrix}}{|A|}$$

where $|A|$ is the determinant of the matrix A, and the numerator is the determinant found from A by replacing the column corresponding to x_i by the right side of the equation. It is evident that x_i is a part of the previous matrix solution

$$X = A^{-1}F$$

4. Inversion of a matrix

Consider the equations

$$a_{11}x_1 + a_{12}x_2 = F_1$$

$$a_{21}x_1 + a_{22}x_2 = F_2$$

By Cramer's rule, the solution is

$$x_1 = \frac{\begin{vmatrix} F_1 & a_{12} \\ F_2 & a_{22} \end{vmatrix}}{|A|} = \frac{a_{22}F_1 - a_{12}F_2}{(a_{11}a_{22} - a_{12}a_{21})}$$

$$x_2 = \frac{\begin{vmatrix} a_{11} & F_1 \\ a_{21} & F_2 \end{vmatrix}}{|A|} = \frac{-a_{21}F_1 + a_{11}F_2}{(a_{11}a_{22} - a_{12}a_{21})}$$

The two equations can then be expressed in the matrix form

$$\begin{Bmatrix} x_1 \\ x_2 \end{Bmatrix} = \frac{1}{|A|} \begin{bmatrix} a_{22} & -a_{12} \\ -a_{21} & a_{11} \end{bmatrix} \begin{Bmatrix} F_1 \\ F_2 \end{Bmatrix}$$

where it is noted that

$$A^{-1}A = \frac{1}{(a_{11}a_{22} - a_{12}a_{21})} \begin{bmatrix} a_{22} & -a_{12} \\ -a_{21} & a_{11} \end{bmatrix} \begin{bmatrix} a_{11} & a_{12} \\ a_{21} & a_{22} \end{bmatrix} = \begin{bmatrix} 1 & 0 \\ 0 & 1 \end{bmatrix}$$

In the more general case of an $n \times n$ matrix, the inverse A^{-1} can be determined according to the following rules and illustrated with

$$A = \begin{bmatrix} 1 & 1 & 1 \\ 1 & 2 & 2 \\ 1 & 0 & 3 \end{bmatrix}$$

$$|A| = 3$$

(a) Determine the minor M_{ij} of each element of the matrix A.

i.e., $M_{11} = \begin{vmatrix} 2 & 2 \\ 0 & 3 \end{vmatrix} = 6$ $M_{12} = \begin{vmatrix} 1 & 2 \\ 1 & 3 \end{vmatrix} = 1$

(b) Supply the proper sign to the minors to form the cofactors

$$\text{cofactor } A_{ij} = (-1)^{i+j}M_{ij}$$

The matrix of the cofactors is the adjoint matrix

$$[A_{ij}] = \begin{bmatrix} 6 & -1 & -2 \\ -3 & 2 & 1 \\ 0 & -1 & 1 \end{bmatrix}$$

(c) The inverse matrix of A is the transpose (rows and columns interchanged) of the adjoint matrix divided by the determinant of A.

$$A^{-1} = \frac{[A_{ij}]'}{|A|} = \frac{[A_{ji}]}{|A|}$$

$$= \tfrac{1}{3} \begin{bmatrix} 6 & -3 & 0 \\ -1 & 2 & -1 \\ -2 & 1 & 1 \end{bmatrix}$$

Check:

$$A^{-1}A = \tfrac{1}{3} \begin{bmatrix} 6 & -3 & 0 \\ -1 & 2 & -1 \\ -2 & 1 & 1 \end{bmatrix} \begin{bmatrix} 1 & 1 & 1 \\ 1 & 2 & 2 \\ 1 & 0 & 3 \end{bmatrix} = \tfrac{1}{3} \begin{bmatrix} 3 & 0 & 0 \\ 0 & 3 & 0 \\ 0 & 0 & 3 \end{bmatrix}$$

$$= \begin{bmatrix} 1 & 0 & 0 \\ 0 & 1 & 0 \\ 0 & 0 & 1 \end{bmatrix}$$

REFERENCES

Pipes, L. A., Applied Mathematics for Engineers and Physicists. McGraw-Hill Book Co. New York. 2nd Ed. 1958, Chapter 4.

Fraser, R. A., W. J. Duncan, & A. R. Collar, Elementary Matrices. Cambridge Univ. Press. London, 1957.

Introduction to Laplace Transformation

If $f(t)$ is a known function of t for values of $t > 0$, its Laplace Transform (L.T.) $\bar{f}(s)$ is defined by the equation

$$\bar{f}(s) = \int_0^\infty e^{-st}f(t)\,dt = \mathscr{L}f(t) \tag{1}$$

where s may be real or complex. The integral exists if

$$\lim_{t \to \infty} e^{-st}f(t) = 0 \tag{2}$$

Example Let $f(t) = t$. We find here that

$$\lim_{t \to \infty} e^{-st}t = 0$$

provided the real part of s is greater than zero. Its L.T. is found by integration by parts, letting

$$u = t \qquad\qquad du = dt$$

$$dv = e^{-st}\,dt \qquad v = -\frac{e^{-st}}{s}$$

The result is

$$\mathscr{L}t = -\frac{te^{-st}}{s}\Bigg]_0^\infty + \frac{1}{s}\int_0^\infty e^{-st}\,dt = \frac{1}{s^2} \qquad \mathscr{R}s > 0$$

In taking the L.T. of a differential equation, the L.T. of a derivative is encountered. If for a continuous function $f(t)$, $\mathscr{L}f(t) = \bar{f}(s)$ exists, and if $\lim_{t \to 0} f(t) = f(0)$, then the L.T. of its derivative $f'(t) = \dfrac{d}{dt}f(t)$ is equal to

$$\mathscr{L}f'(t) = s\bar{f}(s) - f(0) \tag{3}$$

Short Table of Laplace Transforms

	$\bar{f}(s)$	$f(t)$
(1)	1	$\delta(t) =$ unit impulse at $t = 0$
(2)	$\dfrac{1}{s}$	$\mathcal{U}(t) =$ unit step function at $t = 0$
(3)	$\dfrac{1}{s^n}\ (n = 1, 2, \cdots)$	$\dfrac{t^{n-1}}{(n-1)!}$
(4)	$\dfrac{1}{s + a}$	e^{-at}
(5)	$\dfrac{1}{(s + a)^2}$	te^{-at}
(6)	$\dfrac{1}{(s + a)^n}\ (n = 1, 2, \cdots)$	$\dfrac{1}{(n-1)!}\,t^{n-1}e^{-at}$
(7)	$\dfrac{1}{s(s + a)}$	$\dfrac{1}{a}(1 - e^{-at})$
(8)	$\dfrac{1}{s^2(s + a)}$	$\dfrac{1}{a^2}(e^{-at} + at - 1)$
(9)	$\dfrac{s}{s^2 + a^2}$	$\cos at$
(10)	$\dfrac{s}{s^2 - a^2}$	$\cosh at$
(11)	$\dfrac{1}{s^2 + a^2}$	$\dfrac{1}{a}\sin at$
(12)	$\dfrac{1}{s^2 - a^2}$	$\dfrac{1}{a}\sinh at$
(13)	$\dfrac{1}{s(s^2 + a^2)}$	$\dfrac{1}{a^2}(1 - \cos at)$
(14)	$\dfrac{1}{s^2(s^2 + a^2)}$	$\dfrac{1}{a^3}(at - \sin at)$
(15)	$\dfrac{1}{(s^2 + a^2)^2}$	$\dfrac{1}{2a^3}(\sin at - at \cos at)$
(16)	$\dfrac{s}{(s^2 + a^2)^2}$	$\dfrac{t}{2a}\sin at$
(17)	$\dfrac{s^2 - a^2}{(s^2 + a^2)^2}$	$t \cos at$
(18)	$\dfrac{1}{s^2 + 2\zeta\omega_0 s + \omega_0^2}$	$\dfrac{1}{\omega_0\sqrt{1 - \zeta^2}}\,e^{-\zeta\omega_0 t}\sin \omega_0\sqrt{1 - \zeta^2}\,t$

which can be easily proved from integration by parts, starting with Eq. (1). Similarly, the L.T. of the second derivative is

$$\mathscr{L}f''(t) = s^2\bar{f}(s) - sf(0) - f'(0)$$

Most ordinary differential equations can be solved by the elementary theory of L.T. The tables on p. 351 give the L.T. of simple functions. The table is also used to establish the inverse L.T., since if

$$\mathscr{L}f(t) = \bar{f}(s)$$

then

$$f(t) = \mathscr{L}^{-1}\bar{f}(s).$$

REFERENCE

Thomson, W. T., *Laplace Transformation*, 2nd Ed. Englewood Cliffs, N.J. Prentice-Hall, Inc. 1960.

Appendix C

Normal Modes of Uniform Beams

We assume the free vibrations of a uniform beam to be governed by Euler's differential equation.

$$EI \frac{\partial^4 y}{\partial x^4} + m \frac{\partial^2 y}{\partial t^2} = 0 \tag{1}$$

To determine the normal modes of vibration, the solution in the form

$$y(x, t) = \phi_n(x) e^{i\omega_n t} \tag{2}$$

is substituted into Eq. (1) to obtain the equation

$$\frac{d^4 \phi_n(x)}{dx^4} - \beta_n^4 \phi_n(x) = 0 \tag{3}$$

where:

$\phi_n(x)$ = characteristic function describing the deflection of the nth mode
m = mass density per unit length
$\beta_n^2 = m\omega_n^2 / EI$
$\omega_n = (\beta_n l)^2 \sqrt{EI/ml^4}$ = natural frequency of the nth mode.

The characteristic functions $\phi_n(x)$ and the normal-mode frequencies ω_n depend on the boundary conditions, and have been tabulated by Young and Felgar. An abbreviated summary taken from this work is presented here.

REFERENCE

Young, D. and R. P. Felgar Jr., *Tables of Characteristic Functions Representing Normal Modes of Vibration of a Beam*. The University of Texas Publication No. 4913, July 1, 1949.

I. Clamped-Clamped Beam

n	$\beta_n l$	$(\beta_n l)^2$	ω_n/ω_1
1	4.7300	22.3733	1.0000
2	7.8532	61.6728	2.7565
3	10.9956	120.9034	5.4039

Table 1
Characteristic Functions and Derivatives
Clamped-Clamped Beam
First Mode

$\dfrac{x}{l}$	ϕ_1	$\phi_1' = \dfrac{1}{\beta_1}\dfrac{d\phi_1}{dx}$	$\phi_1'' = \dfrac{1}{\beta_1^2}\dfrac{d^2\phi_1}{dx^2}$	$\phi_1''' = \dfrac{1}{\beta_1^3}\dfrac{d^3\phi_1}{dx^3}$
0.00	0.00000	0.00000	2.00000	-1.96500
0.04	0.03358	0.34324	1.62832	-1.96285
0.08	0.12545	0.61624	1.25802	-1.94862
0.12	0.26237	0.81956	0.89234	-1.91254
0.16	0.43126	0.95451	0.53615	-1.84732
0.20	0.61939	1.02342	0.19545	-1.74814
0.24	0.81459	1.02986	-0.12305	-1.61250
0.28	1.00546	0.97870	-0.41240	-1.44017
0.32	1.18168	0.87608	-0.66581	-1.23296
0.36	1.33419	0.72992	-0.87699	-0.99452
0.40	1.45545	0.54723	-1.04050	-0.73007
0.44	1.53962	0.33897	-1.15202	-0.44611
0.48	1.58271	0.11478	-1.20854	-0.15007
0.52	1.58271	-0.11478	-1.20854	0.15007
0.56	1.53962	-0.33897	-1.15202	0.44611
0.60	1.45545	-0.54723	-1.04050	0.73007
0.64	1.33419	-0.72992	-0.87699	0.99452
0.68	1.18168	-0.87608	-0.66581	1.23296
0.72	1.00546	-0.97870	-0.41240	1.44017
0.76	0.81459	-1.02986	-0.12305	1.61250
0.80	0.61939	-1.02342	0.19545	1.74814
0.84	0.43126	-0.95451	0.53615	1.84732
0.88	0.26237	-0.81956	0.89234	1.91254
0.92	0.12545	-0.61624	1.25802	1.94862
0.96	0.03358	-0.34324	1.62832	1.96285
1.00	0.00000	0.00000	2.00000	1.96500

Table 1
Characteristic Functions and Derivatives
Clamped-Clamped Beam
Second Mode

$\dfrac{x}{l}$	ϕ_2	$\phi_2' = \dfrac{1}{\beta_2}\dfrac{d\phi_2}{dx}$	$\phi_2'' = \dfrac{1}{\beta_2^2}\dfrac{d^2\phi_2}{dx^2}$	$\phi_2''' = \dfrac{1}{\beta_2^3}\dfrac{d^3\phi_2}{dx^3}$
0.00	0.00000	0.00000	2.00000	−2.00155
0.04	0.08834	0.52955	1.37202	−1.99205
0.08	0.31214	0.86296	0.75386	−1.93186
0.12	0.61058	1.00644	0.16713	−1.78813
0.16	0.92602	0.97427	−0.35923	−1.54652
0.20	1.20674	0.79030	−0.79450	−1.21002
0.24	1.41005	0.48755	−1.11133	−0.79651
0.28	1.50485	0.10660	−1.28991	−0.33555
0.32	1.47357	−0.30736	−1.32106	0.13566
0.36	1.31314	−0.70819	−1.20786	0.57665
0.40	1.03457	−1.05271	−0.96605	0.94823
0.44	0.66150	−1.30448	−0.62296	1.21670
0.48	0.22751	−1.43728	−0.21508	1.35744
0.52	−0.22751	−1.43728	0.21508	1.35744
0.56	−0.66150	−1.30448	0.62296	1.21670
0.60	−1.03457	−1.05271	0.96605	0.94823
0.64	−1.31314	−0.70819	1.20786	0.57665
0.68	−1.47357	−0.30736	1.32106	0.13566
0.72	−1.50485	0.10660	1.28991	−0.33555
0.76	−1.41005	0.48755	1.11133	−0.79651
0.80	−1.20674	0.79030	0.79450	−1.21002
0.84	−0.92602	0.97427	0.35923	−1.54652
0.88	−0.61058	1.00644	−0.16713	−1.78813
0.92	−0.31214	0.86296	−0.75386	−1.93186
0.96	−0.08834	0.52955	−1.37202	−1.99205
1.00	0.00000	0.00000	−2.00000	−2.00155

Table 1
Characteristic Functions and Derivatives
Clamped-Clamped Beam
Third Mode

$\dfrac{x}{l}$	ϕ_3	$\phi_3' = \dfrac{1}{\beta_3}\dfrac{d\phi_3}{dx}$	$\phi_3'' = \dfrac{1}{\beta_3^2}\dfrac{d^2\phi_3}{dx^2}$	$\phi_3''' = \dfrac{1}{\beta_3^3}\dfrac{d^3\phi_3}{dx^3}$
0.00	0.00000	0.00000	2.00000	−1.99993
0.04	0.16510	0.68646	1.12323	−1.97469
0.08	0.54804	0.99303	0.28189	−1.82280
0.12	0.98720	0.95006	−0.45252	−1.48447
0.16	1.34190	0.62285	−0.99738	−0.96698
0.20	1.50782	0.11050	−1.28572	−0.33199
0.24	1.42971	−0.46573	−1.28637	0.32333
0.28	1.10719	−0.98087	−1.01443	0.88956
0.32	0.59186	−1.32694	−0.53145	1.26880
0.36	−0.02445	−1.43171	0.06438	1.39529
0.40	−0.62837	−1.27099	0.65569	1.24912
0.44	−1.10739	−0.87257	1.12747	0.86096
0.48	−1.37174	−0.31031	1.38852	0.30669
0.52	−1.37174	0.31031	1.38852	−0.30669
0.56	−1.10739	0.87257	1.12747	−0.86096
0.60	−0.62837	1.27099	0.65569	−1.24912
0.64	−0.02445	1.43171	0.06438	−1.39529
0.68	0.59186	1.32694	−0.53145	−1.26880
0.72	1.10719	0.98087	−1.01443	−0.88956
0.76	1.42971	0.46573	−1.28637	−0.32333
0.80	1.50782	−0.11050	−1.28572	0.33199
0.84	1.34190	−0.62285	−0.99738	0.96698
0.88	0.98720	−0.95006	−0.45252	1.48447
0.92	0.54804	−0.99303	0.28189	1.82280
0.96	0.16510	−0.68646	1.12323	1.97469
1.00	0.00000	0.00000	2.00000	1.99993

2. Free-Free Beam

The natural frequencies of the free-free beam are equal to those of the clamped-clamped beam. The characteristic functions of the free-free beam are related to those of the clamped-clamped beam as follows.

$$
\begin{array}{ccc}
\text{free-free} & & \text{clamped-clamped} \\
\phi_n & = & \phi_n'' \\
\phi_n' & = & \phi_n''' \\
\phi_n'' & = & \phi_n \\
\phi_n''' & = & \phi_n'
\end{array}
$$

3. Clamped-Free Beam

n	$\beta_n l$	$(\beta_n l)^2$	ω_n/ω_1
1	1.8751	3.5160	1.0000
2	4.6941	22.0345	6.2669
3	7.8548	61.6972	17.5475

4. Clamped-Pinned Beam

n	$\beta_n l$	$(\beta_n l)^2$	ω_n/ω_1
1	3.9266	15.4182	1.0000
2	7.0686	49.9645	3.2406
3	10.2102	104.2477	6.7613

5. Free-Pinned Beam

The natural frequencies of the free-pinned beam are equal to those of the clamped-pinned beam. The characteristic functions of the free-pinned beam are related to those of the clamped-pinned beam as follows.

$$
\begin{array}{ccc}
\text{free-pinned} & & \text{clamped-pinned} \\
\phi_n & = & \phi_n'' \\
\phi_n' & = & \phi_n''' \\
\phi_n'' & = & \phi_n \\
\phi_n''' & = & \phi_n'
\end{array}
$$

Table 2
Characteristic Functions and Derivatives
Clamped-Free Beam
First Mode

$\dfrac{x}{l}$	ϕ_1	$\phi_1' = \dfrac{1}{\beta_1}\dfrac{d\phi_1}{dx}$	$\phi_1'' = \dfrac{1}{\beta_1^2}\dfrac{d^2\phi_1}{dx^2}$	$\phi_1''' = \dfrac{1}{\beta_1^3}\dfrac{d^3\phi_1}{dx^3}$
0.00	0.00000	0.00000	2.00000	−1.46819
0.04	0.00552	0.14588	1.88988	−1.46805
0.08	0.02168	0.28350	1.77980	−1.46710
0.12	0.04784	0.41286	1.66985	−1.46455
0.16	0.08340	0.53400	1.56016	−1.45968
0.20	0.12774	0.64692	1.45096	−1.45182
0.24	0.18024	0.75167	1.34247	−1.44032
0.28	0.24030	0.84832	1.23500	−1.42459
0.32	0.30730	0.93696	1.12889	−1.40410
0.36	0.38065	1.01771	1.02451	−1.37834
0.40	0.45977	1.09070	0.92227	−1.34685
0.44	0.54408	1.15612	0.82262	−1.30924
0.48	0.63301	1.21418	0.72603	−1.26512
0.52	0.72603	1.26512	0.63301	−1.21418
0.56	0.82262	1.30924	0.54408	−1.15612
0.60	0.92227	1.34685	0.45977	−1.09070
0.64	1.02451	1.37834	0.38065	−1.01771
0.68	1.12889	1.40410	0.30730	−0.93696
0.72	1.23500	1.42459	0.24030	−0.84832
0.76	1.34247	1.44032	0.18024	−0.75167
0.80	1.45096	1.45182	0.12774	−0.64692
0.84	1.56016	1.45968	0.08340	−0.53400
0.88	1.66985	1.46455	0.04784	−0.41286
0.92	1.77980	1.46710	0.02168	−0.28350
0.96	1.88988	1.46805	0.00552	−0.14588
1.00	2.00000	1.46819	0.00000	0.00000

Table 2
Characteristic Functions and Derivatives
Clamped-Free Beam
Second Mode

$\dfrac{x}{l}$	ϕ_2	$\phi_2' = \dfrac{1}{\beta_2}\dfrac{d\phi_2}{dx}$	$\phi_2'' = \dfrac{1}{\beta_2^2}\dfrac{d^2\phi_2}{dx^2}$	$\phi_2''' = \dfrac{1}{\beta_2^3}\dfrac{d^3\phi_2}{dx^3}$
0.00	0.00000	0.00000	2.00000	−2.03693
0.04	0.03301	0.33962	1.61764	−2.03483
0.08	0.12305	0.60754	1.23660	−2.02097
0.12	0.25670	0.80428	0.86004	−1.98590
0.16	0.42070	0.93108	0.49261	−1.92267
0.20	0.60211	0.99020	0.14007	−1.82682
0.24	0.78852	0.98502	−0.19123	−1.69625
0.28	0.96827	0.92013	−0.49475	−1.53113
0.32	1.13068	0.80136	−0.76419	−1.33373
0.36	1.26626	0.63565	−0.99384	−1.10821
0.40	1.36694	0.43094	−1.17895	−0.86040
0.44	1.42619	0.19593	−1.31600	−0.59748
0.48	1.43920	−0.06012	−1.40289	−0.32772
0.52	1.40289	−0.32772	−1.43920	−0.06012
0.56	1.31600	−0.59748	−1.42619	0.19593
0.60	1.17895	−0.86040	−1.36694	0.43094
0.64	0.99384	−1.10821	−1.26626	0.63565
0.68	0.76419	−1.33373	−1.13068	0.80136
0.72	0.49475	−1.53113	−0.96827	0.92013
0.76	0.19123	−1.69625	−0.78852	0.98502
0.80	−0.14007	−1.82682	−0.60211	0.99020
0.84	−0.49261	−1.92267	−0.42070	0.93108
0.88	−0.86004	−1.98590	−0.25670	0.80428
0.92	−1.23660	−2.02097	−0.12305	0.60754
0.96	−1.61764	−2.03483	−0.03301	0.33962
1.00	−2.00000	−2.03693	0.00000	0.00000

Table 2
Characteristic Functions and Derivatives
Clamped-Free Beam
Third Mode

$\dfrac{x}{l}$	ϕ_3	$\phi_3' = \dfrac{1}{\beta_3}\dfrac{d\phi_3}{dx}$	$\phi_3'' = \dfrac{1}{\beta_3^2}\dfrac{d^2\phi_3}{dx^2}$	$\phi_3''' = \dfrac{1}{\beta_3^3}\dfrac{d^3\phi_3}{dx^3}$
0.00	0.00000	0.00000	2.00000	−1.99845
0.04	0.08839	0.52979	1.37287	−1.98892
0.08	0.31238	0.86367	0.75558	−1.92871
0.12	0.61120	1.00785	0.16974	−1.78480
0.16	0.92728	0.97665	−0.35563	−1.54286
0.20	1.20901	0.79394	−0.78975	−1.20575
0.24	1.41376	0.49285	−1.10515	−0.79124
0.28	1.51056	0.11405	−1.28189	−0.32872
0.32	1.48203	−0.29711	−1.31055	0.14479
0.36	1.32534	−0.69422	−1.19398	0.58908
0.40	1.05185	−1.03374	−0.94753	0.96533
0.44	0.68568	−1.27881	−0.59802	1.24030
0.48	0.26103	−1.40247	−0.18130	1.39004
0.52	−0.18130	−1.39004	0.26103	1.40247
0.56	−0.59802	−1.24030	0.68568	1.27881
0.60	−0.94753	−0.96533	1.05185	1.03374
0.64	−1.19398	−0.58908	1.32534	0.69422
0.68	−1.31055	−0.14479	1.48203	0.29711
0.72	−1.28189	0.32872	1.51056	−0.11405
0.76	−1.10515	0.79124	1.41376	−0.49285
0.80	−0.78975	1.20575	1.20901	−0.79394
0.84	−0.35563	1.54236	0.92728	−0.97665
0.88	0.16974	1.78480	0.61120	−1.00785
0.92	0.75558	1.92871	0.31238	−0.86367
0.96	1.37287	1.98892	0.08829	−0.52979
1.00	2.00000	1.99845	0.00000	0.00000

Table 3
Characteristic Functions and Derivatives
Clamped-Pinned Beam
First Mode

$\dfrac{x}{l}$	ϕ_1	$\phi_1' = \dfrac{1}{\beta_1}\dfrac{d\phi_1}{dx}$	$\phi_1'' = \dfrac{1}{\beta_1^2}\dfrac{d^2\phi_1}{dx^2}$	$\phi_1''' = \dfrac{1}{\beta_1^3}\dfrac{d^3\phi_1}{dx^3}$
0.00	0.00000	0.00000	2.00000	−2.00155
0.04	0.02338	0.28944	1.68568	−2.00031
0.08	0.08834	0.52955	1.37202	−1.99203
0.12	0.18715	0.72055	1.06060	−1.97079
0.16	0.31214	0.86296	0.75386	−1.93187
0.20	0.45574	0.95776	0.45486	−1.87177
0.24	0.61058	1.00643	0.16712	−1.78812
0.28	0.76958	1.01105	−0.10554	−1.67975
0.32	0.92601	0.97427	−0.35923	−1.54652
0.36	1.07363	0.89940	−0.59009	−1.38932
0.40	1.20675	0.79029	−0.79450	−1.21002
0.44	1.32032	0.65138	−0.96918	−1.01128
0.48	1.41006	0.48755	−1.11133	−0.79652
0.52	1.47245	0.30410	−1.21875	−0.56977
0.56	1.50485	0.10661	−1.28992	−0.33555
0.60	1.50550	−0.09916	−1.32402	−0.09872
0.64	1.47357	−0.30736	−1.32106	0.13566
0.68	1.40913	−0.51224	−1.28180	0.36247
0.72	1.31313	−0.70820	−1.20786	0.57666
0.76	1.18741	−0.88996	−1.10157	0.77340
0.80	1.03457	−1.05270	−0.96606	0.94823
0.84	0.85795	−1.19210	−0.80507	1.09714
0.88	0.66151	−1.30448	−0.62295	1.21670
0.92	0.44974	−1.38693	−0.42455	1.30414
0.96	0.22752	−1.43727	−0.21507	1.35743
1.00	0.00000	−1.45420	0.00000	1.37533

Table 3
Characteristic Functions and Derivatives
Clamped-Pinned Beam
Second Mode

$\dfrac{x}{l}$	ϕ_2	$\phi_2' = \dfrac{1}{\beta_2}\dfrac{d\phi_2}{dx}$	$\phi_2'' = \dfrac{1}{\beta_2^2}\dfrac{d^2\phi_2}{dx^2}$	$\phi_2''' = \dfrac{1}{\beta_2^3}\dfrac{d^3\phi_2}{dx^3}$
0.00	0.00000	0.00000	2.00000	−2.00000
0.04	0.07241	0.48557	1.43502	−1.99300
0.08	0.25958	0.81207	0.87658	−1.94824
0.12	0.51697	0.98325	0.33937	−1.83960
0.16	0.80176	1.00789	−0.15633	−1.65333
0.20	1.07449	0.90088	−0.58802	−1.38736
0.24	1.30078	0.68345	−0.93412	−1.05012
0.28	1.45308	0.38242	−1.17673	−0.65879
0.32	1.51208	0.02894	−1.30380	−0.23724
0.36	1.46765	−0.34350	−1.31068	0.18649
0.40	1.31923	−0.70122	−1.20092	0.58286
0.44	1.07550	−1.01270	−0.98634	0.92349
0.48	0.75348	−1.25090	−0.68631	1.18364
0.52	0.37700	−1.39515	−0.32640	1.34442
0.56	−0.02536	−1.43265	0.06348	1.39438
0.60	−0.42268	−1.35944	0.45136	1.33056
0.64	−0.78413	−1.18058	0.80569	1.15876
0.68	−1.08158	−0.90972	1.09776	0.89319
0.72	−1.29186	−0.56793	1.30395	0.55537
0.76	−1.39858	−0.18205	1.40755	0.17245
0.80	−1.39351	0.21752	1.40010	−0.22494
0.84	−1.27726	0.59923	1.28198	−0.60506
0.88	−1.05919	0.93288	1.06244	−0.93759
0.92	−0.75676	1.19208	0.75879	−1.19604
0.96	−0.39406	1.35629	0.39504	−1.35983
1.00	0.00000	1.41251	0.00000	−1.41592

Table 3
Characteristic Functions and Derivatives
Clamped-Pinned Beam
Third Mode

$\dfrac{x}{l}$	ϕ_3	$\phi_3' = \dfrac{1}{\beta_3}\dfrac{d\phi_3}{dx}$	$\phi_3'' = \dfrac{1}{\beta_3^2}\dfrac{d^2\phi_3}{dx^2}$	$\phi_3''' = \dfrac{1}{\beta_3^3}\dfrac{d^3\phi_3}{dx^3}$
0.00	0.00000	0.00000	2.00000	−2.00000
0.04	0.14410	0.65020	1.18532	−1.97961
0.08	0.48626	0.97168	0.39742	−1.85535
0.12	0.89584	0.98593	−0.30845	−1.57331
0.16	1.25604	0.74002	−0.86560	−1.13046
0.20	1.47476	0.30725	−1.21523	−0.56678
0.24	1.49419	−0.21934	−1.32168	0.04683
0.28	1.29662	−0.73864	−1.18195	0.62397
0.32	0.90489	−1.15556	−0.82867	1.07934
0.36	0.37703	−1.39512	−0.32637	1.34445
0.40	−0.20439	−1.41364	0.23807	1.37996
0.44	−0.74658	−1.20525	0.76897	1.18287
0.48	−1.16223	−0.80234	1.17711	0.78746
0.52	−1.38422	−0.26994	1.39411	0.26005
0.56	−1.37687	0.30522	1.38344	−0.31179
0.60	−1.14194	0.82907	1.14631	−0.83344
0.64	−0.71844	1.21582	0.72134	−1.21873
0.68	−0.17628	1.40210	0.17821	−1.40403
0.72	0.39519	1.35742	−0.39391	−1.35870
0.76	0.90188	1.08924	−0.90103	−1.09010
0.80	1.26035	0.64175	−1.25980	−0.64233
0.84	1.41160	0.08860	−1.41124	−0.08900
0.88	1.33072	−0.47918	−1.33049	0.47891
0.92	1.03098	−0.96820	−1.03085	0.96800
0.96	0.56168	−1.29798	−0.56162	1.29782
1.00	0.00000	−1.41429	0.00000	1.41414

Appendix D | Specifications of Vibration Bounds

Specifications for vibrations are often based on harmonic motion.

$$x = x_0 \sin \omega t$$

The velocity and acceleration are then available from differentiation and the following relationships for the peak values can be written.

$$\dot{x}_0 = 2\pi f x_0$$
$$\ddot{x}_0 = -4\pi^2 f^2 x_0 = -2\pi f \dot{x}_0$$

These equations can be represented on the log-log paper by rewriting them in the form

$$\ln \dot{x}_0 = \ln x_0 + \ln 2\pi f$$

$$\ln \dot{x}_0 = -\ln \ddot{x}_0 - \ln 2\pi f$$

By letting $x_0 =$ constant, the plot of $\ln \dot{x}_0$ against $\ln 2\pi f$ is a straight line of slope equal to $+1$. By letting $\ddot{x}_0 =$ constant, the plot of $\ln \dot{x}_0$ versus $\ln 2\pi f$ is again a straight line of slope -1. These lines are shown graphically in Fig. D-1. The graph is often used to specify bounds for the vibration. Shown in heavy lines are bounds for a maximum acceleration of 10 g, minimum and maximum frequencies of 5 and 500 c.p.s., and an upper limit for the displacement of 0.30 inch.

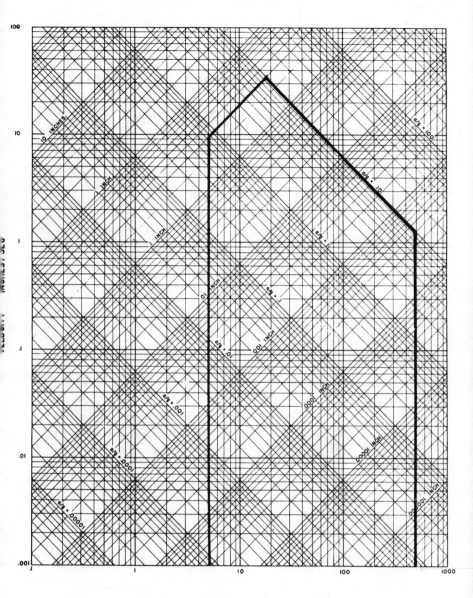

FREQUENCY — CPS

Fig. D-1.

Answers to Selected Problems

Chapter 1

1. 337 c.p.m.

4. 0.159 sec.

7. $f = \dfrac{1}{2\pi} \sqrt{\dfrac{k_1 k_2 (a/l)^2}{m[k_2 + k_1(a/l)^2]}}$

10. $f = \dfrac{1}{2\pi} \sqrt{\dfrac{\pi d^4 G g}{32 W K^2 l}}$

14. 8.0 in.

16. $\omega = \sqrt{\dfrac{T}{m}\left(\dfrac{a+b}{ab}\right)}$

20. 1.99 sec.

25. $\tau = 2\pi \sqrt{\dfrac{J_0 + (W/g)R^2}{ka^2}}$

29. $\omega = \sqrt{\dfrac{gab}{hk^2}}$

33. 5.57 c.p.s.; 2.78 c.p.s.

3. $k_2 = \tfrac{1}{3}k_1$

5. 2.00 lb.; 0.502 lb./in.

8. $f = \dfrac{1}{2\pi} \sqrt{\dfrac{ka^2 - Wb}{J_0}}$

12. 9.37 lb./in./sec^2

15. $f = \dfrac{1}{2\pi} \sqrt{\dfrac{wa \sin \alpha}{J}}$

18. $L = 39.2$ in.; $\theta_0 = 5.53°$

23. $f = \dfrac{1}{2\pi} \sqrt{\dfrac{2g}{l}} \cos \alpha$

27. $\tau = 2\pi \dfrac{L}{a} \sqrt{\dfrac{h}{3g}}$

31. $\omega = \sqrt{\dfrac{48 E I g}{(W + 0.486w)l^3}}$

35. 0.065 lb.

37. $m\ddot{x} + \left(k_1 + \dfrac{T}{l}\right)x + \dfrac{k_2}{2l^2}x^3 = 0$ **40.** $m_{\text{eff.}} = \dfrac{m}{n^2}\left[\dfrac{1}{12} + \left(\dfrac{1}{2} - n\right)^2\right]$

41. $J_{\text{pivot}} = 64.6 \times 10^{-6}$ lb. in. sec.^{2}; $m_{\text{eff}} = 112 \times 10^{-6}$ lb. in.$^{-1}$ sec.^{2}

   $f_{\text{air}} = 2.21$ c.p.s.; $f_{\text{oil}} = 0.0356$ c.p.s.

44. $\omega = 0.00389\sqrt{EI}$ **46.** 0.583 lb.

Chapter 2

1. 0.288 lb. sec./in.

3. $\zeta = 1.45$

5. (a) $\dfrac{x\omega_n}{v_0} = \dfrac{1}{3.46}(e^{-0.268\omega_n t} - e^{-3.73\omega_n t})$

   (b) $\dfrac{x\omega_n}{v_0} = \dfrac{1}{0.865}e^{-0.5\omega_n t}\sin 0.865\,\omega_n t$

   (c) $\dfrac{x\omega_n}{v_0} = \omega_n t e^{-\omega_n t}$

7. $\delta = 0.391$

9. (a) $\zeta = 0.0493$; (b) $\delta = 0.310$; (c) 1.36

11. (a) $\zeta = 0.10$; (b) 3.16 c.p.s.; (c) $\delta = 0.632$; (d) 1.88

13. (a) $\omega_n = \sqrt{\dfrac{k}{m}\left(\dfrac{b}{a}\right)^2 - \left(\dfrac{c}{2m}\right)^2}$; (b) $c_c = 2\dfrac{b}{a}\sqrt{km}$

15. $\eta = 4\pi J\sqrt{f_1^2 - f_2^2}$

Chapter 3

2. 144 per cent

3. $X = 1.40$ in.; $\phi = 51°30'$

5. $c = 0.143c_{cr}$

8. $\dfrac{kx}{F_0} = \dfrac{1}{1 - (\omega/\omega_n)^2}\left[\sin(\omega t - \varphi) - \dfrac{\omega}{\omega_n}\sin\omega_n t \cdot \cos\varphi - \cos\omega_n t \cdot \sin\varphi\right]$

9. $\bar{v} = \dfrac{F_0}{c + i\left(m\omega - \dfrac{k}{\omega}\right)}$; $Z = c + i\left(m\omega - \dfrac{k}{\omega}\right)$

11. $X_1 = \dfrac{(2\zeta\omega/\omega_n)X_2}{\sqrt{[1 - (\omega/\omega_n)^2]^2 + [2\zeta\omega/\omega_n]^2}}$; $\varphi = \tan^{-1}\dfrac{2\zeta\dfrac{\omega}{\omega_n}}{1 - \left(\dfrac{\omega}{\omega_n}\right)^2}$

13. 18.9 lb./in. each

15. (a) No change; (b) reduce to $\dfrac{k_{\text{original}}}{k_{\text{new}}} \times$ previous amplitude

17. (a) 0.0467 in.; (b) 169°22′; (c) 25.9 per cent; 397 lb.; (d) 304°

18. (a) 0.212 in.; (b) 43.1 per cent; (c) 107 lb.

20. $A = 19.05$ sq. in. each; $V_0 = 260$ in.^{3}

21. $X = \dfrac{Y}{1 - (\omega/\omega_n)^2}$; $\omega = 2\pi v/L$; $\omega_n = \sqrt{kg/W}$; $v_c = \dfrac{L}{2\pi} \sqrt{kg/W}$

23. $m\ddot{x}_1 + c\dot{x}_1 + (k + k_0)x_1 = k_0 X_2 \sin \omega t$; $c_c = 2\sqrt{(k + k_0)m}$

27. Semi major axis $= X$

Semi minor axis $= 2\zeta k X \dfrac{\omega}{\omega_n}$

29. $\delta = \dfrac{\pi\gamma}{\sqrt{1 - \frac{1}{4}\gamma^2}}$

31. $c_{eq} = c + \dfrac{4Fc}{\pi\omega X}$

34. 0.0487 in.

35. $\dfrac{Z}{Z_0} = \dfrac{(\omega/\omega_n)^2}{1 - (\omega/\omega_n)^2}$

37. Phase distortion present. Instrument worthless

38. Let $Z = A \sin(\omega_1 t - \varphi_1) + B \sin(\omega_2 t - \varphi_2) + \cdots$

If $\varphi_i = K\dfrac{\omega_i}{\omega_n}$, then

$Z = A \sin(\omega_1 t - \varphi_1) + B \sin \dfrac{\omega_2}{\omega_1}(\omega_1 t - \varphi_1) + \cdots$

$\therefore$ no phase distortion

40. $d = 1.74$ in.; $\theta = 189°27′$

43. 145 lb. each for 1 in. shaft; 26.6 lb. each for $\frac{3}{4}$ in. shaft

45. 0.0102 sec.

46. 2.33 grams at 123°20′

Chapter 4

5. $x = \dfrac{F_0}{k}(1 - \cos \omega t)$ $\quad t < t_0$

$x = \dfrac{F_0}{k}(1 - \cos \omega t) - \dfrac{2F_0}{k}[1 - \cos \omega(t - t_0)]$ $\quad t > t_0$

21. $\omega t_0 = 0.455 = 26.0°$

From Eq. (4.6-9) $X_1 = 0.01003''$

22.

$$\ddot{x} = -2\zeta\omega_n(\dot{x} - \dot{y}) - \omega_n^2(x - y)$$

Integrate with initial conditions $= 0$

$$\dot{x} = -2\zeta\omega_n(x - y) - \omega_n^2\int(x - y)\,dt$$

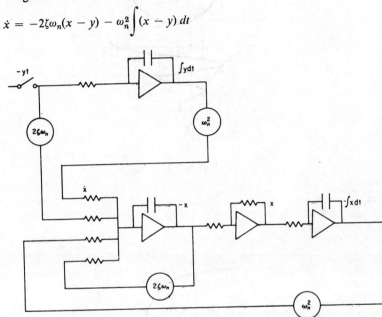

Fig. A.4-22.

Chapter 5

1. $\dfrac{dv}{dx} = -\dfrac{x}{v}, \quad v = \dfrac{\dot{x}}{\omega_n}$

3. $\dfrac{1}{\omega_n}\dfrac{d\dot{x}}{dx} = -2\zeta - \dfrac{x}{\dot{x}/\omega_n}$

Isocline equation

$$v = \frac{-x}{\alpha + 2\zeta}$$

$$\alpha = \frac{dv}{dx} = \text{constant}$$

$$v = \frac{\dot{x}}{\omega_n}, \quad \zeta = \frac{c}{c_c}$$

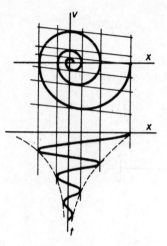

Fig. A.5-3.

4. $\dfrac{dv}{dx} = -2\zeta - \dfrac{x}{v}$

$v = \dfrac{\dot{x}}{\omega_n}$

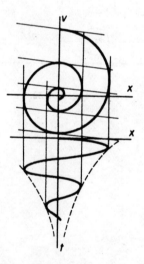

Fig. A.5-4.

5. See Ref. 12 p. A171 Eq. (9)

8. $\dfrac{dv}{dx} = \dfrac{\left(2\zeta v + \dfrac{\mu}{k} x^3\right) + x}{-v}$

9. $\dfrac{dv}{dx} = \dfrac{x + \dfrac{c}{\omega_n^2}\,\mathrm{sgn}\,(v)}{-v} = \dfrac{x + \delta}{-v}$

10.

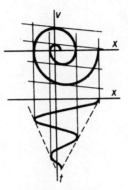

Fig. A.5-10.

11. $\dfrac{d\dot{x}}{dx} = -\dfrac{2T_0}{ml}\dfrac{x}{\dot{x}}\left(1 + k\dfrac{x^2}{l^2}\right), \quad k = \dfrac{1}{2}\left(\dfrac{AE}{T_0} - 1\right)$

14. $\dfrac{dv}{dx} = \dfrac{\sin x}{-v}$

15. $\theta = \theta_0 \cos \omega t + \dfrac{\theta_0^3}{6 \times 32\omega^2}\dfrac{g}{l}\,(\cos \omega t - \cos 3\omega t)$

$\omega^2 = \dfrac{g}{l} - \dfrac{3}{24}\dfrac{g}{l}\,\theta_0^2$

16. $\tau = \dfrac{2\pi}{\sqrt{\dfrac{g}{l}\left(1 - \dfrac{\theta_0^2}{8}\right)}}$

20, 21.

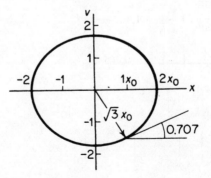

Fig. A.5-20/21.

For $-x_0 < x < x_0$ the curve is a circle of radius $\sqrt{3}\,x_0$

For $|x| > x_0$ the curve is a parabola

22. $\ddot{\theta} + \dfrac{g}{l + R\theta} \sin\theta - \dfrac{R\dot{\theta}^2}{l + R\theta} = 0$

27. $x = A \cos\omega t - \dfrac{T_0 k A^3}{16ml^3\omega^2} (\cos\omega t - \cos 3\omega f)$

$\omega^2 = \dfrac{2T_0}{ml}\left(1 + \dfrac{3}{4}\dfrac{k}{l^2}A^2\right)$

$k = \dfrac{1}{2}\left(\dfrac{AE}{T_0} - 1\right)$

Chapter 6

1. $\begin{bmatrix} J_1 & 0 \\ 0 & J_2 \end{bmatrix} \begin{Bmatrix} \ddot{\theta}_1 \\ \ddot{\theta}_2 \end{Bmatrix} + \begin{bmatrix} (K_1 + K_2) & -K_2 \\ -K_2 & K_2 \end{bmatrix} \begin{Bmatrix} \theta_1 \\ \theta_2 \end{Bmatrix} = \begin{Bmatrix} 0 \\ 0 \end{Bmatrix}$

2. $\omega_1 = 0.616\sqrt{\dfrac{K}{J}} \quad \left(\dfrac{\theta_1}{\theta_2}\right)_1 = 0.62$

$\omega_2 = 1.62\sqrt{\dfrac{K}{J}} \quad \left(\dfrac{\theta_1}{\theta_2}\right)_2 = -1.61$

4. $\omega = \sqrt{\dfrac{K_1 K_2 (J_1 + J_2)}{(K_1 + K_2)J_1 J_2}}$ 　　　　**7.** 22.6 c.p.s.

8. $\begin{bmatrix} m_1 & 0 \\ 0 & m_2 \end{bmatrix} \begin{Bmatrix} \ddot{y}_1 \\ \ddot{y}_2 \end{Bmatrix} + \dfrac{T}{l}\begin{bmatrix} 2 & -1 \\ -1 & 2 \end{bmatrix} \begin{Bmatrix} y_1 \\ y_2 \end{Bmatrix} = \begin{Bmatrix} 0 \\ 0 \end{Bmatrix}$

10. $\omega_1 = 0.796\sqrt{\dfrac{T}{ml}} \quad \left(\dfrac{y_1}{y_2}\right)_1 = 1.365$

$\omega_2 = 1.536\sqrt{\dfrac{T}{ml}} \quad \left(\dfrac{y_1}{y_2}\right)_2 = -0.365$

12. $m\begin{bmatrix} 1 & 2 \\ -2 & 2 \end{bmatrix} \begin{Bmatrix} \ddot{x}_1 \\ \ddot{x}_2 \end{Bmatrix} + k\begin{bmatrix} 1 & 1 \\ -14 & 3 \end{bmatrix} \begin{Bmatrix} x_1 \\ x_2 \end{Bmatrix} = \begin{Bmatrix} 0 \\ 0 \end{Bmatrix}$

15. $\dfrac{m}{k}\begin{bmatrix} 1 & 3 \\ -1 & 1 \end{bmatrix} \begin{Bmatrix} \ddot{x}_1 \\ \ddot{x}_2 \end{Bmatrix} + \begin{bmatrix} 2 & 2 \\ -6 & 2 \end{bmatrix} \begin{Bmatrix} x_1 \\ x_2 \end{Bmatrix} = \begin{Bmatrix} 0 \\ 0 \end{Bmatrix}$

$\dfrac{m\omega^2}{k} = 3 \pm \sqrt{5}$

17.
$$\begin{bmatrix} m & 0 \\ 0 & J \end{bmatrix} \begin{Bmatrix} \ddot{x} \\ \ddot{\theta} \end{Bmatrix} + \begin{bmatrix} (k_1 + k_2) & -\left(\dfrac{k_1}{4} - \dfrac{k_2}{2}\right)l \\ -\left(\dfrac{k_1}{4} - \dfrac{k_2}{2}\right)l & \left(\dfrac{k_1^2}{16} + \dfrac{k_2^2}{4}\right)l^2 \end{bmatrix} \begin{Bmatrix} x \\ \theta \end{Bmatrix} = \begin{Bmatrix} 0 \\ 0 \end{Bmatrix}$$

19. $k_1 = 2k_2$

20. $f_1 = 0.963$ c.p.s.;   node 10.9 ft. behind c.g.

$f_2 = 1.33$ c.p.s.;   node 1.48 ft. ahead of c.g.

23. $\left(\dfrac{x_1}{x_2}\right)^{(1)} = 0.414$, $\left(\dfrac{x_1}{x_2}\right)^{(2)} = -2.41$

26. $k = 4600$ lb./in.; $X = 0.040$ in.

27. $X_1 = \dfrac{m\omega^2 e[k_2 - M_2\omega^2 + i\omega c]}{[(k_1 + k_2) - M_1\omega^2 + i\omega c][k_2 - M_2\omega^2 + i\omega c] - [k_2 + i\omega c]^2}$

$\dfrac{X_2}{X_1} = \dfrac{[k_2 - i\omega c]}{[k_2 - M_2\omega^2 + i\omega c]}$

29. $d_2 = \frac{1}{2}$ in.   **31.** $W = 11.6$ lb.; $k = 17.7$ lb./in.

32. $T = 2\pi\mu\omega R^3 \left[\dfrac{1}{4}\left(R - \dfrac{R_0^4}{R^3}\right) + b\right]$

33. $\omega = \sqrt{\dfrac{g}{l} + \dfrac{k}{ml^2}(1 \pm 1)}$;   beat period = 52.3 sec.

34. $\omega^2 = \left(1 + \dfrac{2}{3}\dfrac{m}{M}\right)\dfrac{g}{l}$; $\dfrac{X_1}{X_2} = -\dfrac{2}{3}\dfrac{m}{M}$; X_1 = displ. of cylinder
X_2 = displ. of pendulum bob.

37. $\zeta_0 = 0.344$; $\dfrac{\omega}{\omega_n} = 0.943$ **38.** $\dfrac{\theta_\zeta = 0.10}{\theta_\zeta = 0.344} \cong 3$

41. $x_1 = \frac{8}{9}\cos\omega_1 t + \frac{1}{9}\cos\omega_2 t$
$x_2 = \frac{4}{9}\cos\omega_1 t - \frac{1}{9}\cos\omega_2 t$

48. $X_1 = \dfrac{1}{D}\begin{vmatrix} F_0 & -k_1 \\ 0 & (k_1 + k_2 - m_2\omega^2) \end{vmatrix}$

$X_2 = \dfrac{1}{D}\begin{vmatrix} (k_1 - m_1\omega^2) & F_0 \\ -k_1 & 0 \end{vmatrix}$

$D = \begin{vmatrix} (k_1 - m_1\omega^2) & -k_1 \\ -k_1 & (k_1 + k_2 - m_2\omega^2) \end{vmatrix}$

57. $F_y = T_i = M_x = 0$;   completely balanced

Chapter 7

1. $\begin{Bmatrix} \theta_1 \\ \theta_2 \end{Bmatrix} = \omega^2 \begin{bmatrix} \left(\dfrac{1}{K_1} + \dfrac{1}{K_2}\right) & \dfrac{1}{K_2} \\ \dfrac{1}{K_2} & \dfrac{1}{K_2} \end{bmatrix} \begin{bmatrix} J_1 & 0 \\ 0 & J_2 \end{bmatrix} \begin{Bmatrix} \theta_1 \\ \theta_2 \end{Bmatrix}$

2. $a_{11} = \dfrac{k_2 + k_3}{\sum k_i k_j}, \quad a_{21} = a_{12} = \dfrac{k_2}{\sum k_i k_j}, \quad a_{22} = \dfrac{k_1 + k_2}{\sum k_i k_j}$

3. $\omega^2 = \dfrac{k}{m}\left(\dfrac{4 - 6n + 5n^2}{1 + 2n^2}\right); \quad n = \dfrac{X_2}{X_1}$

6. $a_{11} = a_{21} = a_{12} = a_{31} = a_{13} = \dfrac{l_1}{(m_1 + m_2 + m_3)g}$

$a_{22} = a_{23} = a_{32} = \dfrac{l_1}{(m_1 + m_2 + m_3)g} + \dfrac{l_2}{(m_2 + m_3)g}$

$a_{33} = \dfrac{l_1}{(m_1 + m_2 + m_3)g} + \dfrac{l_2}{(m_2 + m_3)g} + \dfrac{l_3}{m_3 g}$

11. $\omega = \sqrt{\dfrac{3EI}{Ml^3}\left(1 + \dfrac{n}{2}\right)}; \quad \dfrac{y_1}{y_2} = -\dfrac{n}{2}$ **13.** $\omega_1 = 2.90\sqrt{\dfrac{gEI}{Wl^3}}$

15. $f_{11} = 498$ c.p.s.   **18.** $\begin{Bmatrix} x_1 \\ x_2 \\ x_3 \end{Bmatrix} = \dfrac{m\omega^2}{3k} \begin{bmatrix} 4 & 2 & 1 \\ 4 & 8 & 4 \\ 4 & 8 & 7 \end{bmatrix} \begin{Bmatrix} x_1 \\ x_2 \\ x_3 \end{Bmatrix}$

20. $\omega_1 = 4.93\sqrt{\dfrac{gEI}{Wl^3}}$

24. $\theta_1 = 0.00618$ rad., $\theta_2 = -0.000778$ rad., $\theta_3 = -0.00338$ rad.

26. $\theta_1 = 0.0798$ rad., $0°$; $\theta_2 = 0.0558$ rad., $0°$; $\theta_3 = 0.0204$ rad., $42°$; $\theta_4 = 0.0442$ rad., $128°8'$

28. 22.8 c.p.s.   **29.** 22.5 c.p.s.; 52.5 c.p.s.

31. $J_1 = J_2 = J_3 = J_4 = 0.0442$ lb. in. sec.^{2}

35. $\omega_1 = 0.585\sqrt{\dfrac{EI}{ml^3}}; \quad \omega_2 = 3.88\sqrt{\dfrac{EI}{ml^3}}$

38. $\begin{vmatrix} u_{12} & u_{14} \\ u_{32} & u_{34} \end{vmatrix} = 0$ **41.** $\begin{vmatrix} u_{32} & u_{34} \\ u_{42} & u_{44} \end{vmatrix} = 0$

43. $\begin{vmatrix} u_{11} & u_{12} \\ u_{21} & u_{22} \end{vmatrix} = 0$ Where the u's are determined by multiplying out the 3 matrices of the form given by Eq. 7.9-12.

44. Start with Eq. 7.9-14

$$
\begin{bmatrix} y \\ \theta \\ M \\ V \\ \varphi \\ T \end{bmatrix}
\rightarrow
\begin{bmatrix} 0 \\ 0 \\ M \\ V \\ 0 \\ T \end{bmatrix}_0
=
\begin{bmatrix} \dfrac{l^2}{2EI} M_0 + \dfrac{l^3}{6EI} V_0 \\[2mm] \dfrac{l}{EI} M_0 + \dfrac{l^2}{2EI} V_0 \\[2mm] M_0 + V_0 l \\[2mm] V_0 \\[2mm] hT_0 \\[2mm] T_0 \end{bmatrix}
$$

46. $\omega_k = 2\sqrt{\dfrac{K}{J}} \sin \dfrac{(2k-1)\pi}{2(2N+1)}$ $k = 1, 2, 3, \cdots N$

47. $Y_{r+1} - 2\left(1 - \dfrac{\omega^2 ml}{2T}\right) Y_r + Y_{r-1} = 0$

$\omega_k = 2\sqrt{\dfrac{T}{ml}} \sin \dfrac{k\pi}{2(N+1)}, \quad k = 1, 2, 3, \cdots N$

51. β is found from $2\cos\beta \left(N + \dfrac{1}{2}\right) \sin\dfrac{\beta}{2} = -\dfrac{K_N}{k}\sin\beta N$

and substituted into $\omega = 2\sqrt{\dfrac{k}{m}}\sin\dfrac{\beta}{2}$

53. $m\ddot{y}_n = k(y_{n+1} - 2y_n + y_{n-1})$ boundary eqs. $m\ddot{y}_N = -k(y_N - y_{N-1} + h\theta)$

$$\sum_{n=1}^{N} nh\,(m\ddot{y}_n) - K_\theta\theta = (N+1)m\rho^2\ddot{\theta}$$

ρ = radius of gyration of each floor about its center of mass.

Chapter 8

2. $f = \dfrac{n}{2l}\sqrt{\dfrac{T}{l}}, \quad n = 1, 2, 3, \cdots$ **3.** $\tan\dfrac{\omega l}{c} = \dfrac{(T/kl)(\omega l/c)}{(mc^2/kl^2)(\omega l/c)^2 - 1}$

5. 16,600 ft./sec.

11. $\omega_n = (2n+1)\dfrac{\pi}{l}\sqrt{\dfrac{Gg}{\rho}}, \quad n = 0, 1, 2, \cdots$

12. $\tan\dfrac{\omega l}{c} = \dfrac{2(J_0/J_s)(\omega l/c)}{(J_0/J_s)^2(\omega l/c)^2 - 1}$

14. $b = \dfrac{2}{\pi}$,   node at $x = 0.219l$ from ends

16. Same f_n as free-free bar

18. $\omega = n^2 \sqrt{\dfrac{gEIl}{W_b}}$ where n is determined from

$$(1 + \cosh nl \cdot \cos nl) = nl \frac{W_0}{W_b} (\sinh nl \cdot \cos nl - \cosh nl \cdot \sin nl)$$

20. $\tan \omega \sqrt{\dfrac{\rho}{Gg}}\, l = -\omega \dfrac{I_p G}{K} \sqrt{\dfrac{\rho}{Gg}}$

Chapter 9

3. $K_i = \dfrac{p_0}{l} \displaystyle\int_0^l \phi_i(x)\, dx$

8. $y(x, t) = \dfrac{4p_0}{\pi M \omega_2^2} \sin \dfrac{2\pi x}{l} (1 - \cos \omega_2 t)$

10. Modes absent are 2nd, 5th, 8th, etc.

11. $K_n = \sqrt{2} \cos (2n - 1) \dfrac{\pi}{6}$, $D_n = (1 - \cos \omega_n t)$

$$u = \frac{2F_0 l}{AE} \left[\frac{\cos (\pi/6) \cos (\pi/2)(x/l)}{(\pi/2)^2} D_1 + \frac{\cos (5\pi/6) \cos (5\pi/2)(x/l)}{(5\pi/2)^2} D_2 + \cdots \right]$$

14. $K_1 = \dfrac{1}{l} \displaystyle\int_0^l \phi_1(x)\, dx = 0.784$

$K_2 = \dfrac{1}{l} \displaystyle\int_0^l \phi_2(x)\, dx = 0.434$

$K_3 = \dfrac{1}{l} \displaystyle\int_0^l \phi_3(x)\, dx = 0.254$

19. $\left\{ 1 + \dfrac{K\varphi_1'^2(0)}{M\omega_1^2[1 - (\omega/\omega_1)^2]} \right\} \left\{ 1 + \dfrac{K\varphi_2'^2(0)}{M\omega_2^2[1 - (\omega/\omega_2)^2]} \right\}$

$$= \left\{ \frac{K\varphi_1'(0)\varphi_2'(0)}{M\omega_1^2[1 - (\omega/\omega_1)^2]} \right\} \left\{ \frac{K\varphi_1'(0)\varphi_2'(0)}{M\omega_2^2[1 - (\omega/\omega_2)^2]} \right\}$$

$$\varphi_1 = \sqrt{2} \sin \frac{\pi x}{l}, \quad \varphi_1' = \frac{\pi}{l} \sqrt{2} \cos \frac{\pi x}{l}, \text{ etc.}$$

One mode approximation gives

$$\left(\frac{\omega}{\omega_1} \right)^2 = 1 + \frac{2K}{M\omega_1^2} \left(\frac{\pi}{l} \right)^2, \quad \omega_1 = \pi^2 \sqrt{\frac{EI}{Ml^3}}$$

20. One mode approximation

$$\left(\frac{\omega}{\omega_1} \right)^2 = 1 + \frac{4K}{M\omega_1^2} \left(\frac{\pi}{l} \right)^2$$

21. Using one free-free mode and translation mode of M_0

$$\left(\frac{\omega}{\omega_1}\right)^2 = \frac{M_1}{M_1 + M_0\varphi_1^2(0) - [M_0^2\varphi_1^2(0)/(M_0 + 2ml)]}$$

where $M_1 = \int \varphi_1^2(x)m\,dx = 2ml, \quad \omega_1 = 22.4\sqrt{\dfrac{EI}{m(2l)^4}}$

Chapter 10

4. $f = \dfrac{\omega_n}{2\pi}\sqrt{1 \pm \dfrac{1}{Q}}$

5. $\overline{f^2} = C_0^2 + 2\sum_{n=1}^{\infty}\mid C_n\mid^2$

6. $\overline{x^2} = \left(\dfrac{C_0}{k}\right)^2 + 2\sum_{n=1}^{\infty}\left|\dfrac{C_n}{k}\right|^2\mid H(\omega_n)\mid^2$

9.

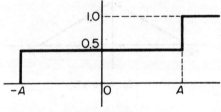

Fig. A.10-9.

10. Instantaneous value 4.6 per cent; Peak value 13.5 per cent

11. $\sigma = 1.99$; r.m.s. = 3.60

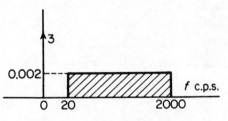

Fig. A.10-11.

12. $\dfrac{C_n C_n^*}{2}\delta(\omega - n\omega_0) = \dfrac{8}{\pi^2}\left\{\delta(\omega - \omega_0) + \dfrac{1}{9}\delta(\omega - 3\omega_0) + \dfrac{1}{25}\delta(\omega - 5\omega_0) + \cdots\right\}$

Sine wave $\dfrac{C_n C_n^*}{2}\delta(\omega - n\omega_0) = \dfrac{1}{2}\delta(\omega - \omega_0) + \dfrac{1}{2}\delta(\omega + \omega_0)$

13. $y = \dfrac{2A}{i\pi} \displaystyle\sum_{n=-\infty}^{\infty} \dfrac{e^{in\omega_0 t}}{n}$ n odd

16. period $= 2T$

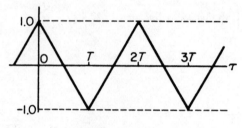

Fig. A.10-16.

17. pulse duration $= 2T$

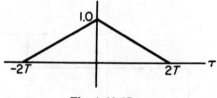

Fig. A.10-17.

Index

GEORGE ALLEN & UNWIN LTD

Head Office:
40 *Museum Street, London* WC1
Telephone: 01-405 8577

Sales, Distribution and Accounts Departments:
Park Lane, Hemel Hempstead, Hertfordshire
Telephone: 0442 3244

Argentina: Defensa 681-5J, Buenos Aires
Australia: Cnr. Bridge Road and Jersey Street, Hornsby, N.S.W. 2077
Bangladesh: Empire Buildings, Victoria Park South, Dacca
Canada: 2330 Midland Avenue, Agincourt, Ontario
Greece: 3 Mitropoleous Street, Athens 118
India: 103/5 Walchand Hirachand Marg, Post Box 21, Bombay 1BR
285J Bepin Behari Ganguli Street, Calcutta 12
2/18 Mount Road, Madras 2
4/21-22B Asaf Ali Road, New Delhi 1
Japan: 29/13 Hongo 5 Chome, Bunkyo Ku, Tokyo 113
Kenya: P.O. Box 30583, Nairobi
Malaysia: 54 Jalan-Pudu, Room 303, Kuala Lumpur
New Zealand: 46 Lake Road, Northcote, Auckland 9
Nigeria: P.O. Box 62 Ibadan
Pakistan: Karachi Chambers, McLeod Road, Karachi 2
22 Falettis' Hotel, Egerton Road, Lahore
Philippines: U.P., P.O. Box 10, Quezon City, D-505
Singapore and Hong Kong: 53L Anson Road, 11th Floor Anson Centre, Singapore 2
South Africa: P.O. Box 31487, Braamfontein, Johannesburg
Thailand: P.O. Box 6/1, Bangkok
West Indies: Rockley New Road, St. Lawrence 4, Barbados